SECOND EDITION

SELF-SUFFICIENCY

A COMPLETE GUIDE TO BAKING,
CARPENTRY, CRAFTS, ORGANIC GARDENING,
PRESERVING YOUR HARVEST, RAISING ANIMALS, AND MORE!

EDITED BY ABIGAIL R. GEHRING

Skyhorse Publishing

Skyhorse Publishing books may be purchased in bulk at special discounts for sales promotion, corporate gifts, fund-raising, or educational purposes. Special editions can also be created to specifications. For details, contact the Special Sales Department, Skyhorse Publishing, 307 West 36th Street, 11th Floor, New York, NY 10018 or info@skyhorsepublishing.com.

Skyhorse® and Skyhorse Publishing® are registered trademarks of Skyhorse Publishing, Inc.®, a Delaware corporation.

Visit our website at www.skyhorsepublishing.com

10 9 8 7 6 5 4 3 2

Library of Congress Cataloging-in-Publication Data is available on file.

ISBN: 978-1-63220-280-2
e-book ISBN: 978-1-63450-022-7

Printed in China

Contents

Introduction   1

Part One: The Family Garden   3

Part Two: The Country Kitchen   95

Part Three: Canning and Preserving   155

Part Four: Country Crafts   237

Part Five: The Barnyard   291

Part Six: The Workshop   341

Appendix 1: Alternative Energy   411

Appendix 2: Food Co-op Directory   421

Sources   433

Acknowledgments   439

Index   441

Introduction

More and more families are being drawn toward a lifestyle that is greener, cleaner, more genuine, and more aware. We want to know where our food is coming from, to the point of touching the dirt that it springs out of, if possible. We want our children (or nieces or nephews or godchildren) to understand that eggs come from chickens—not just from cardboard cartons on supermarket shelves. We love the idea of building things with our own hands, of picking our own berries, of making fresh bread and spreading it with homemade butter. We are, in short, longing for self-sufficiency.

"Self-sufficiency" as a term is somewhat misleading. "The good life" that most of us are seeking in our varied ways does not involve cutting off ties from those who surround us. Complete independence is not possible and, at least for most people, would not bring much satisfaction anyway. The early settlers banded together whenever they could, knowing their lives would be made easier and better by the community's support. In similar ways, we benefit from those who have ventured into back-to-basics living before us, and we would be wise to share ideas, tools, and experiences with those on similar paths around us now. But we do not need to be trapped by dependency on anyone or any group—or any idea, for that matter. We can be responsible for growing or raising at least a portion of what we consume; we can find ways to fix things rather than running to the store to buy replacements; we can teach our children ourselves, rather than leaving the burden entirely on public or private schools.

People and experience are the best teachers when it comes to learning things like how to plant a garden or milk a cow. But sometimes you don't have a neighbor to call on for advice and trial and error will result in more error than the trial is worth. That's where this book comes in. You'll find instructions and tips for everything from growing tomatoes to canning jams and jellies to constructing a chicken coop. Scattered throughout are fun projects for "The Junior Homesteader" and "Homeschooling Hints," which can be used to supplement your children's education, whether or not you choose to participate in a traditional schooling system. You'll also find plenty of photographs and illustrations to add clarity and interest to the written directions. Let these pages inspire and direct you as you discover what self-sufficiency means for you.

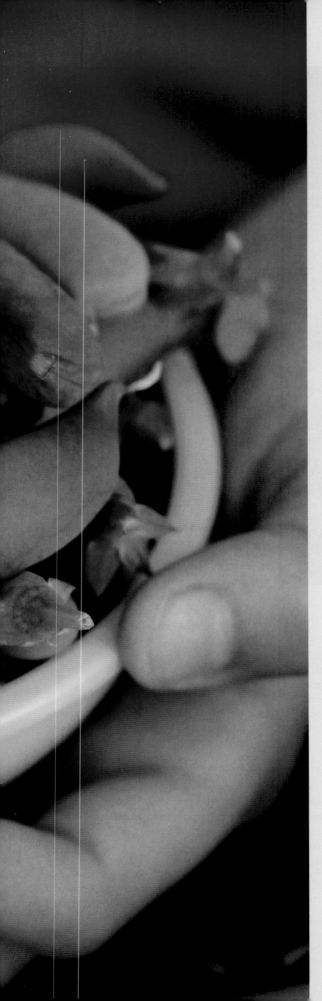

Part 1

The Family Garden

Planning a Garden	4
Improving Your Soil	13
Planting Your Garden	21
Conserving Water	25
Mulching in Your Garden and Yard	29
Organic Gardening	32
Saving Seeds	35
Terracing	38
Start Your Own Vegetable Garden	41
Start Your Own Flower Garden	44
Growing Fruit Bushes and Trees	49
Growing and Threshing Grains	55
Planting Trees for Shade or Shelter	58
Container Gardening	62
Rooftop Gardens	67
Raised Beds	69
Growing Plants Without Soil	71
Pest and Disease Management	76
Harvesting Your Garden	81
Community Gardens	84
Farmers' Markets	90
Attracting Birds, Butterflies, and Bees to Your Garden	93

Planning a Garden

A Plant's Basic Needs

Before you start a garden, it's helpful to understand what plants need in order to thrive. Some plants, like dandelions, are tolerant of a wide variety of conditions, while others, such as orchids, have very specific requirements in order to grow successfully. Before spending time, effort, and money attempting to grow a new plant in a garden, learn about the conditions that particular plant needs in order to grow properly.

Environmental factors play a key role in the proper growth of plants. Some of the essential factors that influence this natural process are as follows:

1. Length of Day

The amount of time between sunrise and sunset is the most critical factor in regulating vegetative growth, blooming, flower development, and the initiation of dormancy. Plants utilize increasing day length as a cue to promote their growth in spring, while decreasing day length in fall prompts them to prepare for the impending cold weather. Many plants require specific day length conditions in order to bloom and flower.

2. Light

Light is the energy source for all plants. Cloudy, rainy days or any shade cast by nearby plants and structures can significantly reduce the amount of light available to the plant. In addition, plants adapted to thrive in shady spaces cannot tolerate full sunlight. In general, plants will only be able to survive where adequate sunlight reaches them at levels they are able to tolerate.

3. Temperature

Plants grow best within an optimal range of temperatures. This temperature range may vary drastically depending on the plant species. Some plants thrive in environments where the temperature range is quite wide; others can only survive within a very narrow temperature variance. Plants can only survive where temperatures allow them to carry on life-sustaining chemical reactions.

4. Cold

Plants differ by species in their ability to survive cold temperatures. Temperatures below 60°F injure some tropical plants. Conversely, arctic species can tolerate temperatures well below zero. The ability of a plant to withstand cold is a function of the degree of dormancy present in the plant and its general health. Exposure to wind, bright sunlight, or rapidly changing temperatures can also compromise a plant's tolerance to the cold.

5. Heat

A plant's ability to tolerate heat also varies widely from species to species. Many plants that evolved to grow in arid, tropical regions are naturally very heat tolerant, while subarctic and alpine plants show very little tolerance for heat.

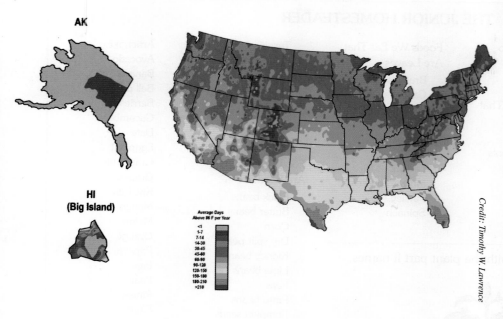

AK

HI
(Big Island)

Credit: Timothy W. Lawrence

Average Days
Above 86 F per Year
<1
1-7
7-14
14-30
30-45
45-60
60-90
90-120
120-150
150-180
180-210
>210

⌃ This map shows the average number of days each year that an area experiences temperatures over 86 degrees ("heat days"). Zone 1 has less than one heat day and Zone 12 has more than 210 heat days. Most plants begin to suffer when it gets any hotter than 86 degrees, though different plants have different levels of tolerance.

6. Water

Different types of plants have different water needs. Some plants can tolerate drought during the summer but need winter rains in order to flourish. Other plants need a consistent supply of moisture to grow well. Careful attention to a plant's need for supplemental water can help you to select plants that need a minimum of irrigation to perform well in your garden. If you have poorly drained, chronically wet soil, you can select lovely garden plants that naturally grow in bogs, marshlands, and other wet places.

7. Soil pH

A plant root's ability to take up certain nutrients depends on the pH—a measure of the acidity or alkalinity—of your soil. Most plants grow best in soils that have a pH between 6.0 and 7.0. Most ericaceous plants, such as azaleas and blueberries, need acidic soils with pH below 6.0 to grow well. Lime can be used to raise the soil's pH, and materials containing sulfates, such as aluminum sulfate and iron sulfate, can be used to lower the pH. The solubility of many trace elements is controlled by pH, and plants can only use the soluble forms of these important micronutrients.

A BASIC PLANT GLOSSARY

Annual—a plant that completes its life cycle in one year or season.

Arboretum—a landscaped space where trees, shrubs, and herbaceous plants are cultivated for scientific study or educational purposes, and to foster appreciation of plants.

Axil—the area between a leaf and the stem from which the leaf arises.

Bract—a leaflike structure that grows below a flower or cluster of flowers and is often colorful. Colored bracts attract pollinators, and are often mistaken for petals. Poinsettia and flowering dogwood are examples of plants with prominent bracts.

Cold hardy—capable of withstanding cold weather conditions.

Conifers—plants that predate true flowering plants in evolution; conifers lack true flowers and produce separate male and female strobili, or cones. Some conifers, such as yews, have fruits enclosed in a fleshy seed covering.

Cultivar—a cultivated variety of a plant selected for a feature that distinguishes it from the species from which it was selected.

Deciduous—having leaves that fall off or are shed seasonally to withstand adverse weather conditions, such as cold or drought.

Herbaceous—having little or no woody tissue. Most plants grown as perennials or annuals are herbaceous.

Hybrid—a plant, or group of plants, that results from the interbreeding of two distinct cultivars, varieties, species, or genera.

Inflorescence—a floral axis that contains many individual flowers in a specific arrangement; also known as a flower cluster.

Native plant—a plant that lives or grows naturally in a particular region without direct or indirect human intervention.

Panicle—a pyramidal, loosely branched flower cluster; a panicle is a type of inflorescence.

Perennial—persisting for several years, usually dying back to a perennial crown during the winter and initiating new growth each spring

Shrub—a low-growing, woody plant, usually less than 15 feet tall, that often has multiple stems and may have a suckering growth habit (the tendency to sprout from the root system).

Taxonomy—the study of the general principles of scientific classification, especially the orderly classification of plants and animals according to their presumed natural relationships.

Tree—a woody perennial plant having a single, usually elongated main stem, or trunk, with few or no branches on its lower part.

Wildflower—a herbaceous plant that is native to a given area and is representative of unselected forms of its species.

Woody plant—a plant with persistent woody parts that do not die back in adverse conditions. Most woody plants are trees or shrubs.

THE JUNIOR HOMESTEADER

We eat lots of
different plant parts!

**Foods We Eat That
Are Roots**

Beet
Carrot
Onion
Parsnip
Potato
Radish
Rutabaga

Sweet potato
Turnip
Yam

**Foods We Eat That
Are Stems**

Asparagus
Bamboo shoots
Bok choy
Broccoli
Celery
Rhubarb

**Foods We Eat That
Are Leaves**

Brussels sprouts
Cabbage
Chard
Collards
Endive
Kale
Lettuce
Mustard greens
Parsley
Spinach

Turnip greens
Watercress

**Foods We Eat That
Are Flowers**

Broccoli
Cauliflower

**Foods We Eat That
Are Seeds**

Black beans
Butter beans
Corn
Dry split peas
Kidney beans
Lima beans
Peas
Pinto beans
Pumpkin seeds
Sunflower seeds

**Foods We Eat That
Are Fruits**

Apple
Apricot

Artichoke
Avocado
Banana
Bell pepper
Berries
Cucumber
Date
Eggplant
Grapefruit
Grapes
Kiwifruit
Mango
Melon
Orange
Papaya
Peach
Pear
Pineapple
Plum
Pomegranate
Pumpkin
Squash
Strawberry
Tangerine
Tomato

Draw a line connecting the word with the plant part it names.

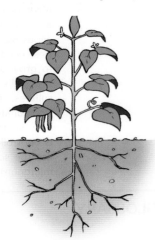

seed
stem
root
leaf
flower
fruit

Bean Plant

THE JUNIOR HOMESTEADER

You can structure your plants to double as playhouses for the kids. Here are a few possibilities:

Bean teepees are the best way to support pole bean plants, and they also make great hiding places for little gardeners. Drive five or six poles that are 7 to 8 feet tall into the ground in a circle with a 4-foot diameter. Bind the tops of the poles together with baling twine or a similar sturdy string. Plant your beans at the bottoms of the poles so they'll grow up and create a tent of vines.

Vine tunnels can be made out of poles and any trailing vines—gourds, cucumbers, or morning glories are a few options. Drive several 7- to 8-foot poles (bamboo works well) into the ground in two parallel lines so that they create a pathway. The poles should be at least 3 feet apart from each other. Then lash horizontal poles to the vertical ones at 2-, 4-, and 6-foot heights. You can also lash poles across the top of the tunnel to connect the two sides and create a roof. Plant your trailing vines at the bases of the poles and watch your tunnel fill in as the weeks go by.

Wigwams and huts are easily fashioned by planting your sunflowers or corn in a circular or square shape. To make a whole house, plant "walls" that are a few rows thick and create several "rooms," leaving gaps for doors.

Choosing a Site for Your Garden

Choosing the best spot for your garden is the first step toward growing the vegetables, fruits, and herbs that you want. You do not need a large space to get started—in fact, often it's wise to start small so that you don't get overwhelmed. A normal garden that is about 25 feet square will provide enough produce for a family of four, and with a little ingenuity (utilizing pots, hanging gardens, trellises, etc.) you can grow more than that in an even smaller space.

« When planning out your garden, first sketch a diagram of what you want your garden to look like. What sorts of plants to you want to grow? Do you want a garden purely for growing vegetables or do you want to mix in some fruits, herbs, and wildflowers? Choosing the appropriate plants to grow next to each other will help your garden grow well and will provide you with ample produce throughout the growing season (see the charts on page 11). If you live in the northern hemisphere, plant taller plants at the north end of your garden so that they won't block sunlight from reaching the smaller plants. If you live in the southern hemisphere, this is reversed.

Five Factors to Consider When Choosing a Garden Site

1. Sunlight

Sunlight is crucial for the growth of vegetables and other plants. For your garden to grow, your plants will need at least six hours of direct sunlight per day. In order to make sure your garden receives an ample amount of sunlight, don't select a garden site that will be in the shade of trees, shrubs, houses, or other structures. Certain vegetables, such as broccoli and spinach, grow just fine in shadier spots, so if your garden does receive some shade, make sure to plant those types of vegetables in the shadier areas. However, on the whole, if your garden does not receive at least six hours of intense sunlight per day, it will not grow as efficiently or successfully.

2. Proximity

Think about convenience as you plot out your garden space. If your garden is closer to your house and easy to reach, you will be more likely to tend it on a regular basis and to harvest the produce at its peak of ripeness. You'll find it a real boon to be able to run out to the garden in the middle of making dinner to pull up a head of lettuce or snip some fresh herbs.

3. Soil Quality

Your soil does not have to be perfect to grow a productive garden. However, it is best to have soil that is fertile, is full of organic materials that provide nutrients to the plant roots, and is easy to dig and till. Loose, well-drained soil is ideal. If there is a section of your yard where water does not easily drain after a good, soaking rain, this is not the spot for your garden; the excess water can easily drown your plants. Furthermore, soils that are of a clay or sandy consistency are not as effective in growing plants. To make these types of soils more nutrient-rich and fertile, add in organic materials (such as compost or manure).

4. Water Availability

Water is vital to keeping your garden green, healthy, and productive. A successful garden needs around 1 inch of water per week to thrive. Rain and irrigation systems are effective in maintaining this 1-inch-per-week quota. Situating your garden near a spigot or hose is ideal, allowing you to keep the soil moist and your plants happy.

« If your garden is not close to your house, you may want to construct a small potting shed in which to keep your tools.

5. Elevation

Your garden should not be located in an area where air cannot circulate or where frost quickly forms. Placing your garden in a low-lying area, such as at the base of a slope, should be avoided. Lower areas do not warm as quickly in the spring, and will easily collect frost in the spring and fall. Your garden should, if at all possible, be on a slightly higher elevation. This will help protect your plants from frost and you'll be able to start your garden growing earlier in the spring and harvest well into the fall.

Tools of the Trade

Gardening tools don't need to be high-tech, but having the right ones on hand will make your life much easier. You'll need a spade or digging fork to dig holes for seeds or seedlings (or, if the soil is loose enough, you can just use your hands). Use a trowel, rake, or hoe to smooth over the garden surface. A measuring stick is helpful when spacing your plants or seeds (if you don't have a measuring stick, you can use a precut string to measure). If you are planting seedlings or established plants, you may need stakes and string to tie them up so they don't fall over in inclement weather or when they start producing fruits or vegetables. Finally, if you are interested in installing an irrigation system for your garden, you will need to buy the appropriate materials for this purpose.

Companion Planting

Plants have natural substances built into their structures that repel or attract certain insects and can have an effect on the growth rate and even the flavor of the other plants around them. Thus, some plants aid each other's growth when planted in close proximity and others inhibit each other. Smart companion planting will help your garden remain healthy, beautiful, and in harmony, while deterring certain insect pests and other factors that could be potentially detrimental to your garden plants.

Here is a chart that lists various types of garden vegetables, herbs, and flowers and their respective companion and "enemy" plants.

VEGETABLES

Type	Companion plant(s)	Avoid
Asparagus	Tomatoes, parsley, basil	Onion, garlic, potatoes
Beans	Eggplant	Tomatoes, onion, kale
Beets	Mint	Runner beans
Broccoli	Onions, garlic, leeks	Tomatoes, peppers, mustard
Cabbage	Onions, garlic, leeks	Tomatoes, peppers, beans
Carrots	Leeks, beans	Radish
Celery	Daisies, snapdragons	Corn, aster flower
Corn	Legumes, squash, cucumbers	Tomatoes, celery
Cucumbers	Radish, beets, carrots	Tomatoes
Eggplant	Marigolds, mint	Runner beans
Leeks	Carrots	Legumes
Lettuce	Radish, carrots	Celery, cabbage, parsley
Melon	Pumpkin, squash	None
Peppers	Tomatoes	Beans, cabbage, kale
Onion	Carrots	Peas, beans
Peas	Beans, corn	Onions, garlic
Potatoes	Horseradish	Tomatoes, cucumbers
Tomatoes	Carrots, celery, parsley	Corn, peas, potatoes, kale

HERBS

Type	Companion Plant(s)	Avoid
Basil	Chamomile, anise	Sage
Chamomile	Basil, cabbage	Other herbs (it will become oily)
Cilantro	Beans, peas	None
Chives	Carrots	Peas, beans
Dill	Cabbage, cucumbers	Tomatoes, carrots
Fennel	Dill	Everything else
Garlic	Cucumbers, peas, lettuce	None
Oregano	Basil, peppers	None
Peppermint	Broccoli, cabbage	None
Rosemary	Sage, beans, carrots	None
Sage	Rosemary, beans	None
Summer savory	Onion, green beans	None

FLOWERS

Types	Companion Plant(s)	Avoid
Geraniums	Roses, tomatoes	None
Marigolds	Tomatoes, peppers, most plants	None
Petunia	Squash, asparagus	None
Sunflower	Corn, tomatoes	None
Tansy	Roses, cucumbers, squash	None

Shade-Loving Plants

Most plants thrive on several hours of direct sunlight every day, but certain plants actually prefer the shade. When buying seedlings from your local nursery or planting your own seeds, read the accompanying label or packet before planting to make sure your plants will thrive in a shadier environment.

Flowering plants that do well in partial and full shade include:

Bee balm   Impatiens
Bellflower   Leopardbane
Bleeding heart   Lily of the valley
Cardinal flower   Meadow rue
Coleus   Pansy
Columbine   Periwinkle
Daylilies   Persian violet

Dichondra   Primrose
Ferns   Rue anemone
Forget-me-not   Snapdragon
Globe daisy   Sweet alyssum
Golden bleeding heart Thyme

Vegetable plants that can grow in partial shade include:

Arugula   Kale
Beans   Leaf lettuce
Beets   Peas
Broccoli   Radish
Brussels sprouts   Spinach
Cauliflower   Swiss chard
Endive

When planting gardens against your home, choose shade-loving plants.

Bleeding heart plants thrive in shady areas.

THE JUNIOR HOMESTEADER

Kids as young as toddlers will enjoy being involved in a family garden. Encourage very young children to explore by touching and smelling dirt, leaves, and flowers. Just be careful they don't taste anything non-edible. If space allows, assign a small plot for older children to plant and tend all on their own. An added bonus is that kids are more likely to eat vegetables they've grown themselves!

Improving Your Soil

Nutrient-rich, fertile soil is essential for growing the best and healthiest plants—plants that will supply you with quality fruits, vegetables, and flowers. Sometimes soil loses its fertility (or has minimum fertility based on the region in which you live), so measures must be taken in order to improve your soil and, subsequently, your garden.

Soil Quality Indicators

Soil quality is an assessment of how well soil performs all of its functions now and how those functions are being preserved for future use. The quality of soil cannot just be determined by measuring row or garden yield, water quality, or any other single outcome, nor can it be measured directly. Thus, it is important to look at specific indicators to better understand the properties of soil. Plants can provide us with clues about how well the soil is functioning—whether a plant is growing and producing quality fruits and vegetables, or failing to yield such things, is a good indicator of the quality of the soil it's growing in.

In short, indicators are measurable properties of soil or plants that provide clues about how well the soil can function. Indicators can be physical, chemical, and biological properties, processes, or characteristics of soils. They can also be visual features of plants.

Useful indicators of soil quality:
- are easy to measure
- measure changes in soil functions
- encompass chemical, biological, and physical properties
- are accessible to many users
- are sensitive to variations in climate and management

Indicators can be assessed by qualitative or quantitative techniques, such as soil tests. After measurements are collected, they can be evaluated by looking for patterns and comparing results to measurements taken at a different time.

Examples of soil quality indicators:
1. Soil Organic Matter—promotes soil fertility, structure, stability, and nutrient retention and helps combat soil erosion.
2. Physical Indicators—these include soil structure, depth, infiltration and bulk density, and water hold capacity. Quality soil will retain and transport

water and nutrients effectively; it will provide habitat for microbes; it will promote compaction and water movement; and, it will be porous and easy to work with.

3. Chemical Indicators—these include pH, electrical conductivity, and extractable nutrients. Quality soil will be at its threshold for plant, microbial, biological, and chemical activity; it will also have plant nutrients that are readily available.

4. Biological Indicators—these include microbial biomass, mineralizable nitrogen, and soil respiration. Quality soil is a good repository for nitrogen and other basic nutrients for prosperous plant growth; it has a high soil productivity and nitrogen supply; and there is a good amount of microbial activity.

Soil and Plant Nutrients

Nutrient Management

There are twenty nutrients that all plants require. Six of the most important nutrients, called macronutrients, are: calcium, magnesium, nitrogen, phosphorous, potassium, and sulfur. Of these, nitrogen, phosphorus, and potassium are essential to healthy plant growth and are required in relatively large amounts. Nitrogen is associated with lush vegetative growth, phosphorus is required for flowering and fruiting, and potassium is necessary for durability and disease resistance. Calcium, sulfur, and magnesium are also required in comparatively large quantities and aid in the overall health of plants.

The other nutrients, referred to as micronutrients, are required in very small amounts. These include such elements as copper, zinc, iron, and boron. While both macro- and micronutrients are required for good plant growth, over-application of these nutrients can be as detrimental to the plant as a deficiency of them. Over-application of plant nutrients may not only impair plant growth, but may also contaminate groundwater by penetrating through the soil or may pollute surface waters.

Soil Testing

Testing your soil for nutrients and pH is important in order to provide your plants with the proper balance of nutrients (while avoiding over-application). If you are establishing a new lawn or garden, a soil test is strongly recommended. The cost of soil testing is minor in comparison to the cost of plant materials and labor. Correcting a problem before planting is much simpler and cheaper than afterwards.

Once your garden is established, continue to take periodic soil samples. While many people routinely lime their soil, this can raise the pH of the soil too high. Likewise, since many fertilizers tend to lower the soil's pH, it may drop below desirable levels after several years, depending on fertilization and other soil factors, so occasional testing is strongly encouraged.

Home tests for pH, nitrogen, phosphorus, and potassium are available from most garden centers. While these may give you a general idea of the nutrients in your soil, they are not as reliable as tests performed by the Cooperative Extension Service at land grant universities. University and other commercial testing services will provide more detail, and you can request special tests for micronutrients if you suspect a problem. In addition to the analysis of nutrients in your soil, these services often provide recommendations for the application of nutrients or how best to adjust the pH of your soil.

The test for soil pH is very simple. pH is a measure of how acidic or alkaline your soil is. A pH of 7 is considered neutral. Below 7 is acidic and above 7 is alkaline. Since pH greatly influences plant nutrients, adjusting the pH will often correct a nutrient problem. At a high pH, several of the micronutrients become less available for plant uptake. Iron deficiency is a common problem, even at a neutral pH, for such plants as rhododendrons and blueberries. At a very low soil pH, other micronutrients may be too available to the plant, resulting in toxicity.

Phosphorus and potassium are tested regularly by commercial testing labs. While there are soil tests for nitrogen, these may be less reliable. Nitrogen is present in the soil in several forms that can change rapidly. Therefore, a precise analysis of nitrogen is more difficult to obtain. Most university soil test labs do not routinely test for nitrogen. Home testing kits often contain a test for nitrogen that may give you a general, though not necessarily completely accurate, idea of the presence of nitrogen in your garden soil.

Organic matter is often part of a soil test. Organic matter has a large influence on soil structure and so is highly desirable for your garden soil. Good soil structure improves aeration, water movement, and retention. This encourages increased microbial activity and root growth, both of which influence the availability of nutrients for plant growth. Soils high in organic matter tend to have a greater supply of plant nutrients compared to many soils low in organic matter. Organic matter tends to bind up some soil pesticides, reducing their effectiveness, and so this should be taken into consideration if you are planning to apply pesticides in your garden.

Tests for micronutrients are usually not performed unless there is reason to suspect a problem. Certain plants have greater requirements for specific micronutrients and may show deficiency symptoms if those nutrients are not readily available.

STEPS FOR TAKING A SOIL TEST

1. If you intend to send your sample to the land grant university in your state, contact the local Cooperative Extension Service for information and sample bags. If you intend to send your sample to a private testing lab, contact them for specific details about submitting a sample.

2. Follow the directions carefully for submitting the sample. The following are general guidelines for taking a soil sample:
   - Sample when the soil is moist but not wet.
   - Obtain a clean pail or similar container.
   - Clear away the surface litter or grass.
   - With a spade or soil auger, dig a small amount of soil to a depth of 6 inches.
   - Place the soil in the clean pail.
   - Repeat steps 3 through 5 until the required number of samples has been collected.
   - Mix the samples together thoroughly.
   - From the mixture, take the sample that will be sent for analysis.
   - Send immediately. Do not dry before sending.

3. If you are using a home soil testing kit, follow the above steps for taking your sample. Follow the directions in the test kit carefully so you receive the most accurate reading possible.

Enriching Your Soil

Organic and Commercial Fertilizers and Returning Nutrients to Your Soil

Once you have the results of the soil test, you can add nutrients or soil amendments as needed to alter the pH. If you need to raise the soil's pH, use lime. Lime is most effective when it is mixed into the soil; therefore, it is best to apply before planting (if you apply lime in the fall, it has a better chance of correcting any soil acidity problems for the next growing season). For large areas, rototilling is most effective. For small areas or around plants, working the lime into the soil with a spade or cultivator is preferable. When working around plants, be careful not to dig too deeply or roughly so that you damage plant roots. Depending on the form of lime and the soil conditions, the change in pH may be gradual. It may take several months before a significant change is noted. Soils high in organic matter and clay tend to take larger amounts of lime to change the pH than do sandy soils.

If you need to lower the pH significantly, especially for plants such as rhododendrons, you can use aluminum sulfate. In all cases, follow the soil test or manufacturer's recommended rates of application. Again, mixing well into the soil is recommended.

There are numerous choices for providing nitrogen, phosphorus, and potassium, the nutrients your plants need to thrive. Nitrogen (N) is needed for healthy, green growth and regulation of other nutrients. Phosphorus (P) helps roots and seeds properly develop and resist disease. Potassium (K) is also important in root development and disease resistance. If your soil is of adequate fertility, applying compost may be the best method of introducing additional nutrients. While compost is relatively low in nutrients compared to commercial fertilizers, it is especially beneficial in improving the condition of the soil

and is nontoxic. By keeping the soil loose, compost allows plant roots to grow well throughout the soil, helping them to extract nutrients from a larger area. A loose soil enriched with compost is also an excellent habitat for earthworms and other beneficial soil microorganisms that are essential for releasing nutrients for plant use. The nutrients from compost are also released slowly, so there is no concern about "burning" the plant with an over-application of synthetic fertilizer.

Manure is also an excellent source of plant nutrients and is an organic matter. Manure should be composted before applying, as fresh manure may be too strong and can injure plants. Be careful when composting manure. If left in the open, exposed to rain, nutrients may leach out of the manure and the runoff can contaminate nearby waterways. Make sure the manure is stored in a location away from wells and any waterways and that any runoff is confined or slowly released into a vegetated area. Improperly applied manure also can be a source of pollution. If you are not composting your own manure, you can purchase some at your local garden store.

For best results, work composted manure into the soil around the plants or in your garden before planting.

If preparing a bed before planting, compost and manure may be worked into the soil to a depth of 8 to 12 inches. If adding to existing plants, work carefully around the plants so as not to harm the existing roots.

Green manures are another source of organic matter and plant nutrients. Green manures are crops that are grown and then tilled into the soil. As they break down, nitrogen and other plant nutrients become available. These manures may also provide additional benefits of reducing soil erosion. Green manures, such as rye and oats, are often planted in the fall after the crops have been harvested. In the spring, these are tilled under before planting.

With all organic sources of nitrogen, whether compost or manure, the nitrogen must be changed to an inorganic form before the plants can use it. Therefore, it is important to have well-drained, aerated soils that provide the favorable habitat for the soil microorganisms responsible for these conversions.

There are also numerous sources of commercial fertilizers that supply nitrogen, phosphorus, and potassium, though it is preferable to use organic fertilizers, such as compost and manures. However, if you choose to use a commercial fertilizer, it is important to know how to read the amount of nutrients contained in each bag. The first number on the fertilizer analysis is the percentage of nitrogen; the second number is phosphorus; and the third number is the potassium content. A fertilizer that has a 10-20-10 analysis contains twice as much of each of the nutrients as a 5-10-5. How much of each nutrient you need depends on your soil test results and the plants you are fertilizing.

As mentioned before, nitrogen stimulates vegetative growth while phosphorus stimulates flowering. Too much nitrogen can inhibit flowering and fruit production. For many flowers and vegetables, a fertilizer higher in phosphorus than nitrogen is preferred, such as a 5-10-5. For lawns, nitrogen is usually required in greater amounts, so a fertilizer with a greater amount of nitrogen is more beneficial.

Fertilizer Application

Commercial fertilizers are normally applied as a dry, granular material or mixed with water and poured onto the garden. If using granular materials, avoid spilling on sidewalks and driveways because these materials are water soluble and can cause pollution problems if rinsed into storm sewers. Granular fertilizers are a type of salt, and if applied too heavily, they have the capability of burning the plants. If using a liquid fertilizer, apply directly to or around the base of each plant and try to contain it within the garden only.

In order to decrease the potential for pollution and to gain the greatest benefits from fertilizer, whether it's a commercial variety, compost, or other organic materials, apply it when the plants have the greatest need for the nutrients. Plants that are not actively growing do not have a high requirement for nutrients; thus, nutrients applied to dormant plants, or plants growing slowly due to cool temperatures, are more likely to be wasted. While light applications of nitrogen may be recommended for lawns in the fall, generally, nitrogen fertilizers should not be applied to most plants in the fall in regions of the country that experience cold winters. Since nitrogen encourages vegetative growth, if it is applied in the fall it may reduce the plant's ability to harden properly for winter.

In some gardens, you can reduce fertilizer use by applying it around the individual plants rather than broadcasting it across the entire garden. Much of the phosphorus in fertilizer becomes unavailable to the plants once spread on the soil. For better plant uptake, apply the fertilizer in a band near the plant. Do not apply directly to the plant or in contact with the roots, as it may burn and damage the plant and its root system.

Rules of Thumb for Proper Fertilizer Use

It is easiest to apply fertilizer before or at the time of planting. Fertilizers can either be spread over a large area or confined to garden rows, depending on the condition of your soil and the types of plants you will be growing. After spreading, till the fertilizer into the soil about 3 to 4 inches deep. Only spread about one half of the fertilizer this way and then dispatch the rest 3 inches to the sides of each row and also a little below each seed or established plant. This method, minus the spreader, is used when applying fertilizer to specific rows or plants by hand.

Composting in Your Backyard

Composting is nature's own way of recycling yard and household wastes by converting them into valuable fertilizer, soil organic matter, and a source of plant nutrients. The result of this controlled decomposition of organic matter—a dark, crumbly, earthy-smelling material—works wonders on all kinds of soil by providing vital nutrients,

Soil Test Reading	What to Do
High pH	Your soil is alkaline. To lower pH, add elemental sulfur, gypsum, or cottonseed meal. Sulfur can take several months to lower your soil's pH, as it must first convert to sulfuric acid with the help of the soil's bacteria.
Low pH	Your soil is too acidic. Add lime or wood ashes.
Low nitrogen	Add manure, horn or hoof meal, cottonseed meal, fish meal, or dried blood.
High nitrogen	Your soil may be over-fertilized. Water the soil frequently and don't add any fertilizer.
Low phosphorus	Add cottonseed meal, bonemeal, fish meal, rock phosphate, dried blood, or wood ashes.
High phosphorous	Your soil may be over-fertilized. Avoid adding phosphorous-rich materials and grow lots of plants to use up the excess.
Low potassium	Add potash, wood ashes, manure, dried seaweed, fish meal, or cottonseed meal.
High potassium	Continue to fertilize with nitrogen and phosphorous-rich soil additions, but avoid potassium-rich fertilizers for at least two years.
Poor drainage or too much drainage	If your soil is a heavy, clay-like consistency, it won't drain well. If it's too sandy, it won't absorb nutrients as it should. Mix in peat moss or compost to achieve a better texture.

HOW TO PROPERLY APPLY FERTILIZER TO YOUR GARDEN

Apply fertilizer when the soil is moist, and then water lightly. This will help the fertilizer move into the root zone where its nutrients are available to the plants, rather than staying on top of the soil where it can be blown or washed away.

Watch the weather. Avoid applying fertilizer immediately before a heavy rain system is predicted to arrive. Too much rain (or sprinkler water) will take the nutrients away from the lawn's root zone and could move the fertilizer into another water system, contaminating it.

Use the minimum amount of fertilizer necessary and apply it in small, frequent applications. An application of two pounds of fertilizer, five times per year, is better than five pounds of fertilizer twice a year.

If you are spreading the fertilizer by hand in your garden, wear gardening gloves and be sure not to damage the plant or roots around which you are fertilizing.

and contributing to good aeration and moisture-holding capacity, to help plants grow and look better.

Composting can be as simple or as involved as you would like, depending on how much yard waste you have, how fast you want results, and the effort you are willing to invest. Since all organic matter eventually decomposes, composting speeds up the process by providing an ideal environment for bacteria and other decomposing micro-organisms. The composting season coincides with the growing season, when conditions are favorable for plant growth, so those same conditions work well for biological activity in the compost pile. However, since compost generates heat, the process may continue later into the fall or winter. The final product—called humus or compost—looks and feels like fertile garden soil.

Compost Preparation

While a multitude of organisms, fungi, and bacteria are involved in the overall process, there are four basic ingredients for composting: nitrogen, carbon, water, and air.

A wide range of materials may be composted because anything that was once alive will naturally decompose. The starting materials for composting, commonly referred to as feed stocks, include leaves, grass clippings, straw, vegetable and fruit scraps, coffee grounds, livestock manure, sawdust, and shredded paper. However, some materials that always should be avoided include diseased plants, dead animals, noxious weeds, meat scraps that may attract animals, and dog or cat manure, which can carry disease. Since adding kitchen wastes to compost may attract flies and insects, make a hole in the center of your pile and bury the waste.

THE JUNIOR HOMESTEADER

Compost Lasagna

Watch your produce scraps decompose! And see how some materials don't.

Things You'll Need:

1 2-liter clear plastic bottle
2 cups fruit and vegetable scraps
1 cup grass clippings and leaves
2 cups soil
Newspaper clippings or shredded paper
Styrofoam packing peanuts
Magic marker

Layer all your ingredients, just like you'd make a lasagna. Start with a couple inches of soil, then add the produce scraps, then more dirt, then the grass clippings and leaves, more dirt, the Styrofoam, more dirt, the shredded paper, and top it all off with a little more dirt.

Use the magic marker to mark the top of the top layer. Then place the bottle upright in a windowsill or another sunny spot. If there's a lot of condensation in the bottle, open the top to let it air out.

Once a week for four weeks check on the bottle and notice how the level of the dirt has changed. Mark it with the marker.

At the end of four weeks, dump the bottle out in a garden spot that hasn't been planted, or add it to your compost pile. Notice which items decomposed the most. Remove the items that didn't decompose and discard them in the trash.

For best results, you will want an even ratio of green, or wet, material, which is high in nitrogen, and brown, or dry, material, which is high in carbon. Simply layer or mix landscape trimmings and grass clippings, for example, with dried leaves and twigs in a pile or enclosure. If there is not a good supply of nitrogen-rich material, a handful of general lawn fertilizer or barnyard manure will help even out the ratio.

Though rain provides the moisture, you may need to water the pile in dry weather or cover it in extremely wet weather. The microorganisms in the compost pile function best when the materials are as damp as a wrung-out sponge—not saturated with water. A moisture content of 40 to 60 percent is preferable. To test for adequate moisture, reach into your compost pile, grab a handful of material, and squeeze it. If a few drops of water come out, it probably has enough moisture. If it doesn't, add water by putting a hose into the pile so that you aren't just wetting the top, or, better yet, water the pile as you turn it.

Air is the only part that cannot be added in excess. For proper aeration, you'll need to punch holes in the pile so it has many air passages. The air in the pile is usually used up faster than the moisture, and extremes of sun or rain can adversely affect this balance, so the materials must be turned or mixed up often with a pitchfork, rake, or other garden tool to add air that will sustain high temperatures, control odor, and yield faster decomposition.

Over time, you'll see that the microorganisms, which are small forms of plant and animal life, will break down the organic material. Bacteria are the first to break down plant tissue and are the most numerous and effective compost-makers in your compost pile. Fungi and protozoans soon join the bacteria and, somewhat later in the cycle, centipedes, millipedes, beetles, sow bugs, nematodes, worms, and numerous others complete the composting process. With the right ingredients and favorable weather conditions, you can have a finished compost pile in a few weeks.

How to Make Your Own Backyard Composting Heap

1. Choose a level, well-drained site, preferably near your garden.
2. Decide whether you will be using a bin after checking on any local or state regulations for composting in urban areas, as some communities require rodent-proof bins. There are numerous styles of compost bins available, depending on your needs, ranging from a moveable bin formed by wire mesh to a more substantial wooden structure consisting of several compartments. You can easily make your own bin using chicken wire or scrap wood. While a bin will help contain the pile, it is not absolutely necessary, as you can build your pile directly on the ground. To help with aeration, you may want to place some woody material on the ground where you will build your pile.
3. Ensure that your pile will have a minimum dimension of 3 feet all around, but is no taller than 5 feet, as not enough air will reach the microorganisms at the center if it is too tall. If you don't have this amount at one time, simply stockpile your materials until a sufficient quantity is available for proper mixing. When composting is completed, the total volume of the original materials is usually reduced by 30 to 50 percent.
4. Build your pile by using either alternating equal layers of high-carbon and high-nitrogen material or by mixing equal parts of both together and then heaping it into a pile. If you choose to alternate layers, make each layer 2 to 4 inches thick. Some composters find that mixing the two together is more effective than layering. Adding a few shovels of soil will also help get the pile off to a good start because soil adds commonly found, decomposing organisms to your compost.
5. Keep the pile moist but not wet. Soggy piles encourage the growth of organisms that can live without oxygen and cause unpleasant odors.
6. Punch holes in the sides of the pile for aeration. The pile will heat up and then begin to cool. The most efficient decomposing bacteria thrive in temperatures between 110 and 160 degrees Fahrenheit. You can track this with a compost thermometer, or you can simply reach into the pile to determine if it is uncomfortably hot to the touch. At these temperatures, the pile kills most weed seeds and plant diseases. However, studies have shown that compost produced at these temperatures has less ability to suppress diseases in the soil, since these temperatures may kill some of the beneficial bacteria necessary to suppress disease.
8. Check your bin regularly during the composting season to assure optimum moisture and aeration are present in the material being composted.
9. Move materials from the center to the outside of the pile and vice versa. Turn every day or two and you should get compost in less than four weeks. Turning every other week will make compost in one to three months. Finished compost will smell sweet and be cool and crumbly to the touch.

COMMON COMPOSTING MATERIALS

Cardboard
Coffee grounds
Corn cobs
Corn stalks
Food scraps
Grass clippings
Hedge trimmings
Livestock manure
Newspapers
Old potting soil
Plant stalks
Pine needles
Sawdust
Seaweed
Shredded paper
Straw
Tea bags
Telephone books
Tree leaves and twigs
Vegetable scraps
Weeds without seed heads
Wood chips
Woody brush

Avoid using:

Bread and grains
Cooking oil
Dairy products
Dead animals
Diseased plant material
Dog or cat manure
Grease or oily foods
Meat or fish scraps
Noxious or invasive weeds
Weeds with seed heads

Other Types of Composting

Cold or Slow Composting

Cold composting allows you to pile just organic material on the ground or in a bin. This method requires no maintenance, but it will take several months to a year or more for the pile to decompose, though the process is faster in warmer climates than in cooler areas. Cold composting works well if you are short on time needed to tend to the compost pile at least every other day, have little yard waste, and are not in a hurry to use the compost.

For this method, add yard waste as it accumulates. To speed up the process, shred or chop the materials by running over small piles of trimmings with your lawn mower, because the more surface area the microorganisms have to feed on, the faster the materials will break down.

Cold composting has been shown to be better at suppressing soil-borne diseases than hot composting and also leaves more non-decomposed bits of material, which can be screened out if desired. However, because of the low temperatures achieved during decomposition, weed seeds and disease-causing organisms may not be destroyed.

Vermicomposting

Vermicomposting uses worms to compost. This takes up very little space and can be done year-round in a basement or garage. It is an excellent way to dispose of kitchen wastes.

Here's how to make your own vermicomposting pile:

1. Obtain a plastic storage bin. One bin measuring 1 foot by 2 feet by 3½ feet will be enough to meet the needs of a family of six.
2. Drill eight to ten holes about ¼ inch in diameter in the bottom of the bin for drainage.
3. Line the bottom of the bin with a fine nylon mesh to keep the worms from escaping.
4. Put a tray underneath to catch the drainage.
5. Rip shredded newspaper into pieces to use as bedding and pour water over the strips until they are thoroughly moist. Place these shredded bits on one side of your bin. Do not let them dry out.
6. Add worms to your bin. It's best to have about two pounds of worms (roughly 2,000 worms) per one pound of food waste. You may want to start with less food waste and increase the amount as your worm population grows. Redworms are recommended for best composting, but other species can be used. Redworms are the common, small worms found in most gardens and lawns. You can collect them from under a pile of mulch or order them from a garden catalog.
7. Provide worms with food wastes such as vegetable peelings. Do not add fat or meat products. Limit their feed, as too much at once may cause the material to rot.

« Any large bucket can be turned into a compost barrel. You can cut out a piece of the barrel for easy access to the compost, as shown here, or simply access the compost through the lid. Drilling holes in the sides and lid of the bucket will increase air circulation and speed up the process. Leave your bucket in the sun and shake it, roll it, or stir the contents regularly.

8. Keep the bin in a dark location away from extreme temperatures.
9. Wait about three months and you'll see that the worms have changed the bedding and food wastes into compost. At this time, open your bin in a bright light and the worms will burrow into the bedding. Add fresh bedding and more food to the other side of the bin. The worms should migrate to the new food supply.
10. Scoop out the finished compost and apply to your plants or save to use in the spring.

Common Problems

Composting is not an exact science. Experience will tell you what works best for you. If you notice that nothing is happening, you may need to add more nitrogen, water, or air; chip or grind the materials; or adjust the size of the pile.

If the pile is too hot, you probably have too much nitrogen and need to add additional carbon materials to reduce the heating.

A bad smell may indicate not enough air or too much moisture. Simply turn the pile or add dry materials to the wet pile to get rid of the odor.

Uses for Compost

Compost contains nutrients, but it is not a substitute for fertilizers. Compost holds nutrients in the soil until plants can use them, loosens and aerates clay soils, and retains water in sandy soils.

To use as a soil amendment, mix 2 to 5 inches of compost into vegetable and flower gardens each year before planting. In a potting mixture, add one part compost to two parts commercial potting soil, or make your own mixture by using equal parts of compost and sand or Perlite.

As a mulch, spread an inch or two of compost around annual flowers and vegetables, and up to 6 inches around trees and shrubs. Studies have shown that compost used as mulch, or mixed with the top 1-inch layer of soil, can help prevent some plant diseases, including some of those that cause damping of seedlings.

As a top dressing, mix finely sifted compost with sand and sprinkle evenly over lawns.

Planting Your Garden

Once you've chosen a spot for your garden (as well as the size you want to make your garden bed), and prepared the soil with compost or other fertilizer, it's time to start planting. Find seeds at your local garden center, browse through seed catalogs, and order seeds that will do well in your area. Alternatively, you can start with bedding plants (or seedlings) available at nurseries and garden centers.

Read the instructions on the back of the seed package or on the plastic tag in your plant pot. You may have to ask experts when to plant the seeds if this information is not stated on the back of the package. Some seeds (such as tomatoes) should be started indoors in small pots or seed trays before the last frost, and only transplanted outdoors when the weather warms up. For established plants or seedlings, be sure to plant as directed on the plant tag or consult your local nursery about the best planting times.

Seedlings

If you live in a cooler region with a shorter growing period, you will want to start some of your plants indoors. To do this, obtain plug flats (trays separated into many small cups or "cells") or make your own small planters by poking holes in the bottoms of paper cups. Fill the cups two-thirds full with potting soil or composted soil. Bury the seed at the recommended depth, according to the instructions on the package. Tamp down the soil lightly and water. Keep the seedlings in a warm, well-lit place, such as the kitchen, to encourage germination.

Once the weather begins to warm up and you are fairly certain you won't be getting any more frosts (you can contact your local extension office to find out the "frost free" date for your area) you can begin to acclimate your seedlings to the great outdoors. First, place them in a partially shady spot outdoors that is protected from strong wind. After a couple of days, move them into direct sunlight, and finally, transplant them to the garden.

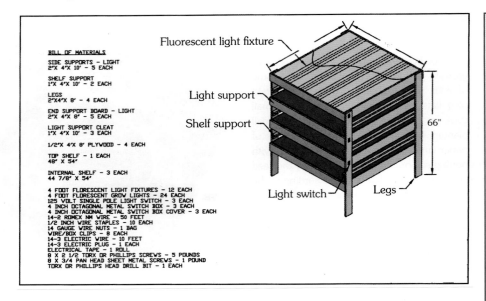

BILL OF MATERIALS
SIDE SUPPORTS - LIGHT
2"X 4"X 10' - 5 EACH

SHELF SUPPORT
1"X 4"X 10' - 2 EACH

LEGS
2"X4"X 8' - 4 EACH

END SUPPORT BOARD - LIGHT
2"X 4"X 8' - 5 EACH

LIGHT SUPPORT CLEAT
1"X 4"X 10' - 3 EACH

1/2"X 4"X 8' PLYWOOD - 4 EACH

TOP SHELF - 1 EACH
48" X 54"

INTERNAL SHELF - 3 EACH
44 7/8" X 54"

4 FOOT FLORESCENT LIGHT FIXTURES - 12 EACH
4 FOOT FLORESCENT GROW LIGHTS - 24 EACH
125 VOLT SINGLE POLE LIGHT SWITCH - 3 EACH
4 INCH OCTAGONAL METAL SWITCH BOX - 3 EACH
4 INCH OCTAGONAL METAL SWITCH BOX COVER - 3 EACH
14-2 ROMEX NM WIRE - 50 FEET
1/2 INCH WIRE STAPLES - 10 EACH
14 GAUGE WIRE NUTS - 1 BAG
WIRE/BOX CLIPS - 8 EACH
14-3 ELECTRIC WIRE - 10 FEET
14-3 ELECTRIC PLUG - 1 EACH
ELECTRICAL TAPE - 1 ROLL
8 X 2 1/2 TORX OR PHILLIPS SCREWS - 5 POUNDS
8 X 3/4 PAN HEAD SHEET METAL SCREWS - 1 POUND
TORX OR PHILLIPS HEAD DRILL BIT - 1 EACH

Fluorescent light fixture

Light support

Shelf support

66"

Light switch

Legs

≈ Follow these plans to make your own propagation rack for starting seeds indoors. Though a propagation rack is not necessary, it will help to ensure your seedlings receive the light and warmth they need to stay strong and healthy.

SPROUTING SEEDS FOR EATING

Seeds can be sprouted and eaten on sandwiches, salads, or stir-fries any time of the year. They are delicious and full of vitamins and proteins. Mung beans, soybeans, alfalfa, wheat, corn, barley, mustard, clover, chickpeas, radishes, and lentils all make good sprouts. Find seeds for sprouting from your local health food store or use dried peas, beans, or lentils from the grocery store. Never use seeds intended for planting unless you've harvested the seeds yourself—commercially available planting seeds are often treated with a poisonous chemical fungicide.

To grow sprouts, thoroughly rinse and strain the seeds, then place in a glass jar, cover with cheesecloth secured with a rubber band, and soak overnight in cool water. You'll need about four times as much water as you have seeds. Drain the seeds by turning the jar upside down and allowing the water to escape through the cheesecloth. Keep the seeds at 60 to 80°F and rinse twice a day, draining them thoroughly after every rinse. Once sprouts are 1 to 1 ½ inches long (generally after three to five days), they are ready to eat.

Germination Temperatures of Selected Vegetable Plants			
Broccoli 77°F	Eggplant 85°F	Onion 70°F	Summer Squash 80°F
Cabbage 86°F	Herbs 65°F	Pepper 85°F	Tomato 85°F
Cucumber 86°F	Melon 90°F	Pumpkin 85°F	Winter Squash 80°F

HOMESCHOOL HINT

How do different treatments change how fast seeds sprout?

Experiment with sprouting seeds under different temperatures, or after being soaked for different times or in different liquids. Or, see how one kind of treatment affects different types of seeds. What makes seeds sprout the fastest? Do some look healthier than others? Write up a lab report to document your findings.

How Best to Water Your Soil

After your seeds or seedlings are planted, the next step is to water your soil. Different soil types have different watering needs. You don't need to be a soil scientist to know how to water your soil properly. Here are some tips that can help to make your soil moist and primed for gardening:

1. Loosen the soil around plants so water and nutrients can be quickly absorbed.
2. Use a 1- to 2-inch protective layer of mulch on the soil surface above the root area. Cultivating and mulching help reduce evaporation and soil erosion.
3. Water your plants at the appropriate time of day. Early morning or night is the best time for watering, as evaporation is less likely to occur at these times.
4. Do not water your plants when it is extremely windy outside. Wind will prevent the water from reaching the soil where you want it to go.

Types of Soil and Their Water Retention

Knowing the type of soil you are planting in will help you best understand how to properly water and grow your garden plants. Three common types of soil

RECOMMENDED PLANTS TO START AS SEEDLINGS

CROP [s] small seed [l] large seed (planting cell size)	WEEKS BEFORE TRANS-PLANTING	SEED PLANTING DEPTH (Inches)	TRANS-PLANT SPACING	WITHIN ROW/ BETWEEN ROW
Broccoli [s]	(1)	4–6	¼–½	8–10"/18–24"
Cabbage [s]	(1)	4–6	¼–½	18–24"/30"
Cucumber [l]	(2)	4–5	½	2'/5–6'
Eggplant [s]	(2)	8	¼	18"/18–24"
Herbs [s]	(1)	4	¼	4–6"/12–18"
Lettuce [s]	(2)	4–5	¼	12"/12"
Melon [l]	(3)	4–5	¼	2–3'/6'
Onion [s]	(8–12)	8	¼	4"/12"
Pepper [s]	(2)	8	¼	12–18"/2–3'
Pumpkin [l]	(3)	2–4	1	5–6'/5–6'
Summer Squash [l]	(3)	2–4	¾–1	18"/2–3'
Tomato [s]	(3)	8	¼	18–24"/3'
Watermelon [l]	(3)	4–5	½–¾	3–4'/3–4'
Winter Squash [l]	(3)	2–4	1	3–4'/4–5'

and their various abilities to absorb water are listed below:

1. Clay Soil

In order to make this type of soil more loamy, add organic materials, such as compost, peat moss, and well-rotted leaves, in the spring before growing and also in the fall after harvesting your vegetables and fruits. Adding these organic materials allows this type of soil to hold more nutrients for healthy plant growth. Till or spade to help loosen the soil.

Since clay soil absorbs water very slowly, water only as fast as the soil can absorb the water.

2. Sandy Soil

As with clay soil, adding organic materials in the spring and fall will help supplement the sandy soil and promote better plant growth and water absorption.

Left on its own (with no added organic matter), the water will run through sandy soil so quickly that plants won't be able to absorb it through their roots and will fail to grow and thrive.

3. Loam Soil

This is the best kind of soil for gardening. It's a combination of sand, silt, and clay. Loam soil is fertile, deep, easily crumbles, and contains organic matter. It will help promote the growth of quality fruits and vegetables, as well as flowers and other plants.

Loam absorbs water readily and stores it for plants to use. Water as frequently as the soil needs to maintain its moisture and to promote plant growth.

Conserving Water

Wise use of water for hydrating your garden and lawn not only helps protect the environment, but saves money and also provides optimum growing conditions for your plants. There are simple ways of reducing the amount of water used for irrigation, such as growing xeriphytic species (plants that are adapted to dry conditions), mulching, adding water-retaining organic matter to the soil, and installing windbreaks and fences to slow winds and reduce evapotranspiration.

You can conserve water by watering your plants and lawn in the early morning, before the sun is too intense. This helps reduce the amount of water lost due to evaporation. Furthermore, installing rain gutters and collecting water from downspouts—in collection bins such as rain barrels—also helps reduce water use.

a process called transpiration. Transpiration, along with evaporation from the soil's surface, accounts for most of the moisture lost from the soil and subsequently from the plants.

When there is a lack of water in the plant tissue, the stomata close to try to limit excessive water loss. If the tissues lose too much water, the plant will wilt. Plants adapted to dry conditions have developed numerous mechanisms for reducing water loss—they typically have narrow, hairy leaves and thick, fleshy stems and leaves. Pines, hemlocks, and junipers are also well adapted to survive extended periods of dry conditions—an environmental factor they encounter each winter when the frozen soil prevents the uptake of water. Cacti, which have thick stems and leaves reduced to spines, are the best

example of plants well adapted to extremely dry environments.

Choosing Plants for Low Water Use

You are not limited to cacti, succulents, or narrow-leafed evergreens when selecting plants adapted to low water requirements. Many plants growing in humid environments are well adapted to low levels of soil moisture. Numerous plants found growing in coastal or mountainous regions have developed mechanisms for dealing with extremely sandy, excessively well-drained soils or rocky, cold soils in which moisture is limited for months at a time. Try alfalfa, aloe,

How Plants Use Water

Water is a critical component of photosynthesis, the process by which plants manufacture their own food from carbon dioxide and water in the presence of light. Water is one of the many factors that can limit plant growth. Other important factors include nutrients, temperature, and amount and duration of sunlight.

Plants take in carbon dioxide through their stomata—microscopic openings on the undersides of the leaves. The stomata are also the place where water is lost, in

---

THE JUNIOR HOMESTEADER

See what happens when a plant (or part of a plant) doesn't get any light!

1. Cut three paper shapes about 2 inches by 2 inches. Circles and triangles work well, but you can experiment with other shapes, too. Clip them to the leaves of a plant, preferably one with large leaves. Either an indoor or an outdoor plant will do. Be very careful not to damage the plant.
2. Leave one paper cutout on for one day, a second on for two days, and a third on for a week. How long does it take for the plant to react? How long does it take for the plant to return to normal?

Photosynthesis means to "put together using light." Plants use sunlight to turn carbon dioxide and water into food. Plants need all of these to remain healthy. When the plant gets enough of these things, it produces a simple sugar, which it uses immediately or stores in a converted form of starch. We don't know exactly how this happens. But we do know that chlorophyll, the green substance in plants, helps it to occur.

⌄ Succulent plants retain water and can therefore thrive in arid environments.

⌄ Garlic plants

THE JUNIOR HOMESTEADER

Did you ever wonder how water gets from a plant's roots to its leaves? The name for this is "capillary action."

Things You'll Need:

4 same-size stalks of fresh celery with leaves
4 cups or glasses
Red and blue food coloring
A measuring cup
4 paper towels
A vegetable peeler
A ruler
Some old newspapers

What to Do:

1. Lay the four pieces of celery in a row on a cutting board or counter so that the place where the stalks and the leaves meet matches up.
2. Cut all four stalks of celery four inches (about 10 centimeters) below where the stalks and leaves meet.
3. Put the four stalks in four separate cups of purple water (use 10 drops of red and 10 drops of blue food coloring for each half cup of water).
4. Label four paper towels in the following way: "2 hours," "4 hours," "6 hours," and "8 hours." (You may need newspapers under the towels.)

5. Every two hours from the time you put the celery into the cups, remove one of the stalks and put it onto the correct towel. (Notice how long it takes for the leaves to start to change.)
6. Each time you remove a stalk from the water, carefully peel the rounded part with a vegetable peeler to see how far up the stalk the purple water has traveled.
7. What do you observe? Notice how fast the water climbs the celery. Does this change as time goes by? In what way?
8. Measure the distance the water has traveled and record this amount.
9. Make a list of other objects around your house or in nature that enable liquids to climb by capillary action. Look for paper towels, sponges, old sweat socks, brown paper bags, and flowers. What other items can you find?

Capillary action happens when water molecules are more attracted to the surface they travel along than to each other. In paper towels, the molecules move along tiny fibers. In plants, they move through narrow tubes that are actually called capillaries. Plants couldn't survive without capillaries because they use the water to make their food.

---

artichokes, asparagus, blue hibiscus, chives, columbine, eucalyptus, garlic, germander, lamb's ear, lavender, ornamental grasses, prairie turnip, rosemary, sage, sedum, shrub roses, thyme, yarrow, yucca, and verbena.

Trickle Irrigation Systems

Trickle irrigation and drip irrigation systems help reduce water use and successfully meet the needs of most plants. With these systems, very small amounts of water

are supplied to the bases of the plants. Since the water is applied directly to the soil—rather than onto the plant—evaporation from the leaf surfaces is reduced, thus allowing more water to effectively reach the roots. In these types of systems, the water is not wasted by being spread all over the garden; rather, it is applied directly to the appropriate source.

Installing Irrigation Systems

An irrigation system can be easy to install, and there are many different products available for home irrigation systems. The simplest system consists of a soaker hose that is laid out around the plants and connected to an outdoor spigot. No installation is required, and the hose can be moved as needed to water the entire garden.

A slightly more sophisticated system is a slotted pipe system. Here are the steps needed in order to install this type of irrigation system in your garden:

1. Sketch the layout of your garden so you know what materials you will need. If you intend to water a vegetable garden, you may want one pipe next to every row or one pipe between every two rows.
2. Depending on the layout and type of garden, purchase the required lengths

of pipe. You will need a length of solid pipe for the width of your garden, and perforated pipes that are the length of your lateral rows (and remember to buy one pipe for each row or two).
3. Measure the distances between rows and cut the solid pipe to the proper lengths.
4. Place T-connectors between the pieces of solid pipe.
5. In the approximate center of the solid pipe, place a T-connector to which a hose connector will be fitted.
6. Cut the perforated pipe to the length of the rows.
7. Attach the perforated pipes to the T-connectors so that the perforations are facing downward. Cap the ends of the pipes.

⩗ A simple trickle irrigation system.

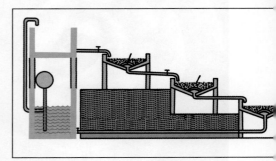

⩗ A slotted pipe irrigation system.

8. Connect a garden hose to the hose connector on the solid pipe. Adjust the pressure of the water flowing from the spigot until the water slowly emerges from each of the perforated pipes. And now you have a slotted pipe irrigation system for your garden.

Rain Barrels

Another very efficient and easy way to conserve water—and save money—is to buy or make your own rain barrel. A rain barrel is a large bin that is placed beneath a downspout and collects rainwater runoff from a roof. The water collected in the rain barrel can then be routed through a garden hose and used to water your garden and lawn.

Rain barrels can be purchased from specialty home and garden stores, but a simple rain barrel is also quite easy to make. Below are simple instructions on how to make your own rain barrel.

How to Make a Simple Rain Barrel

Instructions:

1. Obtain a suitable plastic barrel, a large plastic trash can with a lid, or a wooden barrel (e.g., a wine barrel) that has not been stored dry for too many seasons, since it can start to leak. Good places to find plastic barrels include suppliers of dairy products, metal plating companies, and bulk food suppliers. Just be sure that nothing toxic or harmful to plants and animals (including you!) was stored in the barrel. A wine barrel can be obtained through a winery. Barrels that allow less light to penetrate through will minimize the risk of algae growth and the establishment of other microorganisms.

2. Once you have your barrel, find a location for it under or near one of your home's downspouts. In order for the barrel to fit, you will probably need to shorten the downspout by a few feet. You can do this by removing the screws or rivets located at a joint of the downspout, or by simply cutting off the last few feet with a hacksaw or other cutter. If your barrel will not be able to fit underneath the downspout, you can purchase a flexible downspout at your local home improvement store. These flexible tubes will direct the water from the downspout into the barrel. An alternative, and aesthetically appealing, option is to use a rain chain—a large, metal chain that water can run down.

3. Create a level, stable platform for your rain barrel to sit on by raking the dirt under the spout, adding gravel to smooth out lawn bumps, or using bricks or concrete blocks to make a low platform. Keep in mind that a barrel full of water is very heavy, so if you decide to build a platform, make sure it is sturdy enough to hold such heavy weight.

4. If your barrel has a solid top, you'll need to make a good-sized hole in it for the downspout to pour into. You can do this using a hole-cutting attachment on a power drill or by drilling a series of smaller holes close together and then cutting out the remaining material with a hacksaw blade or a scroll saw.

5. Mosquitoes are drawn to standing water, so to reduce the risk of breeding these insects, and to also keep debris from entering the barrel, fasten a piece of window screen to the underside of the top so it covers the entire hole.

6. Next, drill a hole so the hose bib you'll attach to the side of the barrel fits snugly. Place the hose bib as close to the bottom of the barrel as possible, so you'll be able to gain access to the maximum amount of water in the

barrel. Attach the hose bib using screws driven into the barrel. You'll probably need to apply some caulking, plumber's putty, or silicon sealant around the joint between the barrel and the hose bib to prevent leaks, depending on the type of hardware you're using and how snugly it fits in the hole you drilled.

7. Attach a second hose bib to the side of the barrel near the top, to act as an overflow drain. Attach a short piece of garden hose to this hose bib and route it to a flowerbed, lawn, or another nearby area that won't be damaged by some running water if your barrel gets too full. (Or, if you want to have a second rain barrel for excess water, you can attach it to another hose bib on a second barrel. If you are chaining multiple barrels together, one of them should have a hose attached to drain off the overflow.)

8. Attach a garden hose to the lower hose bib and open the valve to allow collected rain water to flow to your plants. The lower bib can also be used to connect multiple rain barrels together for a larger water reservoir.

9. Consider using a drip irrigation system in conjunction with the rain barrels. Rain barrels don't achieve anything near the pressure of city water supplies, so you won't be able to use microsprinkler attachments, and you will need to use button attachments that are intended to deliver four times the amount of city-supplied water as you need.

10. Now wait for a heavy downpour and start enjoying your rain barrel!

Things to Consider

- Spray some water in the barrel from a garden hose once everything is in place and any sealants have had time to thoroughly dry. The first good downpour is *not* the time to find out there's a leak in your barrel.
- If you don't own the property on which you are thinking of installing a rain barrel, be sure to get permission before altering the downspouts.
- If your barrel doesn't already have a solid top, cover it securely with a circle of painted plywood, an old trash-can lid screwed to the walls of the barrel,

or a heavy tarp secured over the top of the barrel with bungee cords. This will protect children and small animals from falling into the barrel and drowning.

- As stated before, stagnant water is an excellent breeding ground for mosquitoes, so it would be a good idea to take additional steps to keep them out of your barrel by sealing all the openings into the barrel with caulk or putty. You might also consider adding enough non-toxic oil (such as vegetable cooking oil) to the barrel to form a film on top of the water that will prevent mosquito larvae from hatching.

Always double check to make sure the barrel you're using (particularly if it is from a food distribution center or other recycled source) did not contain pesticides, industrial chemicals, weed killers, or other toxins or biological materials that could be harmful to you, your plants, or the environment. If you are concerned about this, it is best to purchase a new barrel or trash can so there is no doubt about its safety.

Mulching in Your Garden and Yard

Mulching is one of the simplest and most beneficial practices you can use in your garden. Mulch is simply a protective layer of material that is spread on top of the soil to enrich the soil, prevent weed growth, and help provide a better growing environment for your garden plants and flowers.

Mulches can either be organic—such as grass clippings, bark chips, compost, ground corncobs, chopped cornstalks, leaves, manure, newspaper, peanut shells, peat moss, pine needles, sawdust, straw, and wood shavings—or inorganic—such as stones, brick chips, and plastic. Both organic and inorganic mulches have numerous benefits, including:

1. Protecting the soil from erosion
2. Reducing compaction from the impact of heavy rains
3. Conserving moisture, thus reducing the need for frequent watering
4. Maintaining a more even soil temperature
5. Preventing weed growth
6. Keeping fruits and vegetables clean
7. Keeping feet clean and allowing access to the garden even when it's damp
8. Providing a "finished" look to the garden

Organic mulches also have the benefit of improving the condition of the soil. As these mulches slowly decompose, they provide organic matter to help keep the soil loose. This improves root growth, increases the infiltration of water, improves the water-holding capacity of the soil, provides a source of plant nutrients, and establishes an ideal environment for earthworms and other beneficial soil organisms.

While inorganic mulches have their place in certain landscapes, they lack the soil-improving properties of organic mulches. Inorganic mulches, because of their permanence, may be difficult to remove if you decide to change your garden plans at a later date.

COMMON ORGANIC MULCHING MATERIALS

Bark chips
Chopped cornstalks
Compost
Grass clippings
Ground corncobs
Hay
Leaves
Manure
Newspaper
Peanut shells
Peat moss
Pine needles
Sawdust
Straw
Wood shavings

≫ Wood chips or shavings make inexpensive and effective mulch.

Where to Find Mulch Materials

You can find mulch materials right in your own backyard. They include:

1. Lawn clippings. They make an excellent mulch in the vegetable garden if spread immediately to avoid heating and rotting. The fine texture allows them to be spread easily, even around small plants.

2. Newspaper. As a mulch, newspaper works especially well to control weeds. Save your own newspapers and only use the text pages, or those with black ink, as color dyes may be harmful to soil microflora and fauna if composted and used. Use three or four sheets together, anchored with grass clippings or other mulch material to prevent them from blowing away.

3. Leaves. Leaf mold, or the decomposed remains of leaves, gives the forest floor its absorbent, spongy structure. Collect leaves in the fall and chop with a lawn-mower or shredder. Compost leaves over winter, as some studies have indicated that freshly chopped leaves may inhibit the growth of certain crops.

4. Compost. The mixture makes wonderful mulch—if you have a large supply—as it not only improves the soil structure but also provides an excellent source of plant nutrients.

5. Bark chips and composted bark mulch. These materials are available at garden centers, and are sometimes used with landscape fabric or plastic that is spread atop the soil and beneath the mulch to provide additional protection against weeds. However, the barrier between the soil and the mulch also prevents any improvement in the soil condition and makes planting additional plants more

difficult. Without the barrier, bark mulch makes a neat finish to the garden bed and will eventually improve the condition of the soil. It may last for one to three years or more, depending on the size of the chips or how well composted the bark mulch is. Smaller chips are easier to spread, especially around small plants.

6. Hay and straw. These work well in the vegetable garden, although they may harbor weed seeds.

7. Seaweed mulch, ground corncobs, and pine needles. Depending on where you live, these materials may be readily available and also can be used as mulch. However, pine needles tend to increase the acidity of the soil, so they work best around acid-loving plants, such as rhododendrons and blueberries.

When choosing a mulch material, think of your primary objective. Newspaper and grass clippings are great for weed control, while bark mulch gives a perfect, finishing touch to a front-yard perennial garden. If you're looking for a cheap solution, consider using materials found in your own yard or see if your community offers chipped wood or compost to its residents.

If you want the mulch to stay in place for several years around shrubs, for example, you might want to consider using inorganic mulches. While they will not provide organic matter to the soil, they will be more or less permanent.

When to Apply Mulch

Time of application depends on what you hope to achieve by mulching. Mulches, by providing an insulating barrier between the soil and the air, moderate the soil temperature. This means that a mulched soil in the summer will be cooler than an adjacent, unmulched soil; while in the winter, the mulched soil may not freeze as deeply. However, since mulch acts as an insulating layer, mulched soils tend to warm up more slowly in the spring and cool down more slowly in the fall than unmulched soils.

If you are using mulches in your vegetable or flower garden, it is best to apply or add additional mulch after the soil has warmed up in the spring. Organic mulches reduce the soil temperature by 8 to 10 degrees Fahrenheit during the summer, so if they are applied to cold garden soils, the soil will warm up more slowly and plant maturity will be delayed.

Mulches used to help moderate winter temperatures can be applied late in the fall after the ground has frozen, but before the coldest temperatures arrive. Applying mulches before the ground has frozen may attract rodents looking for a warm over-wintering site. Delayed applications of mulch should prevent this problem.

Mulches used to protect plants over the winter should be composed of loose material, such as straw, hay, or pine boughs that will help insulate the plants without compacting under the weight of snow and ice. One of the benefits from winter applications of mulch is the reduction in the freezing and thawing of the soil in the late winter and early spring. These repeated cycles of freezing at night and then thawing in the warmth of the sun cause many small or shallow-rooted plants to be heaved out of the soil. This leaves their root systems exposed and results in injury, or death, of the plant. Mulching helps prevent these rapid fluctuations in soil temperature and reduces the chances of heaving.

General Guidelines

Mulch is measured in cubic feet, so, for example, if you have an area measuring 10 feet by 10 feet, and you wish to apply 3 inches (¼ foot) of mulch, you would need 25 cubic feet to do the job correctly.

While some mulch can come from recycled material in your own yard, it can also be purchased bagged or in bulk from a garden center. Buying in bulk may be cheaper if you need a large volume and have a way to haul it. Bagged mulch is often easier to handle, especially for smaller projects, as most bagged mulch comes in 3-cubic-foot bags.

To start, remove any weeds. Begin mulching by spreading the materials in your garden, being careful not to apply mulch to the plants themselves. Leave an inch or so of space next to the plants to help prevent diseases from flourishing in times of excess humidity.

How Much Do I Apply?

The amount of mulch to apply to your garden depends on the mulching material used. Spread bark mulch and wood chips 2 to 4 inches deep, keeping it an inch or two away from tree trunks.

Scatter chopped and composted leaves 3 to 4 inches deep. If using dry leaves, apply about 6 inches.

Grass clippings, if spread too thick, tend to compact and rot, becoming quite slimy and smelly. They should be applied 2 to 3 inches deep, and additional layers should be added as clippings decompose. Make sure not to use clippings from lawns treated with herbicides.

Sheets of newspaper should only be ¼ inch thick, and covered lightly with grass clippings or other mulch material to anchor them. If other mulch materials are not available, cover the edges of the newspaper with soil.

If using compost, apply 3 to 4 inches deep, as it's an excellent material for enriching the soil.

Organic Gardening

"Organically grown" food is food grown and processed using no synthetic fertilizers or pesticides. Pesticides derived from natural sources (such as biological pesticides—compost and manure) may be used in producing organically grown food.

Organic gardeners grow the healthiest, highest quality foods and flowers—all without the addition of chemical fertilizers, pesticides, or herbicides. Organic gardening methods are healthier, environmentally friendly, safe for animals and humans, and are typically less expensive, since you are working with natural materials. It is easy to grow and harvest organic foods in your backyard garden and typically, organic gardens are easier to maintain than gardens that rely on chemical and unnatural components to help them grow effectively.

Organic production is not simply the avoidance of conventional chemical inputs, nor is it the substitution of natural inputs for synthetic ones. Organic farmers apply techniques first used thousands of years ago, such as crop rotations and the use of composted animal manures and green manure crops, in ways that are economically sustainable in today's world.

Organic farming entails:

- Use of cover crops, green manures, animal manures, and crop rotations to fertilize the soil, maximize biological activity, and maintain long-term soil health.

- Use of biological control, crop rotations, and other techniques to manage weeds, insects, and diseases.
- An emphasis on biodiversity of the agricultural system and the surrounding environment.
- Reduction of external and off-farm inputs and elimination of synthetic pesticides and fertilizers and other materials, such as hormones and antibiotics.
- A focus on renewable resources, soil and water conservation, and management practices that restore, maintain, and enhance ecological balance.

How to Start Your Own Organic Garden

Step One: Choose a Site for Your Garden

1. Think small, at least at first. A small garden takes less work and materials than a large one. If done well, a 4 x 4-foot garden will yield enough vegetables and fruit for you and your family to enjoy.

2. Be careful not to over-plant your garden. You do not want to end up with

HOMESCHOOL HINT

How do microorganisms in the soil affect plants?

Take a sample of fertile soil from a field or garden and divide it into two portions. Bake one in an oven at 350°F for half an hour (to destroy the microorganisms). Leave the other portion alone as a control. Plant the same number of seeds in each soil sample. Remember to treat both samples the same while the plants are growing. Make sure all the plants receive the same amounts of water and light, and are kept at the same temperature. How do the plants differ as they grow?

Next, discover how some microorganisms and plants form mutually beneficial partnerships. For example, certain bacteria make a natural nitrogen fertilizer for plants in the family called legumes, which includes peas, alfalfa, and soybeans. The nitrogen-fixing bacteria are available from garden supply stores and by mail order. Grow both legumes and non-legume plants with and without the bacteria. Are there differences in how well the plants grow?

too many vegetables that will end up over-ripening or rotting in your garden.

3. You can even start a garden in a window box if you are unsure of your time and dedication to a larger bed.

Step Two: Make a Compost Pile

Compost is the main ingredient for creating and maintaining rich, fertile soil. You can use most organic materials to make compost that will provide your soil with essential nutrients. To start a compost pile, all you need are fallen leaves, weeds, grass clippings, and other vegetation that is in your yard. (See the Improving Your Soil chapter for more details on how to make compost.)

Step Three: Add Soil

In order to have a thriving organic garden, you must have excellent soil. Adding organic material (such as that in your compost pile) to your existing soil will only make it better. Soil containing copious amounts of organic material is very good for your garden. Organically rich soil:

- Nourishes your plants without any chemicals, keeping them natural
- Is easy to use when planting seeds or seedlings and it also allows for weeds to be more easily picked

- Is softer than chemically treated soil, so the roots of your plants can spread and grow deeper
- Helps water and air find the roots

Step Four: Weed Control

1. Weeds are invasive to your garden plants and thus must be removed in order for your organic garden to grow efficiently. Common weeds that can invade your garden are ivy, mint, and dandelions.
2. Using a sharp hoe, go over each area of exposed soil frequently to keep weeds from sprouting. Also, plucking off the green portions of weeds will deprive them of the nutrients they need to survive.
3. Gently pull out weeds by hand to remove their root systems and to stop continued growth. Be careful when weeding around established plants so you don't uproot them as well.
4. Mulch unplanted areas of your garden so that weeds will be less likely to grow. You can find organic mulches, such as wood chips and grass clippings, at your local garden store. These mulches will not only discourage weed growth but will also eventually break down and help enrich the soil. Mulching also

helps regulate soil temperatures and helps in conserving water by decreasing evaporation. (See the Mulching In Your Garden and Yard chapter for more on mulching.)

Step Five: Be Careful of Lawn Fertilizers

If you have a lawn and your organic garden is situated in it, be mindful that any chemicals you place on your lawn may find their way into your organic garden. Therefore, refrain from fertilizing your lawn with chemicals and, if you wish to return nutrients to your grass, simply let your cut grass clippings remain in the yard to decompose naturally and enrich the soil beneath.

Things to Consider

- "Organic" means that you don't use any kinds of chemicals or materials, such as paper or cardboard, that contain chemicals, and especially not fertilizer or pesticides. Make sure that these products do not find their way into your garden or compost pile.
- If you are adding grass clippings to your compost pile, make sure they don't come from a lawn that has been treated with chemical fertilizer.
- If you don't want to start a compost pile, simply add leaves and grass clippings directly to your garden bed. This will act

like a mulch, deter weeds from growing, and will eventually break down to help return nutrients to your soil.
- If you find insects attacking your plants, the best way to control them is by picking them off by hand. Also practice crop rotation (planting different types of plants in a given area from year to year), which will hopefully reduce your pest problem. For some insects, just a strong stream of water is effective in removing them from your plants.
- Shy away from using bark mulch. It robs nitrogen from the soil as it decomposes and can also attract termites.

Saving Seeds

If you have a unique heirloom variety plant that you want to preserve or if you don't want to buy new seeds every year, you can save seeds from your healthy plants. Saving

seeds is relatively easy for dry plants, like beans, where the seeds are easily distinguishable from the vegetable or fruit. In these cases, simply scrape the seeds from the vegetable, place them in a single layer in a glass dish, and leave them near a sunny window to dry for one week.

For some plants, like tomatoes, the seeds are surrounded by a wet pulp. For these plants, remove the seeds from the flesh of the fruit or vegetable with your fingers and then rinse thoroughly in a wire mesh strainer. You may need to soak them for a while to remove all residue. Then dry as described above.

Once seeds are thoroughly dried, store them in labeled envelopes in a cool, dry place.

Keep in mind that plants often naturally cross-pollinate, especially when different types of plants are near each other in a garden, resulting in a hybrid seed. Hybrid seeds are unpredictable and often grow into inferior plants. Also, most seeds that you buy today are already hybrids. If you plan to save your seeds, invest in heirloom variety or open-pollinated seeds.

Hybrid vs. Heirloom Seeds

Plants are like any other living thing in that there are male and female parts and it

Saving Seeds

takes both to create offspring. Some plants contain both the male and female parts within their own flowers (self-pollinators), and others have separate male plants and female plants. With the latter type, bees or birds carry the pollen from male plants to the ovule of female plants. Thus, plants can be bred to have certain characteristics and qualities by ensuring that the desired male and female plants are in close proximity and that undesirable potential "parents" are kept at a distance. Nowadays, seeds can also be artificially crossbred and genetically altered.

Seed manufacturers frequently breed seeds to be high yield, often at the expense of disease resistance, since the majority of plants now are grown with pesticides anyway. These hybrid seeds are high-maintenance; they require special fertilizers, they're less hardy, and they are more susceptible to disease. However, with the right supports (pesticides, herbicides, and irrigation) they will produce a greater volume of plants. Other seeds are bred for other characteristics, such as size of the fruit or vegetable.

There are several concerns regarding the popularity of hybrid seeds. One is that it creates too much dependence on the major seed producers, as well as suppliers of pesticides and other inorganic gardening products. Since hybrid plants do not produce reliable seeds, farmers must return to the seed supplier every year before they begin planting and then often depend on pesticides to keep their plants healthy. This is an especially serious issue in poorer countries where the people are at the mercy of major seed supply companies.

Heirloom varieties are much more diverse than hybrids. Not only does this mean that by using them you'll be harvesting more interesting (and often more flavorful) produce, but you'll also be helping to prevent a potential food shortage disaster. Because major seed suppliers are breeding seeds for specific purposes, they're narrowing down the varieties of seeds they provide to only those that best meet their needs. This will become a major problem if a disease attacks those plants.

⌃ **Close-up of an ear of heirloom corn.**

If there are many varieties, some will resist the disease. If there are only a few varieties available, they might all be wiped out, as happened with the Irish potato famine of the 1840s.

Heirloom seeds are generally more expensive than hybrids, but you only have to buy them once, since you can save their seeds at the end of every growing season to plant the following spring. Thus, it makes sense to incorporate as many heirloom varieties into your garden as you can.

⌄ **Heirloom tomatoes are often unusual shapes and colors. They can be as beautiful as they are delicious.**

Terracing

Terraces can create several mini-gardens in your backyard. On steep slopes, terracing can make planting a garden possible. Terraces also prevent erosion by shortening a long slope into a series of shorter, more level steps. This allows heavy rains to soak into the soil rather than to run off and cause erosion and poor plant growth.

Materials Needed for Terraces

Numerous materials are available for building terraces. Treated wood is often used in terrace building and has several advantages: it is easy to work with, it blends well with plants and the surrounding environment, and it is often less expensive than other materials. There are many types of treated wood available for terracing—railroad ties and landscaping timbers are just two examples. These materials will last for years, which is crucial if you are hoping to keep your terraced garden intact for quite a while. There has been some concern about using these treated materials around plants, but studies by Texas A&M University and the Southwest Research Institute concluded that these materials are not harmful to gardens or people when used as recommended.

Other materials for terraces include bricks, rocks, concrete blocks, and similar masonry materials. Some masonry materials are made specifically for walls and terraces and can be more easily installed by a homeowner than other materials. These include fieldstone and brick. One drawback is that most stone or masonry products tend to be more expensive than wood, so if you are looking to save money, treated wood will make a sufficient terrace wall.

How High Should the Terrace Walls Be?

The steepness of the slope on which you wish to garden often dictates the appropriate height of the terrace wall. Make the terraces in your yard high enough so the land between them is fairly level. Be sure the terrace material is strong enough and anchored well to stay in place through freezing and thawing, and during heavy rainstorms. Do not underestimate the pressure of waterlogged soil behind a wall—it can be enormous and will cause improperly constructed walls to bulge or collapse.

Many communities have building codes for walls and terraces. Large projects will most likely need the expertise of a professional landscaper to make sure the walls can stand up to water pressure in the soil. Large terraces also need to be built with adequate drainage and to be tied back into the slope properly. Because of the expertise and equipment required to do this correctly, you will probably want to restrict terraces you build on your own to no more than a foot or two high.

« Some ferns, decorative grasses, mosses, and other plants grow well on slopinglandscapes and don't require any terracing in order to thrive.

Building Your Own Terrace

The safest way to build a terrace is by using the cut and fill method. With this method, little soil is disturbed, giving you protection from erosion should a sudden storm occur while the work is in progress. This method will also require little, if any, additional soil. Here are the steps needed to build your own terrace:

1. Contact your utility companies to identify the location of any buried utility lines and pipes before starting to dig.
2. Determine the rise and run of your slope. The rise is the vertical distance from the bottom of the slope to the top. The run is the horizontal distance between the top and the bottom. This will allow you to determine how many terraces you will need. For example, if your run is 20 feet and the rise is 8 feet, and you want each bed to be 5 feet wide, you will need four beds. The rise of each bed will be 2 feet.
3. Start building the beds at the bottom of your slope. You will need to dig a trench in which to place your first tier. The depth and width of the trench will

⌃ Terracing is used extensively for growing rice in Indonesia.

vary depending on how tall the terrace will be and the specific building materials you are using. Follow the manufacturer's instructions carefully when using masonry products, as many of these have limits on the number of tiers or the height that can safely be built. If you are using landscape timbers and your terrace is low (less than 2 feet), you only need to bury the timber to about half its thickness or less. The width of the trench should be slightly wider than your timber. Make sure the bottom of the trench is firmly packed and completely level, and then place your timbers into the trench.

4. For the sides of your terrace, dig a trench into the slope. The bottom of this trench must be level with the bottom of the first trench. When the depth of the trench is one inch greater than the thickness of your timber, you have reached the back of the terrace and can stop digging.

⌄ Periwinkle

5. Cut a piece of timber to the correct length and place it into the trench.

6. Drill holes through your timbers and pound long spikes, or pipes, through the holes and into the ground. A minimum of 18 inches of pipe length is recommended, and longer pipes may be needed in higher terraces for added stability.

7. Place the next tier of timbers on top of the first, overlapping the corners and joints. Pound a spike through both tiers to fuse them together.

8. Move the soil from the back of the bed to the front of the bed until the surface is level. Add another tier as needed.

9. Repeat, starting with step 2, to create the remaining terraces. In continuously connected terrace systems, the first timber of the second tier will also be the back wall of your first terrace.

10. The back wall of the last bed will be level with its front wall.

11. When finished, you can start to plant and mulch your terraced garden.

Other Ways to Make Use of Slopes in Your Yard

If terraces are beyond the limits of your time or money, you may want to consider other options for backyard slopes. If you have a slope that is hard to mow, consider using ground-covers on the slope rather than grass. There are many plants adapted to a wide range of light and moisture conditions that require little care (and do not need mowing) and provide soil erosion protection. These include:

- Juniper
- Wintercreeper
- Periwinkle
- Cotoneaster
- Potentilla
- Heathers and heaths

Strip-cropping is another way to deal with long slopes in your yard. Rather than terracing to make garden beds level, plant perennial beds and strips of grass across the slope. Once established, many perennials are effective in reducing erosion. Adding mulch also helps reduce erosion. If erosion does occur, it will be basically limited to the gardened area. The grass strips will act as filters to catch much of the soil that may run off the beds. Grass strips should be wide enough to mow easily, as well as wide enough to reduce erosion effectively.

⌃ Heather will grow easily on sloped landscapes and will help prevent erosion.

Start Your Own Vegetable Garden

If you want to start your own vegetable garden, just follow these simple steps and you'll be on your way to growing your own yummy vegetables—right in your own backyard!

Steps to Making Your Own Vegetable Garden

1. Select a site for your garden.
   - Vegetables grow best in well-drained, fertile soil (loamy soils are the best).
   - Some vegetables can cope with shady conditions, but most prefer a site with a good amount of sunshine—at least six hours a day of direct sunlight.
2. Remove all weeds in your selected spot and dispose of them. If you are using compost to supplement your garden soil, do not put the weeds on the compost heap, as they may germinate once again and cause more weed growth among your vegetable plants.
3. Prepare the soil by tilling it. This will break up large soil clumps and allow you to see and remove pesky weed roots. This would also be the appropriate time to add organic materials (such as compost) to the existing soil to help make it more fertile. The tools used for tilling will depend on the size of your garden. Some examples are:
   - Shovel and turning fork—using these tools is hard work, requiring strong upper body strength.
   - Rotary tiller—this will help cut up weed roots and mix the soil.
4. After the soil has been tilled you are ready to begin planting. If you would like straight rows in your garden, a guide can be made from two wooden stakes and a bit of rope.
5. Vegetables can be grown from seeds or transplanted:
   - If your garden has problems with pests such as slugs, it's best to transplant older plants, as they are more likely to survive attacks from these organisms.
   - Transplanting works well for vegetables like tomatoes and onions, which usually need a head start to mature within a shorter growing season. These can be germinated indoors on seed trays on a windowsill before the growing season begins.

6. Follow these basic steps to grow vegetables from seeds:
   - Information on when and how deep to plant vegetable seeds is usually printed on seed packages or on various Web sites. You can also contact your local nursery or garden center to inquire after this information.
   - Measure the width of the seed to determine how deep it should be planted. Take the width and multiply by two. That is how deep the seed should be placed in the hole. As a general rule, the larger the seed, the deeper it should be planted.
7. Water the plants and seeds well to insure a good start. Make sure they receive water at least every other day, especially if there is no rain in the forecast.

Things to Consider

In the early days of a vegetable garden, all your plants are vulnerable to attack by insects and animals. It is best to plan multiples of the same plant in order to ensure that some survive. Placing netting and fences around your garden can help keep out certain animal pests. Coffee grains or slug traps filled with beer will also help protect your plants against insect pests.

If sowing seed straight onto your bed, be sure to obtain a photograph of what your seedlings will look like so you don't mistake the growing plant for a weed.

Weeding early on is very important to the overall success of your garden. Weeds steal water, nutrients, and light from your vegetables, which will stunt their growth and make it more difficult for them to thrive.

⌄ Label your garden rows as soon as you plant the seeds or seedlings so you'll remember what you planted where.

RADISH

Hoophouses

Hoophouses are small, semi-portable structures that can be used as a small greenhouse structure for starting seedlings outdoors and for growing heat-loving vegetables. A hoophouse provides frost protection, limited insect protection, and season extension. Hoophouse structures are easy to construct and will last many years. You can make them any size, but a structure 4 feet x 10 feet is generally adequate. These dimensions allow easy side access for weeding and adequate hoop arch strength relative to span.

Seeds can be started in flats in the hoophouse. Temperature is regulated by varying the size of the end openings and/or lifting the side wall plastic. After seed trays are removed (to be planted in an outdoor garden), heat-loving plants such as tomatoes, peppers, and melons can be grown directly in the soil in the hoophouse. Plastic can be left in place to keep hoophouse temperatures warm until outside temperatures will support active plant growth or until plant vegetation outgrows the confines of the box. Plants requiring staking, such as tomatoes, can be planted near the edge of the box near a hoophouse support. After the plastic and hoops are removed, a rigid stick or dowel can be inserted into the plastic hoop retainer. Tomato plants can then be tied and supported by the rigid upright stake.

Hoophouse Pros and Cons

Advantages of Using Hoophouses:

1. Hoophouses allow for earlier soil warming and protects from frost, lengthening the growing season.

HOMESCHOOL HINT

Root Vegetable "Magic"

Cut off the top 1 inch of a carrot, turnip, or beet. Put the top on a saucer, cut side down. Add just enough water to make the bottom of the vegetable top wet. Keep the saucer in a sunny window, add water every day so the bottom of the vegetable stays wet. Watch new leaves and roots grow!

Seeds that grow well in containers:

- Tomatoes
- Radishes
- Cucumbers
- Peppers
- Leaf lettuce
- Herbs

THE JUNIOR HOMESTEADER

Starting a Mini Garden

You don't need a big field or even a backyard to grow some of your own food. You can grow some on a windowsill, balcony, porch, deck, or doorstep! Follow these steps to create your own mini garden:

Things You'll Need:

- Container, such as milk carton, bleach jug, coffee can, ice cream tub, or ceramic pot
- Seeds
- Soil
- Plant fertilizer
- Tray or plate
- Water

1. Select seeds to plant. See "Seeds that grow well in containers" for ideas of seeds to select.
2. Select a container. Match the container to the size of the plant. For example, tomatoes require a much bigger container than herbs. Rinse the container. Punch holes in the bottom, if there are none.
3. In a bucket, combine soil with water until the soil is damp. Fill your container with the damp soil to ½ inch from the top.
4. Read the seed packet to see how far apart and how deep to plant seeds. Cover seeds gently with soil.
5. Keep the seed bed watered well. The seeds need a lot of water, but don't add it all at once. Pour some on, let it sink in, and pour more on. Stop pouring when you see water coming out the bottom of the container. Keep a plate or tray under the plant container so the container will not leak. Keep the soil moist, but not sopping wet.
6. Place container(s) in a sunny location.
7. Once a week, add fertilizer following directions on the label.
8. Turn the containers often, so that sunlight reaches all sides of the growing plants.
9. As the plants grow larger, use scissors to trim the leaves of side-by-side plants, so they do not touch each other.

⌃ To grow lima beans indoors, first soak several lima beans in room temperature water overnight. If the beans do not sprout automatically, place them on damp paper towel until they begin to send out a thin shoot. Then fill a pot with soil and make a small hole in the center. Place the seed in the hole, sprout end down, and cover with soil. Place the pot in a sunny spot and water daily.

HOOPHOUSE PLANS

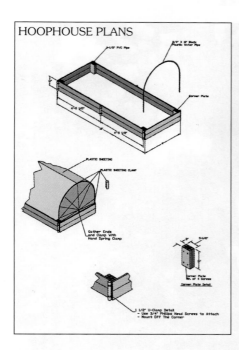

« **Materials for Hoophouse**
¾' x 5½' x 10' treated wood (6 each)
1½' PVC pipes
1½" U-Clamp (24 each)
¾" Black Plastic Water Pipe (35 Lw. Ft.)
Plastic Sheeting (10' x 16')
Hand Spring Clamp (2 ea.)
10 x ¾" Galvanized Phillips Head Screws (24 ea.)
10 x 2 Torx Head Climatek Plated Deck Screws (48 ea.)

2. Small heaters can be used to give additional frost protection.
3. Hoophouses are easily constructed from readily available materials.
4. Hoop/plastic covering can be manipulated and/or removed to control internal temperatures.

Disadvantages of Using Hoophouses:

1. Relatively high cost per square foot of growing space.

2. Internal temperatures can rise quickly on sunny days and kill plants unless the plastic covering is adjusted to allow for adequate ventilation.
3. Hoop covering must be removed at the end of the growing season, as snow load will crush the hoops.
4. Plastic covering will only last one to two years unless more expensive greenhouse plastic is used.

The quality of the soil is critical to the proper functioning of a hoophouse. The hoophouse may be filled with topsoil that is either purchased or acquired on-site. If the latter is used, be prepared to deal with imported weed seed that often are present in the soil.

Hoophouses will last many years if cared for properly. The plastic covering is the only component that needs periodic replacing. Any clear plastic may be used as a covering, although ultraviolet light will tend to break down plastics not designed for outdoor use after one season. Many types of greenhouse plastics are available and will last for three to ten years.

⩔ Hoophouses can be made large or small to suit your needs.

Start Your Own Flower Garden

If you are looking to grow a beautiful garden full of flowers, just follow these simple steps to achieve the perfect beginner's flower garden.

Start with a Small Garden

Gardening takes a lot of work, and so for the beginner gardener, tackling a large garden can be overwhelming. Start with a small flowerbed around 25 square feet. This will provide you with room for about twenty to thirty plants—enough room for three types of annuals and two types of perennials. As your gardening experience grows, so can the size of your garden!

If you want to start even smaller, you can begin your first flower garden in a container, or create a border from treated wood or bricks and stones around your existing bed. That way, when you are ready to expand your garden, all you need to do is remove the temporary border and you'll be all set. Even a small container filled with a few different types of plants can be a wonderful addition to any yard.

Plan Your Flower Garden

Draw up a plan of how you'd like your garden to look, and then dig a flowerbed to fit that plan. Planning your garden before gathering the seeds or plants and beginning the digging can give you a clearer sense of how your garden will be organized and can facilitate the planting process.

Choose a Spot for Your Garden

It is important, when choosing where your flower garden will be located, that you consider an area that receives at least six hours of direct sunlight per day, as this will be adequate for a large variety of garden plants. Be careful that you will not be digging into utility lines or pipes, and that you place your garden at least a short distance away from fences or other structures.

If you live in a part of the country that is quite hot, it might be beneficial for your flowers to be placed in an area that gets some

shade during the hot afternoon sun. Placing your garden on the east side of your home will help your flowers flourish. If your garden will get more than six hours of sunlight per day, it would be wise to choose flowers that thrive in hot, sunny spaces, and make sure to water them frequently.

It is also important to choose a spot that has good, fertile soil in which your flowers can grow. Try to avoid any areas with rocky, shallow soil or where water collects and pools. Make sure your garden is away from large trees and shrubs, as these plants will compete with your flowers for water and nutrients. If you are concerned that your soil may not contain enough nutrients for your flowers to grow properly, you can have a soil test done, which will tell you the pH of the soil. Depending on the results, you can then adjust the types of nutrients needed in your soil by adding organic materials or certain types of fertilizers.

Start Digging

Now that you have a site picked out, mark out the boundaries with a hose or string. Remove the sod and any weed roots that may re-grow. Use your spade or garden fork to dig up the bed at least 8 to 12 inches deep, removing any rocks or debris you come across.

Once your bed is dug, level it and break up the soil with a rake. Add compost or manure if the soil is not fertile. If your soil is sandy, adding peat moss or grass clippings will help it hold more water. Work any additions into the top 6 inches of soil.

Purchase Your Seeds or Plants

Once you've chosen which types of flowers you'd like to grow in your garden, visit your local garden store or nursery and pick out already-established plants or packaged seeds. Follow the planting instructions on the plant tabs or seed packets. The smaller plants should be situated in the front of the bed. Once your plants or seeds are in their holes, pack in the soil around them. Make sure to leave ample space between your seeds or plants for them to grow and spread out (most labels and packets will alert you to how large your flower should be expected to grow, so you can adjust the spacing as needed).

Water Your Flower Garden

After your plants or seeds are first put into the ground, be sure they get a thorough watering. Then, continue to check your garden to see whether or not the soil is drying out. If so, give your garden a good soaking with the garden hose or watering can. The amount of water your garden needs is dependant on the climate you live in, the exposure to the sun, and how much rain your area has received.

Cutting Your Flowers

Once your flowers begin to bloom, feel free to cut them and display the beautiful blooms in your home. Pruning your flower garden (cutting the dead or dying blooms off the plant) will help certain plants to re-bloom. Also, if you have plants that are becoming top heavy, support them with a stake and some string so you can enjoy their blossoms to the fullest.

Things to Consider

- Annuals are plants that you need to replant every year. They are often inexpensive, and many have brightly colored flowers. Annuals can be rewarding for beginner gardeners, as they take little effort and provide lovely color to your garden. The following season, you'll need to replant or start over from seed.
- Perennials last from one year to the next. They, too, will require annual maintenance but not yearly replanting. Perennials may require division, support, and extra care during winter months. Perennials may also need their old blooms and stems pruned and cut back every so often.
- Healthy, happy plants tend not to be as susceptible to pests and diseases. Here, too, it is easier to practice prevention rather than curing existing problems. Do your best to give your plants good soil, nutrients, and appropriate moisture, and choose plants that are well suited to your climate. This way, your garden will be more likely to grow to its maximum potential and your plants will be strong and healthy.

THE JUNIOR HOMESTEADER

Plan a Garden Party

A party in the garden is a great way to celebrate a birthday, last day of school, or other special event. And it will show kids how much fun gardens can be. For invitations, cut large flower shapes out of construction paper and let the kids paste pictures from seed catalogs or gardening magazines onto one side. On the other side, write the date, time, and place of your party. The garden itself will serve as a beautiful backdrop for your celebration, but if you want to go further, choose seasonal decorations such as bouquets of lilacs and apple blossoms in spring, daisy chains in summer, and pumpkins and corn stalks in fall. Once at the party, kids will enjoy decorating their own clay flowerpots by painting them with acrylic paints or, for older kids, using a glue gun to attach stones or beads to the outsides of the pots. Then provide seeds and soil so they can bring a potted plant home and watch it grow. Have a contest to see who can make the most creative faces or sculptures out of fresh produce—and then eat the fun creations! Scavenger hunts are always a hit. Provide a chart showing different types of leaves, grasses, and flowers and see who can find all the items and check them off the chart first. Kids will enjoy coming up with their own games, too. Just spending time together in the garden will create lasting fun memories.

PLANTERS

Barrel Plant Holder

If you have some perennials you want to display in your yard away from your flower garden you can create a planter out of an old barrel. This plant holder is made by sawing an old barrel (wooden or metal) into two pieces and mounting it on short or tall legs—whichever design fits better in your yard. You can choose to either paint it or leave it natural. Filling the planter with good quality soil and compost and planting an array of multi-colored flowers into the barrel planters will brighten up your yard all summer long. If you do not want to mount the barrel on legs, it can be placed on the ground on a smooth and level surface where it won't easily tip over.

Rustic Plant Stand

If you'd like to incorporate a rustic, natural-looking plant stand in your garden or on your patio or deck, one can easily be made from a preexisting wooden box or by nailing boards together. This box should be mounted on legs. To make the legs, saw the piece of wood meant for the leg in half to a length from the top equal to the depth of the box. Then, cross-cut and remove one half. The corner of the box can then be inserted in the middle of the crosscut and the leg nailed to the side of the box.

The plant stand can be decorated to suit your needs and preference. You can nail smaller, alternating twigs or cut branches around the stand to give it a more natural feel or you can simply paint it a soothing, natural color and place it in your yard.

Wooden Window Box

Planting perennial flowers and cascading plants in window boxes is the perfect way to brighten up the front exterior of your home. Making a simple wooden window box to hold your flowers and plants is quite easy. These boxes can be made from preexisting wooden boxes (such as fruit crates) or you can make your own out of simple boards. Whatever method you choose, make sure the boards are stout enough to hold the brads firmly.

The size of your window will ultimately determine the size of your box, but this plan calls for a box roughly 21 x 7 x 7 inches (fig. 14). You can decorate your boxes with waterproof paint or you can nail strips of wood or sticks to the panels. Make sure to cut a few holes in the bottom of the box to allow for water drainage. Figure 16 shows the wedge pieces that are fitted to the stone or cement sill to bring the box completely level. The window box can be kept in position by two metal angle pieces screwed to the wood sill and to the back of the box.

Growing Fruit Bushes and Trees

If you take the time to properly plan and care for your fruit bushes and trees, they'll provide you with delicious, nutrient-dense fruit year after year. Some fruit plants, like strawberries, are easy to grow and will reward you with ripe fruit relatively quickly. Fruit trees, like apple or pear, will require more work and time, but with the right maintenance they will bear fruit for generations.

Think carefully about where you choose to plant and then take time to prepare the site. Most fruit plants need at least six hours a day of sun and require well-drained soil. If the soil is not already cultivated and relatively free of pests, spend the first year preparing the site. Planting a cover crop of rye, wheat, or oats will improve the quality of your soil and reduce weeds that could compete with your fruit plants. The cover crops will die in the late fall and add to the organic matter of the soil. Just leave them to decompose on the surface of the soil and then turn them under the soil come spring.

Testing your soil pH the year ahead of planting will give you time to adjust it if necessary to give your plants the best chance of thriving. Fruit trees, grapes, strawberries, blackberries, and raspberries do best if the soil pH is between 6.0 and 6.5. Blueberries require a more acidic soil, around 4.5.

What plants will thrive will depend largely on where you live, your planting zone (see page 5), your altitude, and your proximity to large bodies of water (since areas close to water tend to be more temperate). Refer to the chart below for hardiness zones for most fruits, but keep in mind that hardiness varies by variety. Refer to seed catalogs or talk to other local gardeners before settling on a particular variety of fruit to plant.

⤊ Grains such as wheat can be used as an inexpensive and simple cover crop to enrich your soil.

FRUIT GROWING CHART

Fruit	Hardiness Zone	Soil pH	Space between Plants	Space between Rows	Bearing Age (in Years)	Potential Yield (in Pounds)	When to Harvest
Apple	5 to 7	6 to 6.5	7 to 18 (depending on variety)	13 to 24 (depending on variety)	3 to 5	60 to 250	August through October
Apricot	4 to 8	6 to 6.5	15	20	4	100	July to August
Cherry, sweet	5 to 7	6 to 6.5	24	30	7	300	July
Cherry, tart	4 to 7	6 to 6.5	18	24	4	100	July
Peach and nectarine	5 to 8	6 to 6.5	15	20	4 to 5	100	August to September
Pear and quince	4 to 7	6 to 6.5	15 to 20	15 to 20	4	100	August to October
Plum	5 to 7	6 to 6.5	10	15	5	75	July to September
Grapes	5 to 10 (depending on variety)	6 to 6.5	8	9	3	10 to 20	September to October
Blackberry	3 to 9 (depending on variety)	6 to 6.5	2	10	2	2 to 3	July to August
Blueberry	3 to 11 (depending on variety)	4 to 5	4 to 5	10	3 to 6	3 to 10	July to September
Currant	2 to 6	5.5 to 7	4	8	2 to 4	6 to 8	July
Elderberry	3 to 9	6 to 6.5	6	10	2 to 4	4 to 8	August to September
Gooseberry	2 to 6	5.5 to 7	4	10	2 to 4	2 to 4	July to August
Raspberry	3 to 8	5.6 to 6.2	2	8	2	1 to 2	July to September
Strawberry	4 to 9	5.5 to 6.5	12 to 18	12 to 18	1 to 2	1 to 3	May to July

Strawberries

Purchase young strawberry plants to plant in the spring after the last frost. Try to find plants that are certified disease-free, since diseases from strawberries can spread through your whole garden. Strawberries thrive with lots of sun and well-drained soil. If you have access to a gentle south-facing slope, this is ideal. Till the top 12 inches of soil. If you planted a cover crop, turn under all the organic matter. If not, be sure to add manure or compost to a reach a rich, slightly acidic soil.

Dig a 5- to 7-inch wide hole for each plant. It should be deep enough to accommodate the root system without squishing it. Place the plant in the hole and fill in the soil, tamping it down gently around the plant. Space plants about 12 inches apart on all sides. The roots will shoot out runners that produce more small plants. To allow the plants to focus their energy on fruit production, snip the runners and transplant or discard any new plants.

An alternate planting method is the matted-row system. This method requires less maintenance but offers a slightly lower quality yield. Space plants about 18 inches apart, allowing the roots to shoot out runners and produce new plants. If planting more than one row, space them three to four feet apart. To aid picking and to keep the plants from competing with each other, prune out the plants on the outer edges of rows by snipping the runners and pulling out the plants.

Strawberry plants should receive at least an inch of water per week. In the first year, snip away or pick blossoms as soon as they develop. You will sacrifice your fruit crop in the first year, but you will have healthier plants and a greater fruit yield for many years afterward.

Brambles and Bush Fruits

Most brambles and bush fruits should be planted in the spring after the last frost. Blackberries and raspberries should have any old or damaged canes removed before planting in a 4-inch deep hole or furrow. Do not fertilize for several weeks after planting and even after that use fertilizer sparingly; brambles are easily damaged by over-fertilizing. If the weather is dry, water bushes in the morning, just after the dew has dried, being careful to avoid getting the foliage very wet.

and thin canes to three or four per foot. Trim down the tops of canes to about 4 to 5 feet high so that you can easily pick the berries. Prune similarly every spring.

Blueberries

Blueberries thrive in acidic soil (around pH 4.8). If your soil is naturally over pH 6.5, you're better off planting your blueberries in container gardens or raised beds, where you can more easily manipulate the soil's pH.

Blueberries should be planted in a hole about 6 inches deep and 20 inches in diameter. The crowns should be right at soil level. Surround the stems with about 6 inches of sawdust mulch or leaves and water the bushes in the morning in dryer climates. Blueberries are very sensitive to drought.

Some varieties of blueberries only require pruning if there are dead or damaged branches that need to be removed. Highbush varieties should be pruned every year starting after the third year after planting. When old canes get twiggy, cut them off at soil level to allow new, stronger canes to emerge.

Currants and Gooseberries

These berries thrive in colder regions in well-drained soils. Plant them in holes 12 inches across and 12 inches deep. They should bear fruit in their second year and will give the highest yield in their in the third year. After that the canes will begin to darken, which is a sign that it's time to prune them by mowing or clipping at soil level. By

Brambles need to be pruned once a year to keep them healthy and to keep them from spreading out of control. How you prune your brambles will depend on the variety. Fall-bearing brambles (primocane-fruiting brambles) produce fruit the same year they are planted. If planted in spring, they'll produce some berries in late summer or early fall and then again (lower on the canes) in early summer of their second year. For these varieties, there are two pruning choices. The first option produces a smaller yield but higher quality berries in late summer. For this pruning method, mow down the bush all the way to the ground in late fall. The second method will produce berries in summer and fall, and is the same method used for flori-cane-fruiting brambles. Allow the canes to grow through the first year and prune them gently in the early spring of the following year. Remove any damaged or diseased parts

« Blackcurrants do best planted in full sun and moist, slightly acidic (pH 6 to 6.5) soil. The berries are tart and perfect for use in jams, pies, and other desserts.

Next, prune away any broken or damaged roots. Pruning shears or a sharp knife work best. Remove roots that are very tangled or too long to fit in the hole. Place the roots into the hole, on top of the dirt mound. If the tree was grafted, it will have a "graft union," a bulge where the roots meet the trunk. For most trees, this mound should be barely visible above the ground. For dwarf trees, it should be just below the soil level.

Most young trees need extra support. Drive a 5- to 6-foot garden stake into the ground a few inches away from the trunk and on the south side. The stake should go about 2 feet deeper than the roots.

Begin filling the hole back in with soil, using your fingers to work the soil around the roots and eliminate any air pockets. Add soil until the hole is filled, pat it down until it's slightly lower than the ground level (to help retain water), and then pour a bucket of water over the soil to pack it down further. Prune away all but the three or four strongest branches and then tie the trunk gently to the stake with a soft rag.

Spread leaves, mulch, or bark around the base of the tree to protect the roots and help retain the water. Because young tree bark is easily injured, wrap the trunk carefully with burlap from the ground to its lowest branches. This will protect the bark from being scalded by sun (even in winter) or damaged by deer or rodents. Leave the wrap on for the first two or three years.

Pruning

Your fruit trees will benefit from gentle pruning once a year. The goals of pruning are to remove dead or damaged branches, to keep branches from crowding each other, and to keep trees from growing so large that they begin to invade each other's space. Pruning, when done properly, will produce healthier trees and more fruit.

Pruning shears or a pruning saw can be used. When removing a whole branch, try to cut flush with the trunk, so that no "stub" is left behind. Stubs soon decay, inviting insects to invade your tree. A cut that is flush with the trunk will heal over quickly (in one growing season). If you're removing part of a branch, cut slightly above a bud and cut at a slant. Try to choose a bud that slants in the direction you want a new branch to grow.

When removing particularly large branches, care should be taken to ensure that the branch doesn't tear away large pieces of bark from the trunk as it falls. To do this, start below the branch and cut upwards about a third of the way through the branch. Then cut down from the top, starting an inch or two further away from the trunk.

If your pruning leaves a wound that is larger than a silver dollar, use a knife to peel away the bark above and below the wound to create a vertical diamond shape. Then cover the wound with shellac or tree wound paint to protect it from decay and insects.

this point, new canes will likely have developed. Currants and gooseberries do not require much pruning, but dead or diseased branches should always be removed.

Fruit Trees

Planting

Once you've decided on which varieties to plant and have planned and prepared the best site, it's time to purchase the trees and plant them. Most young trees come with the roots planted in a container of soil, embedded in a ball of soil and wrapped in burlap (balled-and-burlapped, or B&B), or packed in damp moss or Excelsior. It's best to plant your trees as soon as possible after purchasing, though B&B stock or potted trees can be kept for several weeks in a shady area.

To plant, dig a hole that is about 2 feet deep and wide enough to give the roots plenty of room to spread (about 1 ½ feet wide). As you dig, try to keep the sod, topsoil, and subsoil in separate piles. Once the hole is the right size, loosen the dirt at the bottom and then place the sod into the hole, upside down. Then make a small mountain of topsoil in the center of the hole. The roots will sit on top of the mountain and hang over the edges.

Growing Fruit Bushes and Trees

Grapes

Talk to someone at your local nursery or to other growers in your area to determine which grape variety will work best in your location. All varieties fall under the categories of wine, table, or slipskin. Grapes need full sun to stay healthy and benefit from loose, well-drained, loamy soil.

Before planting your vine cutting, soak it in a bucket of water for at least six hours. Cuttings should never dry out. Grape vines need a trellis or a similar support. Dig a hole near the trellis deep enough to accommodate the roots and douse the hole with water. Place the cutting in the hole and fill in the soil around

it, adding more water as you go, and then tamping it down firmly. Prune away all but the best cane and tie it gently to the support (a stake or the bottom wire of a trellis). Water the vine once a week for at least the first month.

Don't use any fertilizer in the first year as it can actually damage the young vines. If necessary, begin fertilizing the soil in the second year.

Buds will begin to grow after several weeks. After about ten weeks, remove all but the strongest shoots as well as any flower clusters or side shoots. Every year in the late fall, after the last grapes have been picked, remove 90 percent of the new growth from that season. You should be able to harvest fruit in the plant's third year.

Growing and Threshing Grains

Grains are a type of grass and they grow almost as easily as the grass in your yard does. There are many reasons for growing your own grains, including supplying feed for your livestock, providing food for you and your family, or to use as a green manure (a crop that will be plowed back into the soil to enrich it). Growing grains requires much

less work than growing a vegetable garden, though getting the grains from the field to the table requires a bit more work.

Whether your are growing wheat, oats, barley, or another grain, the process is basically the same:

1. Decide which grain to grow. Most cereal grains have a spring variety and a winter variety. Winter grains are often preferred because they are more nutritious than spring varieties and are less affected by weeds in the spring. However, spring wheat is preferred in cold climates as winter wheat may not survive very harsh winters. If you have trouble finding smaller amounts of seeds to purchase from seed supply houses, try health food stores. They often have bins full of grains you can buy in bulk for eating, and they work just as well for planting, as long as you know what variety of grain you're buying. Winter grains should be planted from late September to mid-October, after most insects have disappeared but before the hard frosts set in. Spring wheat should be planted in early spring.

2. Decide how much grain you want to grow. A 10-foot by 10-foot plot of wheat will provide enough flour for about twenty loaves of bread. An acre of corn will provide feed for a pig, a milk cow, a beef steer, and thirty laying hens for an entire year.

3. Prepare the soil. Rototill or use a shovel to turn over the earth, remove any stones or weeds, and make the plot as even as possible using a garden rake.

4. Sprinkle the seeds over the entire plot. How much seed you use will depend on the grain (refer to the chart on page 57). For wheat, use a ratio of around 3 ounces of seed per 100 square feet. Aim to plant about one seed per square inch. Rake over the plot to cover all the seeds with earth.

5. Water the seeds immediately after planting and then about once a month throughout the growing season if there's not adequate rainfall.

6. When the grain is golden with a few streaks of green left, it's ready for harvest. For winter grains, harvest is usually ready in June or July. To cut the grains, use a scythe, machete, or other sharp knife, and cut near the base of the stems. Gather the grains into bundles, tie them with twine, and stand them upright in the plot to finish ripening. Lean three or four bundles together to keep them from falling over. If there is danger of rain, move the

⋩ **An antique thresher.**

sheaves into a barn or other covered area to prevent them from molding. Once all the green has turned to gold, the grains are ready for threshing.

7. The simplest way to thresh is to grasp a bunch of stalks and beat it around the inside of a barrel, heads facing down. The grain will fall right off the stalks. Alternatively, you can lay the stalks down on a hard surface covered by an old sheet and beat the seed heads with a broom or baseball bat. Discard (or compost) the stalks. If there is enough breeze, the chaff will blow away, leaving only the grains. You can also pour the grain and chaff back and forth between two barrels and allow the wind (which can be supplied by a fan if necessary) to blow away the chaff.

8. Store grain in a covered metal trash can or a wooden bin. Be sure it is kept completely dry and that no rodents can get in.

HOW MUCH GRAIN SHOULD YOU GROW TO FEED YOUR FAMILY?

An acre of wheat will supply about 30 bushels of grain, or around 1,800 pounds. The average American consumes about 140 pounds of wheat in a year. The Federal Emergency Management Association (FEMA) recommends the following consumption rates:

- Adult males, pregnant or nursing mothers, active teens ages 14 to 18: 275 lbs./year
- Women, kids ages 7 to 13, seniors: 175 lbs./year
- Children 6 and under: 60 lbs./year

Grain Growing Chart

Type of Grain	Amount of seed per acre (in pounds)	Grain yield per acre (in bushels)	Characteristics and Uses
Amaranth	1	125	Very tolerant of arid environments. High in protein and gluten-free. Use in baking or animal feed.
Barley	100	120 to 140	Tolerates salty and alkaline soils better than most grains. Use in animal feed, soups, as a side dish, and for making beer and malts.
Buckwheat	50	20 to 30	Matures rapidly (sixty to ninety days). Rich, nutty flavor perfect for baking and in pancakes.
Field corn	6 to 8	180 to 190	Use in animal feed, corn starch, hominy, and grits.
Grain sorghum	2 to 8	70 to 100	Drought tolerant. Use in animal feed or in baking.
Oats	80	70 to 100	Thrive in cool, moist climates. High in protein. Use in animal feed, baking, or as a breakfast cereal.
Rye	84	25 to 30	Tolerant of cold, dampness, and drought. Use for animal feed, in baking, to make whiskey, or as a cover crop.
Wheat	75 to 90	40 to 70	Hard red winter wheat is used in bread and is highly nutritious. Soft red winter wheat is good for cakes and pastries. Hard red spring wheat is the most common bread wheat. Durum wheat is best for pasta.

Planting Trees for Shade or Shelter

Trees in your yard can become home to many different types of wildlife. Trees also reduce your cooling costs by providing shade, help clean the air, add beauty and color, provide shelter from the wind and the sun, and add value to your home.

Choosing a Tree

Choosing a tree should be a well thought-out decision. Tree planting can be a significant investment, both in money and time. Selecting the proper tree for your yard can provide you with years of enjoyment, as well as significantly increasing the value of your property. However, a tree that is inappropriate for your property can be a constant maintenance problem, or even a danger to your and others' safety. Before you decide to purchase a tree, take advantage of the many references on gardening at local libraries, universities, arboretums, native plant and gardening clubs, and nurseries. Some questions to consider in selecting a tree include:

1. What purpose will this tree serve?

Trees can serve numerous landscape functions, including beautification, screening of sights and sounds, shade and energy conservation, and wildlife habitat.

2. Is the species appropriate for your area?

Reliable nurseries will not sell plants that are not suitable for your area. However, some mass marketers have trees and shrubs that are not fitted for the environment in which they are sold. Even if a tree is hardy, it may not flower consistently from year to year if the environmental factors are not conducive for it to do so. If you are buying a tree for its spring flowers and fall fruits, consider climate when deciding which species of tree to plant.

Be aware of microclimates. Microclimates are localized areas where weather conditions may vary from the norm. A very sheltered yard may support vegetation not normally adapted to the region. On the other hand, a north-facing slope may be significantly cooler or windier than surrounding areas, and survival of normally adapted plants may be limited.

Select trees native to your area. These trees will be more tolerant of local weather and soil conditions, will enhance natural biodiversity in your neighborhood, and be more beneficial to wildlife than many non-native trees. Avoid exotic trees that can invade other areas, crowd out native plants, and harm natural ecosystems.

3. How big will it get?

When planting a small tree, it is often difficult to imagine that in twenty years it will most likely be shading your entire yard. Unfortunately, many trees are planted and later removed when the tree grows beyond the dimensions of the property.

4. What is the average life expectancy of the tree?

Some trees can live for hundreds of years. Others are considered "short-lived" and may live for only twenty or thirty years. Many short-lived trees tend to be smaller, ornamental species. Short-lived species should not necessarily be ruled out when considering plantings, as they may have other desirable characteristics, such as size, shape, tolerance of shade, or fruit, that would be useful in the landscape. These species may also fill a void in a young landscape, and can be removed as other larger, longer-lived species mature.

5. Does it have any particular ornamental value, such as leaf color or flowers and fruits?

Some species provide beautiful displays of color for short periods in the spring or fall. Other species may have foliage that is

8. Is the tree evergreen or deciduous?

Evergreen trees will provide cover and shade year round. They may also be more effective as wind and noise barriers. On the other hand, deciduous trees will give you summer shade but allow the winter sun to shine in. If planting a deciduous tree, keep these heating and cooling factors in mind when placing the tree in your yard.

Placement of Trees

Proper placement of trees is critical for your enjoyment and for their long-term survival. Check with local authorities about regulations pertaining to placement of trees in your area. Some communities have ordinances restricting placement of trees within a specified distance from a street, sidewalk, streetlight, or other city utilities.

Before planting your tree, consider the tree's potential maximum size. Ask yourself these simple questions:

1. When the tree nears maturity, will it be too close to your or a neighbor's house? An evergreen tree planted on your north side may block the winter sun from your next-door neighbor.
2. Will it provide too much shade for your vegetable and flower gardens? Most vegetables and many flowers require considerable amounts of sun. If you intend to grow these plants in your yard, consider how the placement of trees will affect these gardens.
3. Will the tree obstruct any driveways or sidewalks?
4. Will it cause problems for buried or overhead power lines and utility pipes?

Once you have taken these questions into consideration and have bought the perfect tree for your yard, it is time to start digging!

Planting a Tree

A properly planted and maintained tree will grow faster and live longer than one that is incorrectly planted. Trees can be planted almost any time of the year, as long as the ground is not frozen. Late summer or early fall is the optimum time to plant trees in many areas. By planting during these times, the tree has a chance to establish new roots before winter arrives and the ground freezes. When spring comes, the tree is then ready to grow. Another feasible time for planting

---

THINGS YOU'LL NEED

- Tree
- Shovel
- Watering can or garden hose
- Measuring stick
- Mulch
- Optional: scissors or knife to cut the burlap or container, stakes, and supporting wires

---

trees is late winter or early spring. Planting in hot summer weather should be avoided if possible as the heat may cause the young tree to wilt. Planting in frozen soil during the winter is very difficult, and is tough on tree roots. When the tree is dormant and the ground is frozen, there is no opportunity for the new roots to begin growing.

Trees can be purchased as container-grown, balled and burlapped (B&B), or bare root. Generally, container-grown are the easiest to plant and to successfully establish in any season, including summer. With container-grown stock, the plant has been growing in a container for a period of time. When planting container-grown trees, little damage is done to the roots as the plant is transferred to the soil. Container-grown trees range in size from very small plants in gallon pots up to large trees in huge pots.

B&B trees are dug from a nursery, wrapped in burlap, and kept in the nursery for an additional period of time, giving the roots opportunity to regenerate. B&B plants can be quite large.

Bare root trees are usually extremely small plants. Because there is no soil around the roots, they must be planted when they are dormant to avoid drying out, and the roots must be kept moist until planted. Frequently, bare root trees are offered by seed and nursery mail order catalogs, or in the wholesale trade. Many state-operated nurseries and local conservation districts also sell bare root stock in bulk quantities for only a few cents per plant. Bare root plants are usually offered in the early spring and should be planted as soon as possible.

Be sure to carefully follow the planting instructions that come with your tree. If specific instructions are not available, here are some general tree-planting guidelines:

1. Before starting any digging, call your local utility companies to identify the location of any underground wires or lines. In the United States, you can call

---

reddish or variegated and can add color in your yard year round. Trees bearing fruits or nuts can provide an excellent source of food for many species of wildlife.

6. Does it have any particular insect, disease, or other problem that may reduce its usefulness in the future?

Certain insects and diseases can cause serious problems for some desirable species in certain regions. Depending on the pest, control of the problem may be difficult and the pest may significantly reduce the attractiveness, if not the life expectancy, of the tree. Other species, such as the silver maple, are known to have weak wood that is susceptible to damage in ice storms or heavy winds. All these factors should be kept in mind, as controlling pests or dealing with tree limbs that have snapped in foul weather can be expensive and potentially damaging.

7. How common is this species in your neighborhood or town?

Some species are over-planted. Increasing the natural diversity in your area will provide habitat for wildlife and help limit the opportunity for a single pest to destroy large numbers of trees.

811 to have your utility lines marked for free.

2. Dig a hole twice as wide as, and slightly shallower than, the root ball. Roughen the sides and bottom of the hole with a pick or shovel so that the roots can easily penetrate the soil.

3. With a potted tree, gently remove the tree from the container. To do this, lay the tree on its side with the container end near the planting hole. Hit the bottom and sides of the container until the root ball is loosened. If roots are growing in a circular pattern around the root ball, slice through the roots on a couple of sides of the root ball. With trees wrapped in burlap, remove the string or wire that holds the burlap to the root crown; it is not necessary to remove the burlap completely. Plastic wraps must be completely removed. Gently separate circling roots on the root ball. Shorten exceptionally long roots and guide the shortened roots downward and outward. Root tips die quickly when exposed to light and air, so complete this step as quickly as possible.

4. Place the root ball in the hole. Leave the top of the root ball (where the roots end and the trunk begins) ½ to 1 inch above the surrounding soil, making sure not to cover it unless the roots are exposed. For bare root plants, make a mound of soil in the middle of the hole and spread plant roots out evenly over the mound. Do not set the tree too deep into the hole.

5. As you add soil to fill in around the tree, lightly tap the soil to collapse air pockets, or add water to help settle the soil. Form a temporary water basin around the base of the tree to encourage water penetration, and be sure to water the tree thoroughly after planting. A tree with a dry root ball cannot absorb water; if the root ball is extremely dry, allow water to trickle into the soil by placing the hose at the trunk of the tree.

6. Place mulch around the tree. A circle of mulch, 3 feet in diameter, is common.

7. Depending on the size of the tree and the site conditions, staking the tree in place may be beneficial. Staking supports the tree until the roots are well established to properly anchor it. Staking should allow for some movement of the tree on windy days. After trees are established, remove all supporting wires. If these are not removed, they can girdle the tree, cut into the trunk, and eventually kill the tree.

Maintenance

For the first year or two, especially after a week or so of especially hot or dry weather, watch your tree closely for signs of moisture stress. If you see any leaf wilting or hard, caked soil, water the tree well and slowly enough to allow the water to soak in. This will encourage deep root growth. Keep the area under the tree mulched.

Some species of evergreen trees may need protection against winter sun and wind. A thorough watering in the fall before the ground freezes is recommended.

Fertilization is usually not needed for newly planted trees. Depending on the soil and growing conditions, fertilizer may be beneficial at a later time.

Young trees need protection against rodents, frost cracks, sunscald, lawn mowers, and weed whackers. In the winter months, mice and rabbits frequently girdle small trees by chewing away the bark at the snow level. Since the tissues that transport nutrients in the tree are located just under the bark, a girdled tree often dies in the spring when growth resumes. Weed whackers are also a common cause of girdling. In order to prevent girdling from occurring, use plastic guards, which are inexpensive and easy to control.

Frost cracking is caused by the sunny side of the tree expanding at a different rate than the colder, shaded side. This can cause large splits in the trunk. To prevent this, wrap young trees with paper tree wrap, starting from the base and wrapping up to the bottom branches. Sunscald can occur when a young tree is suddenly moved from a shady spot into direct sunlight. Light-colored tree wraps can be used to protect the trunk from sunscald.

Pruning

Usually, pruning is not needed on newly planted trees. As the tree grows, lower branches may be pruned to provide clearance above the ground, or to remove dead or damaged limbs or suckers that sprout from the trunk. Sometimes larger trees need pruning to allow more light to enter the canopy. Small branches can be removed easily with pruners. Large branches should be removed with a pruning saw. All cuts should be vertical. This will allow the tree to heal quickly without the use of any artificial sealants. Major pruning should be done in late winter or early spring. At this time, the tree is more likely to "bleed," as sap is rising through the plant. This is actually healthy and will help prevent invasion by many disease-carrying organisms.

Under no circumstance should trees be topped (topping is chopping off large top tree branches). Not only does this practice ruin the natural shape of the tree, but it increases its susceptibility to diseases and results in very narrow crotch angles (the angle between the trunk and the side branch). Narrow crotch angles are weaker than wide ones and more susceptible to damage from wind and ice. If a large tree requires major reduction in height or size, contact a professionally trained arborist.

Final Thoughts

Trees are natural windbreaks, slowing the wind and providing shelter and food for wildlife. Trees can help protect livestock, gardens, and larger crops. They also help prevent dust particles from adding to smog over urban areas. Tree plantings are key components of an effective conservation system and can provide your yard with beauty, shade, and rich natural resources.

Container Gardening

An alternative to growing vegetables, flowers, and herbs in a traditional garden is to grow them in containers. While the amount that can be grown in a container is certainly limited, container gardens works well for tomatoes, peppers, cucumbers, herbs, salad greens, and many flowering annuals. Choose vegetable varieties that have been specifically bred for container growing. You can obtain this information online or at your garden center. Container gardening also brings birds and butterflies right to your doorstep. Hanging baskets of fuchsia or pots of snapdragons are frequently visited by hummingbirds, allowing for up-close observation.

Container gardening is an excellent method of growing vegetables, herbs, and flowers, especially if you do not have adequate outdoor space for a full garden bed. A container garden can be placed anywhere—on the patio, balcony, rooftop, or windowsill. Vegetables such as leaf lettuce, radishes, small tomatoes, and baby carrots can all be grown successfully in pots.

How to Grow Vegetables in a Container Garden

Here are some simple steps to follow for growing vegetables in containers:

1. Choose a sunny area for your container plants. Your plants will need at least five to six hours of sunlight a day. Some plants, such as cucumbers, may need more. Select plants that are suitable for container growing. Usually their names will contain words such as "patio," "bush," "dwarf," "toy," or "miniature." Peppers, onions, and carrots are also good choices.
2. Choose a planter that is at least five gallons, unless the plant is very small. Poke holes in the bottom if they don't already exist; the soil must be able to drain in order to prevent the roots from rotting. Avoid terracotta or dark colored pots as they tend to dry out quickly.

3. Fill your container with potting soil. Good potting soil will have a mixture of peat moss and vermiculite. You can make your own potting soil using composted soil. Read the directions on the seed packet or label to determine how deep to plant your seeds.

4. Check the moisture of the soil frequently. You don't want the soil to become muddy, but the soil should always feel damp to the touch. Do not wait until the plant is wilting to water it—at that point, it may be too late.

Things to Consider

- Follow normal planting schedules for your climate when determining when to plant your container garden.
- You may wish to line your container with porous materials such as shredded newspaper or rags to keep

the soil from washing out. Be sure the water can still drain easily.

How to Grow Herbs in a Container

Herbs will thrive in containers if cared for properly. And if you keep them near your kitchen, you can easily snip off pieces to use in cooking. Here's how to start your own herb container garden:

1. If your container doesn't already have holes in the bottom, poke several to allow the soil to drain. Pour gravel into the container until it is about a quarter of the way full. This will help the water drain and help to keep the soil from washing out.

2. Fill your container three-quarters of the way with potting soil or a soil-based compost.

3. It's best to use seedlings when planting herbs in containers. Tease the roots slightly, gently spreading them apart with your fingertips. This will encourage them to spread once planted. Place each herb into the pot and cover the root base with soil. Place herbs that will grow taller in the center of your container, and the smaller ones around the edges. Leave about 4 square inches of space between each seedling.

4. As you gently press in soil between the plants, leave an inch or so between the container's top and the soil. You don't want the container to overflow when you water the herbs.

5. Cut the tops off the taller herb plants to encourage them to grow faster and to produce more leaves.

6. Pour water into the container until it begins to leak out the bottom. Most herbs like to dry out between watering, and over-watering can cause some herbs to rot and die, so only water every few days unless the plants are in a very hot place.

Things to Consider

- Growing several kinds of herbs together helps the plants to thrive. A few exceptions to this rule are oregano, lemon balm, and tea balm. These herbs should be planted on their own because they will overtake the other herbs in your container.
- You may wish to choose your herbs according to color to create attractive

arrangements for your home. Any of the following herbs will grow well in containers:

- Silver herbs: artemisias, curry plants, santolinas
- Golden herbs: lemon thyme, calendula, nasturtium, sage, lemon balm
- Blue herbs: borage, hyssop, rosemary, catnip
- Green herbs: basil, mint, marjoram, thyme, parsley, chives, tarragon
- Pink and purple herbs: oregano (the flowers are pink), lavender
- If you decide to transplant your herbs in the summer months, they will grow quite well outdoors and will give you a larger harvest.

How to Grow Flowers from Seeds in a Container

1. Cover the drainage hole in the bottom of the pot with a flat stone. This will keep the soil from trickling out when the plant is watered.
2. Fill the container with soil. The container should be filled almost to the top. For the best results, use potting soil from your local nursery or garden center.
3. Make holes for the seeds. Refer to the seed packet to see how deep to make the holes. Always save the seed packet for future reference—it most likely has helpful directions about thinning young plants.
4. Place a seed in each hole. Pat the soil gently on top of each seed.
5. Use a light mist to water your seeds, making sure that the soil is only moist and not soaked.
6. Make sure your seeds get the correct amount of sunlight. Refer to the seed packet for the adequate amount of sunlight each seedling needs.
7. Watch your seeds grow. Most seeds take three to seventeen days to sprout. Once the plants start sprouting, be sure to pull out plants that are too close together so the remaining plants will have enough space to establish good root systems.
8. Remember to water and feed your container plants. Keep the soil moist so your plants can grow. And in no time at all, you should have wonderful flowers growing in your container garden.

Preserving Your Container Plants

As fall approaches, frost will soon descend on your container plants and can ultimately destroy your garden. Container plants are particularly susceptible to frost damage, especially if you are growing tropical plants, perennials, and hardy woody plants in a single container garden. There are many ways that you can preserve and maintain your container garden plants throughout the winter season.

Preservation techniques will vary depending on the plants in your container garden. Tropical plants can be over-wintered using methods replicating a dry season, forcing the plant into dormancy; hardy perennials and woody shrubs need a cold dormancy to grow in the spring, so they must stay outside; cacti and succulents prefer their winters warm and dry and must be brought inside, while many annuals can be propagated by stem cuttings or can just be repotted and maintained inside.

Preserving Tropical Bulbs and Tubers

Many tropical plants, such as cannas, elephant ears, and angel's trumpets can be saved from an untimely death by over-wintering them in a dark corner or sunny window of your home, depending on the type of plant. A lot of bulbous and tuberous tropical plants have a natural dry season (analogous to our winter) when their leafy parts die off, leaving the bulb behind. Don't throw the bulbs away. After heavy frosts turn the aboveground plant parts to mush, cut the damaged foliage off about 4 inches above the thickened bulb. Then, dig them up and remove all excess soil from the roots. If a bulb has been planted for several years and it's performance is beginning to decline, it may need dividing. Daffodil's, for example, should generally be divided every three years. If you do divide the bulb, be sure to dust all cut surfaces with a sulfur-based fungicide made for bulbs to prevent the wounds from rotting. Cut the roots back to 1 inch from the bulb and leave to dry out evenly. Rotten bulbs or roots need to be thrown away so infection doesn't spread to the healthy bulbs.

A bulb's or tuber's drying time can last up to two weeks if it is sitting on something absorbent like newspaper and located somewhere shaded and dry, such as a garage or basement. Once clean and dry, bulbs

watered sparingly until they can be placed outside. The emerging leaves will be stunted, but once outside, the plant will replace any spindly leaves with lush, new ones.

Annuals

Many herbaceous annuals can also be saved for the following year. By rooting stem cuttings in water on a sunny windowsill, plants like impatiens, coleus, sweet potato vine cultivars, and purple heart can be held over winter until needed in the spring. Otherwise, the plants can be cut back by half, potted in a peat-based, soilless mix, and placed on a sunny windowsill. With a wide assortment of "annuals" available on the market, some research is required to determine which annuals can be over-wintered successfully. True annuals (such as basils, cockscomb, and zinnias)—regardless of any treatment given—will go to seed and die when brought inside.

Cacti and Succulents

If you planted a mixed dry container this year and want to retain any of the plants for next year, they should be removed from the main container and repotted into a high-sand-content soil mix for cacti and succulents. Keep them in a sunny window and water when dry. Many succulents and cacti

should be stored—preferably at around 50°F—all winter in damp (not soggy) milled peat moss. This prevents the bulbs from drying out any further, which could cause them to die. Many gardeners don't have a perfectly cool basement or garage to keep bulbs dormant. Alternative methods for dry storage include a dark closet with the door cracked for circulation, a cabinet, or underneath a bed in a cardboard box with a few holes punched for airflow. The important thing to keep in mind is that the bulb needs to be kept on the dry side, in the dark, and moderately warm.

If a bulb was grown as a single specimen in its own pot, the entire pot can be placed in a garage that stays above 50°F or a cool basement and allowed to dry out completely. Cut all aboveground plant parts flush with the soil and don't water until the outside temperatures stabilize above 60°F. Often, bulbs break dormancy unexpectedly in this dry pot method. If this happens, pots can be moved to a sunny location near a window and

Container Gardening

do well indoors, either in a heated garage or a moderately sunny corner of a living room.

As with other tropical plants, succulents also need time to adjust to sunnier conditions in the spring. Move them to a shady spot outside when temperatures have stabilized above 60°F and then gradually introduce them to brighter conditions.

Hardy Perennials, Shrubs, and Vines

Hardy perennials, woody shrubs, and vines needn't be thrown away when it's time to get rid of accent containers. Crack-resistant, four-season containers can house perennials and woody shrubs year-round. Below is a list of specific perennials and woody plants that do well in both hot and cold weather, indoors and out:

- Shade perennials, like coral bells, lenten rose, assorted hardy ferns, and Japanese forest grass are great for all weather containers.
- Sun-loving perennials, such as sedges, some salvias, purple coneflower, daylily, spiderwort, and bee blossom are also very hardy and do well in year-round containers. Interplant them with cool growing plants, like kale, pansies, and Swiss chard, for fall and spring interest.
- Woody shrubs and vines—many of which have great foliage interest with four-season appeal—are ideal for container gardens. Red-twigged dogwood cultivars, clematis vine cultivars, and dwarf crape myrtle cultivars are great container additions that can stay outdoors year-round.

If the container has to be removed, hardy perennials and woody shrubs can be temporarily planted in the ground and mulched.

⌃ If your floor space is limited, a ladder can provide additional surface area on which to keep your potted plants.

Dig them from the garden in the spring, if you wish, and replant into a container. Or, leave them in their garden spot and start over with fresh ideas and new plant material for your container garden.

Sustainable Plants and Money in Your Pocket

Over-wintering is a great form of sustainable plant conservation achieved simply and effectively by adhering to each plant's cultural and environmental needs. With careful planning and storage techniques, you'll save money as well as plant material. The beauty and interest you've created in this season's well-grown container garden can also provide enjoyment for years to come.

Rooftop Gardens

If you live in an urban area and don't have a lawn, that does not mean that you cannot have a garden. Whether you live in an apartment building or you're a homeowner without yard space, you can grow your very own garden, right on your roof!

Is Your Roof Suitable for a Rooftop Garden?

Theoretically, any roof surface can be greened—even sloped or curved roofs can support a layer of sod or wildflowers. However, if the angle of your roof is over 30 degrees you should consult with a specialist. Very slanted roofs make it difficult to keep the soil in place until the plants' roots take hold. Certainly, a flat roof, approximating level ground conditions, is the easiest on which to grow a garden, though a slight slant can be helpful in allowing drainage.

Also consider how much weight your roof can bear. A simple, lightweight rooftop garden will weigh between 13 and 30 pounds per square foot. Add to this your own weight—or that of anyone who will be tending or enjoying the garden—gardening tools, and, if you live in a colder climate, the additional weight of snow in the winter.

Will a Rooftop Garden Cause Water Leakage or Other Damage?

No. In fact, planting beds or surfaces are often used to protect and insulate roofs. However, you should take some precautions to protect your roof:

1. Cover your roof with a layer of waterproof material, such as a heavy-duty pond liner. You may want to place an old rug on top of the waterproof material to help it stay in place and to give additional support to the materials on top.
2. Place a protective drainage layer on top of the waterproof material. Otherwise, shovels, shoe heels, or dropped tools could puncture the roof. Use a coarse material such as gravel, pumice, or expanded shale.
3. Place a filter layer on top of the drainage layer to keep soil in place so that it won't clog up your drainage. A lightweight polyester geotextile (an inexpensive, non-woven fabric found at most home improvement stores) is ideal for this. Note that if your roof has an angle greater than 10 degrees, only install the filter layer around the edges of the roof as it can increase slippage.
4. Using moveable planters or containers, modular walkways and surfacing treatment, and compartmentalized planting beds will make it easier to fix leaks should they appear.

BENEFITS OF ROOFTOP GARDENING

- Create more outdoor green space within your urban environment.
- Grow your own fresh vegetables—even in the city.
- Improve air quality and reduce CO_2 emissions.
- Help delay storm water runoff.
- Give additional insulation to building roofs.
- Reduce noise.

How to Make a Rooftop Garden

Preparation

1. Before you begin, find out if it is possible and legal to create a garden on your roof. You don't want to spend lots of time and money preparing for a garden and then find out that it is prohibited.
2. Make sure that the roof is able to hold the weight of a rooftop garden. If so, figure out how much weight it can hold. Remember this when making the garden and use lighter containers and soil as needed.

Setting Up the Garden

1. Install your waterproof, protective drainage, and filter layers, as described above. If your roof is angled, you may want to place a wooden frame around the edges of the roof to keep the layers from sliding off. Be sure to use rot-resistant wood and cut outlets into the frame to allow excess water to drain away. Layer pebbles around the outlets to aid drainage and to keep vegetation from clogging them.
2. Add soil to your garden. It should be 1 to 4 inches thick and will be best if it's a mix

of three-fourths inorganic soil (crushed brick or a similar granular material) and one-fourth organic compost.

Planting and Maintaining the Garden

1. Start planting. You can plant seeds or seedlings, or transplant mature plants. Choose plants that are wind-resistant and won't need a great deal of maintenance. Sedums make excellent rooftop plants as they require very little attention once planted, are hardy, and are attractive throughout most of the year. Most vegetables can be grown in-season on rooftops, though the wind will make taller vegetables (like corn or beans) difficult to grow. If your roof is slanted, plant drought-resistant plant varieties near the peak, as they'll get less water.

2. Water your garden immediately after planting, and then regularly throughout the growing season, unless rain does the work for you.

Things to Consider

1. If you live in a very hot area, you may want to build small wooden platforms to elevate your plants above the hot rooftop. This will help increase the ventilation around the plants.

2. When determining whether or not your roof is strong enough to support a garden, remember that large pots full of water and soil will be very heavy, and if the roof is not strong enough, your garden could cause structural damage.

3. You can use pots or other containers on your rooftop rather than making a full garden bed. You should still first find out how much weight your roof can hold and choose lightweight containers.

4. Consider adding a fence or railing around your roof, especially if children will be helping in the garden.

⌃ In many cases, potted plants make more sense than laying a garden bed when it comes to green roofs.

Raised Beds

If you live in an area where the soil is quite wet (preventing a good vegetable garden from growing the spring), or find it difficult to bend over to plant and cultivate your vegetables or flowers, or if you just want a different look to your backyard garden, you should consider building a raised bed.

A raised bed is an interesting and affordable way to garden. It creates an ideal environment for growing vegetables, since the soil concentration can be closely monitored and, as it is raised above the ground, it reduces the compaction of plants from people walking on the soil.

Raised beds are typically 2 to 6 feet wide and as long as needed. In most cases, a raised bed consists of a "frame" that is filled in with nutrient-rich soil (including compost or organic fertilizers) and is then planted with a variety of vegetables or flowers, depending on the gardener's preference. By controlling the bed's construction and the soil mixture that goes into the bed, a gardener can effectively reduce the amount of weeds that will grow in the garden.

When planting seeds or young sprouts in a raised bed, it is best to space the plants equally from each other on all sides. This will ensure that the leaves will be touching once the plant is mature, thus saving space and reducing the soil's moisture loss.

How to Make a Raised Bed

Step One: Plan Out Your Raised Bed

1. Think about how you'd like your raised bed to look, and then design the shape. A raised bed is not extremely complicated, and all you need to do is build an open-top and open-bottom box (if you are ambitious, you can create a raised bed in the shape of a circle, hexagon, or star). The main purpose of this box is to hold soil.
2. Make a drawing of your raised bed, measure your available garden space, and add those measurements to your drawing. This will allow you to determine how much material is needed. Generally, your bed should be at least 24 inches in height.
3. Decide what kind of material you want to use for your raised bed. You can use lumber, plastic, synthetic wood, railroad ties, bricks, rocks, or a number of other items to hold the dirt. Using lumber is the easiest and most efficient method.
4. Gather your supplies.

Step Two: Build Your Raised Bed

1. Make sure your bed will be situated in a place that gets plenty of sunlight. Carefully assess your placement, as your raised bed will be fairly permanent.
2. Connect the sides of your bed together (with either screws or nails) to form the desired shape of your bed. If you are using lumber, you can use 4-inch x 4-inch posts to serve as the corners of your bed, and then nail or screw the sides to these corner posts. By doing so, you will increase strength of the structure and ensure that the dirt will stay inside.
3. Cut a piece of gardening plastic to fit inside your raised bed. This will significantly reduce the amount of weeds growing in your garden. Lay it out in the appropriate location.
4. Place your frame over the gardening plastic (this might take two people).

Step Three: Start Planting

1. Add some compost into the bottom of the bed and then layer potting soil on top of the compost. If you have soil from other parts of your yard, feel free to use that in

addition to the compost and potting soil. Plan on filling at least one-third of your raised bed with compost or composted manure (available from nurseries or garden centers in 40-pound bags).

2. Mix in dry organic fertilizers (like wood ash, bone meal, and blood meal) while building your bed. Follow the package instructions for how best to mix it in.

3. Decide what you want to plant. Some people like to grow flowers in their raised beds; others prefer to grow vegetables. If you do want to grow food, raised beds are excellent choices for salad greens, carrots, onions, radishes, beets, and other root crops.

Things to Consider

1. To save money, you can dig up and use soil from your yard. Potting soil can be expensive, and yard soil is just as effective when mixed with compost. However, removing grass and weeds

---

THINGS YOU'LL NEED

- Forms for your raised bed (consider using 4-inch x 4-inch posts cut to 24 inches in height for corners, and 2-inch x 12-inch boards for the sides)
- Nails or screws
- Hammer or screwdriver
- Plastic liner (to act as a weed barrier at the bottom of your bed)
- Shovel
- Compost or composted manure
- Soil (either potting soil or soil from another part of your yard)
- Rake (to smooth out the soil once in the bed)
- Seeds or young plants
- Optional: PVC piping and greenhouse plastic (to convert your raised bed to a greenhouse)

---

from the soil before filling your raised beds can be time-consuming.

---

2. Be creative when building your raised planting bed. You can construct a great raised bed out of recycled goods or old lumber.

3. You can convert your raised bed into a greenhouse. Just add hoops to your bed by bending and connecting PVC pipe over the bed. Then clip greenhouse plastic to the PVC pipes, and you have your own greenhouse.

4. Make sure to water your raised bed often. Because it is above ground, your raised bed will not retain water as well as the soil in the ground. If you keep your bed narrow, it will help conserve water.

5. Decorate or illuminate your raised bed to make it a focal point in your yard.

6. If you use lumber to construct your raised bed, keep a watch out for termites.

7. Beware of old pressure-treated lumber, as it may contain arsenic and could potentially leak into the root systems of any vegetables you might grow in your raised bed. Newer pressure-treated lumber should not contain these toxic chemicals.

Growing Plants Without Soil

Hydroponics is the method of growing plants in a container filled with a nutrient-rich bath (water with special fertilizer) and no soil. Plants grown in soilless cultures still need the basic requirements of plant growth, such as temperature, light (if indoors, use a heat-lamp and set the container near or on a windowsill), water, oxygen (you can produce good airflow by using a small, rotating fan indoors), carbon dioxide, and mineral nutrients (derived from solutions). But instead of planting their roots in soil, hydroponically grown plants have their roots either free-floating in a nutrient-rich solution or bedded in a soil-like medium, such as sand, gravel, brick shards, Perlite, or rockwool. These plants do not have to exert as much energy to gather nutrients from the soil and thus they grow more quickly and, usually, more productively.

The Benefits and Drawbacks of Growing Plants in a Hydroponics System

Benefits:

- Plants can be grown in areas where normal plant agriculture is difficult (such as deserts and other arid places, or cities).
- Most terrestrial plants will grow in a hydroponics system.

- There is minimal weed growth.
- The system takes up less space than a soil system.
- It conserves water.
- No fear of contaminated runoff from garden fertilizers.
- There is less labor and cost involved.
- Certain seasonal plants can be raised during any season.
- The quality of produce is generally consistent.
- Old nutrient solution can be used to water houseplants.

Drawbacks:

- Can cause salmonella to grow due to the wet and confined conditions.
- More difficult to grow root vegetables, such as carrots and potatoes.
- If nutrient solution is not regularly changed, plants can become nutrient deficient and thus not grow or produce.

Types of Hydroponics Systems

There are two main types of soilless cultures that can be used in order to grow plants and vegetables. The first is a water culture, in which plants are supplied with mineral nutrients directly from the water solution. The second, called aggregate culture or "sand culture," uses an aggregate (such as sand, gravel, or Perlite) as soil to provide an anchoring support for the plant roots. Both types of hydroponics are effective in growing soilless plants and in providing essential nutrients for healthy and productive plant growth.

Water Culture

The main advantage of using a water culture system is that a significant part of the nutrient solution is always in contact with the plants' roots. This provides an adequate amount of water and nutrients. The main disadvantages of this system are providing sufficient air supply for the roots and providing the roots with proper support and anchorage.

Water culture systems are not extremely expensive, though the cost does depend on the price of the chemicals and water used in the preparation of the nutrient solutions, the size of your container, and whether or not your are using mechanized objects, such as pumps and filters. You can decrease the cost by starting small and using readily available materials.

Materials Needed to Make Your Own Water Culture

A large water culture system will need either a wood or concrete tank 6 to 18 inches deep and 2 to 3 feet wide. If you use a wooden container, be sure there are no knots in the wood and seal the tank with non-creosote or tar asphalt.

For small water culture systems, which are recommended for beginners, glass jars,

earthenware crocks, or plastic buckets will suffice as your holding tanks. If your container is transparent, be sure to paint the outside of the container with black paint to keep the light out (and to keep algae from growing inside your system). Keep a narrow vertical strip unpainted in order to see the level of the nutrient solution inside your container.

The plant bed should be 3 or more inches deep and large enough to cover the container or tank. In order to support the weight of the litter (where your seeds or seedlings are placed), cover the bottom of the bed with chicken wire and then fill the bed with litter (wood shavings, sphagnum moss, peat, or other organic materials that do not easily decay). If you are starting your plants from seeds, germinate the seeds in a bed of sand and then transplant to the water culture bed, keeping the bed moist until the plants get their roots down into the nutrient solution.

Aeration

A difficulty in using water culture is keeping the solution properly aerated. It is important to try to keep enough space between the seed bed and the nutrient solution so the plant's roots can receive proper oxygen. In order to make sure that air can easily flow into the container, either prop up the seed bed slightly to allow air flow or drill a hole in your container just above the highest solution level.

In order to make sure there is sufficient oxygen reaching the plant roots, you can install an aquarium pump in your water culture system. Just make sure that the water is not agitated too much or the roots may be damaged. You can also use an air stone or

perforated pipe to gently introduce air flow into your container.

Water Supply

Your hydroponics system needs an adequate supply of fresh water in order to maintain healthy plant life. Make sure that the natural minerals in your water are not going to adversely affect your hydroponics plants. If there is too much sodium in your water (usually an effect of softened water), it could become toxic to your plants. In general, the minerals in water are typically not harmful to the growth of your plants.

Nutrient Solution

You may add nutrient solution by hand, by a gravity-feed system, or mechanically. In smaller water culture systems, mixing the nutrient solution in a small container and adding it by hand, as needed, is typically adequate.

If you are using a larger setup, a gravity-feed system will work quite well. In this type of system, the nutrient solution is mixed in a vat and then tapped from the vat into your container as needed. You can use a plastic container or larger earthenware jar as the vat.

A pump can also be used to supply your system with adequate nutrient solution. You can insert the pump into the vat and then transfer the solution to your hydroponics system.

When your plants are young, it is important to keep the space between your seed bed and the nutrient solution quite small (that way, the young plant roots can reach the nutrients). As your plants grow, the amount of space between the bed and solution should increase (but do this slowly and keep the level rather consistent).

If the temperature is rather high and there is increased evaporation, it is important to keep the roots at the correct level in the water and change the nutrient solution every day, if needed.

Drain your container every two weeks and then renew the nutrient solution from your vat or by hand. This must be done in a short amount of time, so the roots do not dry out.

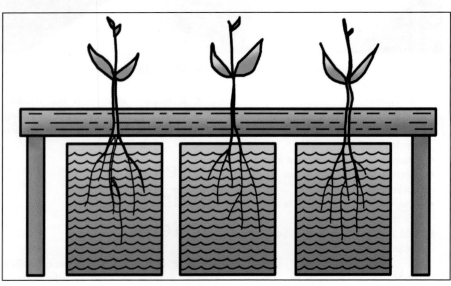

⌃ In a water culture system, the roots are always in contact with the nutrient solution.

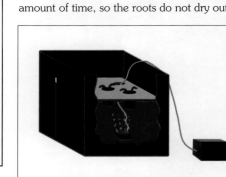

⌃ A simple hydroponics system

Credit: Timothy W. Lawrence

Transplanting

When transplanting your seedlings, it's important to make sure that you are careful with the tiny root systems. Gently work the roots through the support netting and down into the nutrient solution. Then fill in the support netting with litter to help the plant remain upright.

THINGS YOU'LL NEED

- External pump
- Air line or tubing
- Air stones
- Waterproof bin, bucket, or fish tank to use as a reserve
- Styrofoam
- Net pots
- Type of growing medium, such as rockwool or grow rocks
- Hydroponics nutrients, such as grow formula, bloom formula, supplements, and pH
- Black spray paint (this is only required if the reservoir is transparent)
- Knife, box cutter, or scissors
- Tape measure

How to Build a Simple, Homemade Hydroponics System

Steps to Building Your Hydroponics System

1. Find a container to use as a reservoir, such as a fish tank, a bin, or a bucket. The reservoir should be painted black if it is not lightproof (or covered with a thick, black trash bag if you want to reuse the tank at some point), and allowed to dry before moving on to the next step. Allowing light to enter the reservoir will promote the growth of algae. Use a reservoir that is the same dimensions (length and width) from top to bottom.
2. Using a knife or sharp object, score a line on the tank (scratch off some paint in a straight line from top to bottom). This will be your water level meter, which will allow you to see how much water is in the reservoir and will give you a more accurate and convenient view of the nutrient solution level in your tank.
3. Use a tape measure to determine the length and width of your reservoir. Measure the inside of the reservoir from one end to the other. Once you have the dimensions, cut the Styrofoam ¼ inch smaller than the size of the reservoir. For example, if your dimensions are 36 x 20 inches, you should cut the Styrofoam to 35 ¾ x 19¾ inches. The Styrofoam should fit nicely in the reservoir, with just enough room to adjust to any water level changes. If the reservoir tapers off at the bottom (the bottom is smaller in dimension than the top) the floater (Styrofoam) should be 2 to 4 inches smaller than the reservoir, or more if necessary.
4. Do not place the Styrofoam in the reservoir yet. First, you need to cut holes for the net pots. Put the net pots on the Styrofoam where you want to place each plant. Using a pen or pencil, trace around the bottom of each net pot. Use a knife or box cutter to follow the trace lines and cut the holes for pots. On one end of the Styrofoam, cut a small hole for the air line to run into the reservoir.
5. The number of plants you can grow will depend on the size of the garden you build and the types of crops you want to grow. Remember to space plants appropriately so that each receives ample amounts of light.
6. The pump you choose must be strong enough to provide enough oxygen to sustain plant life. Ask for advice choosing a pump at your local hydroponics supply store or garden center.
7. Connect the air line to the pump and attach the air stone to the free end. The air line should be long enough to travel from the pump into the bottom of the reservoir, or at least float in the middle of the tank so the oxygen bubbles can get to the plant roots. It also must be the right size for the pump you choose. Most pumps will come with the correct size air line. To determine the tank's capacity, use a one-gallon bucket or bottle and fill the reservoir. Remember to count how many gallons it takes to fill the reservoir and you will know the correct capacity of your tank.

THE JUNIOR HOME-STEADER

Encourage middle or high school aged kids to come up with their own hydroponic designs. Encourage them to research different methods and find materials around the house. Hydroponic growing systems make great science fair projects!

Setting Up Your Hydroponics System

1. Fill the reservoir with the nutrient solution.
2. Place the Styrofoam into the reservoir.
3. Run the air line through the designated hole or notch.
4. Fill the net pots with growing medium and place one plant in each pot.
5. Put the net pots into the designated holes in the Styrofoam.
6. Plug in the pump, turn it on, and start growing with your fully functional, homemade hydroponics system.

Things to Consider

- A homemade hydroponics system like this is not ideal for large-scale production of plants or for commercial usage. This particular system does not offer a way to conveniently change the nutrient solution. An extra container would be required to hold the floater while you change the solution.
- Lettuce, watercress, tomatoes, cucumbers, and herbs grow especially well hydroponically.

Aggregate Culture

This type of hydroponics system utilizes different mediums that act in the place of soil to stabilize the plant and its roots. The aggregate in the container is flooded with the nutrient solution. The advantage of this type of system is that there is not as much trouble with aerating the roots. Also, aggregate culture systems allow for the easy transplantation of seedlings into the aggregate medium and it is less expensive.

Materials Needed for an Aggregate Culture System

The container should be watertight to help conserve the nutrient solution. Large tanks can be made of concrete or wood, and smaller operations can effectively be done in glass jars, earthenware containers, or plastic buckets. Make sure to paint transparent containers black.

Aggregate materials may differ greatly, depending on what type you choose to use. Silica sand (well washed) is one of the best materials that can be used. Any other type of coarse-textured sand is also effective, but make sure it does not contain lime. Sand holds moisture quite well and it allows for easy transplantation. A mixture of sand and gravel together is also an effective aggregate. Other materials, such as peat moss, vermiculite, wood shavings, and coco peat, are also good aggregates. You can find aggregate materials at your local garden center, home center, or garden-supply house.

Aeration

Aggregate culture systems allow much easier aeration than water culture systems. Draining and refilling the container with nutrient solution helps the air to move in and out of the aggregate material. This brings a fresh supply of oxygen to the plant roots.

Water Supply

The same water requirements are needed for this type of hydroponics system as for a water culture system. Minerals in the water tend to collect in the aggregate material, so it's a good idea to flush the material with fresh water every few weeks.

Nutrient Solution

The simplest way of adding the nutrient solution to aggregate cultures is to pour it over the aggregates by hand. You may also use a manual gravity-feed system with buckets or vats. Attach the vat to the bottom of the container with a flexible hose, raise the vat to flood the container, and lower it to drain it. Cover the vat to prevent evaporation and replenish it with new nutrient solution once every two weeks.

A gravity drip-feed system also works well and helps reduce the amount of work you do. Place the vat higher than the container, and then control the solution drip so it is just fast enough to keep the aggregate moist.

It is important that the nutrient solution is added and drained or raised and lowered at least once a day. In hotter weather, the aggregate material may need more wetting with the solution. Make sure that the material is not drying out the roots. Drenching the aggregate with solution often will not harm the plants but letting the roots dry out could have detrimental effects.

Always replace your nutrient solution after two weeks. Not replacing the solution will cause salts and harmful fertilizer residues to build up, which may ultimately damage your plants.

Planting

You may use either seedlings or rooted cuttings in an aggregate culture system. The aggregate should be flooded and solution drained before planting to create a moist, compacted seed bed. Seeds may also be planted directly into the aggregate material. Do not plant the seeds too deep, and flood the container frequently with water to keep the aggregate moist. Once the seedlings have germinated, you may start using the nutrient solution.

If you are transplanting seedlings from a germination bed, make sure they have germinated in soilless material, as any soil left on the roots may cause them to rot and may hamper them in obtaining nutrients from the solution.

Making Nutrient Solutions

In order for plants to grow properly, they must receive nitrogen, phosphorous, potassium, calcium, magnesium, sulfur, iron, manganese, boron, zinc, copper, molybdenum, and chlorine. There are a wide range of nutrient solutions that can be used. If your plants are receiving inadequate amounts of nutrients, they will show this in different ways. This means that you must proceed with caution when selecting and adding the minerals that will be present in your nutrient solution.

It is important to have pure nutrient materials when preparing the solution. Using fertilizer-grade chemicals is always the best route to go, as it is cheapest. When housing your nutrient solution, be sure the containers are not transparent (if they are, paint them black to keep out the light). Make sure the containers are closed and not exposed to air. Evaporation from solution concentrates the amount of salt, which could harm your plants.

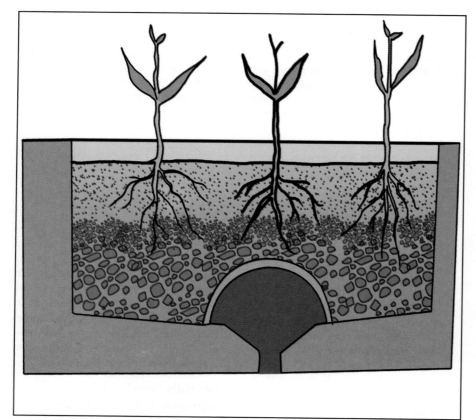

⩓ A simple aggregate culture system

Salt	Grade	Nutrients	Amt. for 25 gallons of solution
Potassium phosphate	Technical	Potassium, phosphorus	½ ounce (1 Tbsp)
Potassium nitrate	Fertilizer	Potassium, nitrogen	2 ounces (4 Tbsp of powdered salt)
Calcium nitrate	Fertilizer	Calcium, nitrogen	3 ounces (7 Tbsp)
Magnesium sulfate	Fertilizer	Magnesium, sulfur	1 ½ ounces (4 Tbsp)

Pre-mixed Chemicals

Many of the essential nutrients needed for hydroponic plant growth are now available already mixed in their correct proportions. You may find these solutions in catalogs or from garden-supply stores. They are typically inexpensive and only small quantities are needed to help your plants grow strong and healthy. Always follow the directions on the container when using pre-mixed chemicals.

Making Your Own Solution

In the event that you want to make your own nutrient solution, here is a formula for a solution that will provide all the major elements required for your plants to grow.

You can obtain all of these chemicals from garden-supply stores or drugstores.

After all the chemicals have been mixed into the solution, check the pH of the solution. A pH of 7.0 is neutral; anything below 7.0 is acidic and anything above is alkaline. Certain plants grow best in certain pHs. Plants that grow well at a lower pH (between 4.5 and 5.5) are azaleas, buttercups, gardenias, and roses; plants that grow well at a neutral pH are potatoes, zinnias, and pumpkins; most plants grow best in a slightly acidic pH (between 5.5 and 6.5).

To determine the pH of your solution, use a pH indicator (these are usually paper strips). The strip will change color when placed in different levels of pH. If you find your pH level to be above your desired range, you can bring it down by adding dilute sulfuric acid in small quantities using an eyedropper. Keep retesting until you reach your desired pH level.

Plant Nutrient Deficiencies

When plants are lacking nutrients, they typically display these deficiencies outwardly. Following is a list of symptoms that might occur if a plant is lacking a certain type of nutrient. If your plants display any of these symptoms, it is imperative that the level of that particular nutrient be increased.

Soil nutrient chart. »

≈ **Potatoes grow best in soil with a neutral pH.**

Deficient Nutrient	Symptoms
Boron	Tip of the shoot dies; stems and petioles are brittle
Calcium	Tip of the shoot dies; tips of the young leaves die; tips of the leaves are hooked
Iron	New upper leaves turn yellow between the veins; edges and tips of leaves may die
Magnesium	Lower leaves are yellow between the veins; leaf margins curl up or down; leaves die
Manganese	New upper leaves have dead spots; leaf might appear netted
Nitrogen	Leaves are small and light green; lower leaves are lighter than upper leaves; weak stalks
Phosphorous	Dark-green foliage; lower leaves are yellow between the veins; purplish color on leaves
Potassium	Lower leaves might be mottled; dead areas near tips of leaves; yellowing at leaf margins and toward the center
Sulfur	Light-green upper leaves; leaf veins are lighter than surrounding area

Pest and Disease Management

Pest management can be one of the greatest challenges to the home gardener. Yard pests include weeds, insects, diseases, and some species of wildlife. Weeds are plants that are growing out of place. Insect pests include an enormous number of species from tiny thrips that are nearly invisible to the naked eye, to the large larvae of the tomato hornworm. Plant diseases are caused by fungi, bacteria, viruses, and other organisms—some of which are only now being classified. Poor plant nutrition and misuse of pesticides also can cause injury to plants. Slugs, mites, and many species of wildlife, such as rabbits, deer, and crows, can be extremely destructive as well.

Identify the Problem

Careful identification of the problem is essential before taking measures to control the issue in your garden. Some insect damage may at first appear to be a disease, especially if no visible insects are present. Nutrient problems may also mimic diseases. Herbicide damage, resulting from misapplication of chemicals, can also be mistaken for other problems. Learning about different types of garden pest is the first step in keeping your plants healthy and productive.

Insects and Mites

All insects have six legs, but other than that they are extremely different depending on the species. Some insects include such organisms as beetles, flies, bees, ants, moths, and butterflies. Mites and spiders have eight legs—they are not, in fact, insects but will be treated as such for the purposes of this section.

Insects damage plants in several ways. The most visible damage caused by insects is chewed plant leaves and flowers. Many pests are visible and can be readily identified, including the Japanese beetle, Colorado potato beetle, and numerous species of caterpillars such as tent caterpillars and tomato hornworms. Other chewing insects, however, such as cutworms (which are caterpillars) come out at night to eat, and burrow into the soil during the day. These are much harder to identify but should be considered likely culprits if young plants seem to disappear overnight or are found cut off at ground level.

Sucking insects are extremely common in gardens and can be very damaging to your vegetable plants and flowers. The most known of these insects are leafhoppers, aphids, mealy bugs, thrips, and mites. These insects insert their mouthparts into the plant tissues and suck out the plant juices. They also may carry diseases that they spread from plant to plant as they move about the yard. You may suspect that these insects are present if you notice misshapen plant leaves or flower petals. Often the younger leaves will appear curled or puckered. Flowers developing from the buds may only partially develop if they've been sucked by these bugs. Look on the undersides of the leaves—that is where many insects tend to gather.

Other insects cause damage to plants by boring into stems, fruits, and leaves, possibly disrupting the plant's ability to transport water. They also create opportunities for disease organisms to attack the plants. You may suspect the presence of boring insects if you see small accumulations of sawdust-like material on plant stems or fruits. Common examples of boring insects include squash vine borers and corn borers.

Integrated Pest Management (IPM)

It is difficult, if not impossible, to prevent all pest problems in your garden every year. If your best prevention efforts have not been entirely successful, you may need to use some control methods. Integrated pest management (IPM) relies on several techniques to keep pests at acceptable population levels without excessive use of chemical controls. The basic principles of IPM include monitoring (scouting), determining tolerable injury levels (thresholds), and applying appropriate strategies and tactics to solve the pest issue. Unlike other methods of pest control where pesticides are applied on a rigid schedule, IPM applies only those controls that are needed, when they are needed, to control pests that will cause more than a tolerable level of damage to the plant.

Monitoring

Monitoring is essential for a successful IPM program. Check your plants regularly. Look for signs of damage from insects and diseases as well as indications of adequate fertility and moisture. Early identification of potential problems is essential.

There are thousands of insects in a garden, many of which are harmless or even beneficial to the plants. Proper identification is needed before control strategies can be adopted. It is important to recognize the different stages of insect development for several reasons. The caterpillars eating your plants may be the larvae of the butterflies you were trying to attract. Any small larva with six spots on its back is probably a young ladybug, a very beneficial insect.

⬙ Japanese beetle.

Thresholds

It is not necessary to kill every insect, weed, or disease organism invading your garden in order to maintain the plants' health. When dealing with garden pests, an economic threshold comes into play and is the point where the damage caused by the pest exceeds the cost of control. In a home garden, this can be difficult to determine. What you are growing and how you intend to use it will determine how much damage you are willing to tolerate. Remember that larger plants, especially those close to harvest, can tolerate more damage than a tiny seedling. A few flea beetles on a radish seedling may warrant control, whereas numerous Japanese beetles eating the leaves of beans close to harvest may not.

If the threshold level for control has been exceeded, you may need to employ control strategies. Effective and safe strategies can be discussed with your local Cooperative Extension Service, garden centers, or nurseries.

⌃ Worm holes are a common problem for apples not sprayed with pesticides.

Mechanical/Physical Control Strategies

Many insects can simply be removed by hand. This method is definitely preferable if only a few, large insects are causing the problem. Simply remove the insect from the plant and drop it into a container of soapy water or vegetable oil. Be aware that some insects have prickly spines or excrete oily substances that can cause injury to humans. Use caution when handling unfamiliar insects. Wear gloves or remove insects with tweezers.

Many insects can be removed from plants by spraying water from a hose or sprayer. Small vacuums can also be used to suck up insects. Traps can be used effectively for some insects as well. These come in a variety of styles depending on the insect to be caught. Many traps rely on the use of pheromones—naturally occurring chemicals produced by the insects and used to attract the opposite sex during mating. They are extremely specific for each species and, therefore, will not harm beneficial species. One caution with traps is that they may actually draw more insects into your yard, so don't place them directly into your garden. Other traps (such as yellow and blue sticky cards) are more generic and will attract numerous species. Different insects are attracted to different colors of these traps. Sticky cards also can be used effectively to monitor insect pests.

Other Pest Controls

Diatomaceous earth, a powder-like dust made of tiny marine organisms called diatoms, can be used to reduce damage from soft-bodied insects and slugs. Spread this material on the soil—it is sharp and cuts or irritates these soft organisms. It is harmless to other organisms. In order to trap slugs, put out shallow dishes of beer.

Biological Controls

Biological controls are nature's way of regulating pest populations. Biological controls rely on predators and parasites to keep organisms under control. Many of our present pest problems result from the loss of predator species and other biological control factors.

Some biological controls include birds and bats that eat insects. A single bat can eat up to 600 mosquitoes an hour. Many bird species eat insect pests on trees and in the garden.

Chemical Controls

When using biological controls, be very careful with pesticides. Most common pesticides are broad spectrum, which means that they kill a wide variety of organisms. Spray applications of insecticides are likely to kill numerous beneficial insects as well as the pests. Herbicides applied to weed species may drift in the wind or vaporize in the heat of the day and injure non-targeted plants. Runoff of pesticides can pollute water. Many pesticides are toxic to humans as well as pets and small animals that may enter your yard. Try to avoid using these types of pesticides at all costs—and if you do use them, read the labels carefully and avoid spraying them on windy days.

Some common, non-toxic household substances are as effective as many toxic pesticides. A few drops of dishwashing detergent mixed with water and sprayed on plants is extremely effective in controlling many soft-bodied insects, such as aphids and whiteflies. Crushed garlic mixed with water may control certain insects. A baking soda solution has been shown to help control some fungal diseases on roses.

Beneficial Insects that Help Control Pest Populations	
Insect	Pest Controlled
Green lacewings	Aphids, mealy bugs, thrips, and spider mites
Ladybugs	Aphids and Colorado potato beetles
Praying mantises	Almost any insect
Ground beetles	Caterpillars that attack trees and shrubs
Seedhead weevils and other beetles	Weeds

≳ **Aphids**

Alternatives to Pesticides and Chemicals

When used incorrectly, pesticides can pollute water. They also kill beneficial as well as harmful insects. Natural alternatives prevent both of these events from occurring and save you money. Consider using natural alternatives to chemical pesticides: Non-detergent insecticidal soaps, garlic, hot pepper spray, 1 teaspoon of liquid soap in a gallon of water, used dishwater, or a forceful stream of water from a hose all work to dislodge insects from your garden plants.

Another solution is to consider using plants that naturally repel insects. These plants have their own chemical defense systems, and when planted among flowers and vegetables, they help keep unwanted insects away.

Plant Diseases

Plant disease identification is extremely difficult. In some cases, only laboratory analysis can conclusively identify some diseases. Disease organisms injure plants in several ways: Some attack leaf surfaces and limit the plant's ability to carry on photosynthesis; others produce substances that clog plant tissues that transport water and nutrients; still other disease organisms produce toxins that kill the plant or replace plant tissue with their own.

Symptoms that are associated with plant diseases may include the presence of mushroom-like growths on trunks of trees; leaves with a grayish, mildewed appearance; spots on leaves, flowers, and fruits; sudden wilting or death of a plant or branch; sap exuding from branches or trunks of trees; and stunted growth.

Natural Pest Repellants

Pest	Repellant
Ant	Mint, tansy, or pennyroyal
Aphids	Mint, garlic, chives, coriander, or anise
Bean leaf beetle	Potato, onion, or turnip
Codling moth	Common oleander
Colorado potato bug	Green beans, coriander, or nasturtium
Cucumber beetle	Radish or tansy
Flea beetle	Garlic, onion, or mint
Imported cabbage worm	Mint, sage, rosemary, or hyssop
Japanese beetle	Garlic, larkspur, tansy, rue, or geranium
Leaf hopper	Geranium or petunia
Mice	Onion
Root knot nematodes	French marigolds
Slugs	Prostrate rosemary or wormwood
Spider mites	Onion, garlic, cloves, or chives
Squash bug	Radish, marigolds, tansy, or nasturtium
Stink bug	Radish
Thrips	Marigolds
Tomato hornworm	Marigolds, sage, or borage
Whitefly	Marigolds or nasturtium

≳ **Cutworms**

Misapplication of pesticides and nutrients, air pollutants, and other environmental conditions—such as flooding and freezing—can also mimic some disease problems. Yellowing or reddening of leaves and stunted growth may indicate a nutritional problem. Leaf curling or misshapen growth may be a result of herbicide application.

Pest and Disease Management Practices

Preventing pests should be your first goal when growing a garden, although it is unlikely that you will be able to avoid all pest problems because some plant seeds and disease organisms may lay dormant in the soil for years.

Diseases need three elements to become established in plants: the disease organism, a susceptible species, and the proper environmental conditions. Some disease organisms can live in the soil for years; other organisms are carried in infected plant material that falls to the ground. Some disease organisms are carried by insects. Good sanitation will help limit some problems with disease. Choosing resistant varieties of plants also prevents many diseases from occurring. Rotating annual plants in a garden can also prevent some diseases.

Plants that have adequate, but not excessive, nutrients are better able to resist attacks from both diseases and insects. Excessive rates of nitrogen often result in extremely succulent vegetative growth and can make plants more susceptible to insect and disease problems, as well as decreasing their winter hardiness. Proper watering and spacing of plants limits the

⩘ Powdery mildew leaf disease.

spread of some diseases and provides good aeration around plants, so diseases that fester in standing water cannot multiply. Trickle irrigation, where water is applied to the soil and not the plant leaves, may be helpful.

Removal of diseased material certainly limits the spread of some diseases. It is important to clean up litter dropped from diseased plants. Prune diseased branches on trees and shrubs to allow for more air circulation. When pruning diseased trees and shrubs, disinfect your pruners between cuts with a solution of chlorine bleach to avoid spreading the disease from plant to plant. Also try to control insects that may carry diseases to your plants.

You can make your own natural fungicide by combining 5 teaspoons each of baking soda and hydrogen peroxide with a gallon of water. Spray on your infected plants. Milk diluted with water is also an effective fungicide, due to the potassium phosphate in it, which boosts a plant's immune system. The more diluted the solution, the more frequently you'll need to spray the plant.

Harvesting Your Garden

It is essential, in order to get the best freshness, flavor, and nutritional benefits from your garden vegetables and fruits, to harvest them at the appropriate time. The vegetable's stage of maturity and the time of day at which it is harvested are essential for good-tasting and nutritious produce. Over-ripe vegetables and fruits will be stringy and coarse. When possible, harvest your vegetables during the cool part of the morning. If you are going to can and preserve your vegetables and fruits, do so as soon as possible. Or, if this process must be delayed, make sure to cool the vegetables in ice water or crushed ice and store them in the refrigerator. Here are some brief guidelines for harvesting various types of common garden produce:

Asparagus—Harvest the spears when they are at least 6 to 8 inches tall by snapping or cutting them at ground level. A few spears may be harvested the second year after crowns are set out. A full harvest season will last four to six weeks during the third growing season.

Beans, snap—Harvest before the seeds develop in the pod. Beans are ready to pick if they snap easily when bent in half.

Beans, lima—Harvest when the pods first start to bulge with the enlarged seeds. Pods must still be green, not yellowish.

Broccoli—Harvest the dark green, compact cluster, or head, while the buds are shut tight, before any yellow flowers appear. Smaller side shoots will develop later, providing a continuous harvest.

Brussels sprouts—Harvest the lower sprouts (small heads) when they are about 1 to 1½ inches in diameter by twisting them off. Removing the lower leaves along the stem will help to hasten the plant's maturity.

Cabbage—Harvest when the heads feel hard and solid.

Cantaloupe—Harvest when the stem slips easily from the fruit with a gentle tug. Another indicator of ripeness is when the netting on the skin becomes rounded and the flesh between the netting turns from a green to a tan color.

Carrots—Harvest when the roots are ¾ to 1 inch in diameter. The largest roots generally have darker tops.

Cauliflower—When preparing to harvest, exclude sunlight when the curds (heads) are 1 to 2 inches in diameter by loosely tying the outer leaves together above the curd with a string or rubber band. This process is known as blanching. Harvest the curds when they are 4 to 6 inches in diameter but still compact, white, and smooth. The head should be ready ten to fifteen days after tying the leaves.

Collards—Harvest older, lower leaves when they reach a length of 8 to 12 inches. New leaves will grow as long as the central growing point remains, providing a continuous harvest. Whole plants may be harvested and cooked if desired.

Corn, sweet—The silks begin to turn brown and dry out as the ears mature. Check a few ears for maturity by opening the top of the ear and pressing a few kernels with your thumbnail. If the exuded liquid is milky rather than clear, the ear is ready for harvesting. Cooking a few ears is also a good way to test for maturity.

Cucumbers—Harvest when the fruits are 6 to 8 inches in length. Harvest when the color is deep green and before yellow color appears. Pick four to five times per week to encourage continuous production. Leaving mature cucumbers on the vine will stop the production of the entire plant.

Eggplant—Harvest when the fruits are 4 to 5 inches in diameter and their color is a glossy, purplish black. The fruit is getting too ripe when the color starts to dull or become bronzed. Because the stem is woody, cut—do not pull—the fruit from the plant. A short stem should remain on each fruit.

Kale—Harvest by twisting off the outer, older leaves when they reach a length of 8 to 10 inches and are medium green in color. Heavy, dark green leaves are overripe and are likely to be tough and bitter. New leaves will grow, providing a continuous harvest.

Lettuce—Harvest the older, outer leaves from leaf lettuce as soon as they are 4 to 6

⌃ Only the eggplant vegetable is edible. Do not eat the leaves, stem, roots, or flowers.

Pods are getting too old when they lose their brightness and turn light or yellowish green.

Peppers—Harvest sweet peppers with a sharp knife when the fruits are firm, crisp, and full size. Green peppers will turn red if left on the plant. Allow hot peppers to attain their bright red color and full flavor while attached to the vine; then cut them and hang them to dry.

Potatoes (Irish)—Harvest the tubers when the plants begin to dry and die down. Store the tubers in a cool, high-humidity location with good ventilation, such as the basement or crawl space of your house. Avoid exposing the tubers to light, as greening, which denotes the presence of dangerous alkaloids, will occur even with small amounts of light.

Pumpkins—Harvest pumpkins and winter squash before the first frost. After the vines dry up, the fruit color darkens and the skin surface resists puncture from your thumbnail. Avoid bruising or scratching the fruit while handling it. Leave a 3- to 4-inch portion of the stem attached to the fruit and store it in a cool, dry location with good ventilation.

Radishes—Harvest when the roots are ½ to 1½ inches in diameter. The shoulders of radish roots often appear through the soil surface when they are mature. If left in the ground too long, the radishes will become tough and woody.

Rutabagas—Harvest when the roots are about 3 inches in diameter. The roots may

inches long. Harvest heading types when the heads are moderately firm and before seed stalks form.

Mustard—Harvest the leaves and leaf stems when they are 6 to 8 inches long; new leaves will provide a continuous harvest until they become too strong in flavor and tough in texture due to temperature extremes.

Okra—Harvest young, tender pods when they are 2 to 3 inches long. Pick the okra at least every other day during the peak growing season. Overripe pods become woody and are too tough to eat.

Onions—Harvest when the tops fall over and begin to turn yellow. Dig up the onions and allow them to dry out in the open sun for a few days to toughen the skin. Then remove the dried soil by brushing the onions lightly. Cut the stem, leaving 2 to 3 inches attached, and store in a net-type bag in a cool, dry place.

Peas—Harvest regular peas when the pods are well rounded; edible-pod varieties should be harvested when the seeds are fully developed but still fresh and bright green.

be stored in the ground and used as needed, if properly mulched.

Spinach—Harvest by cutting all the leaves off at the base of the plant when they are 4 to 6 inches long. New leaves will grow, providing additional harvests.

Squash, summer—Harvest when the fruit is soft, tender, and 6 to 8 inches long. The skin color often changes to a dark, glossy green or yellow, depending on the variety. Pick every two to three days to encourage continued production.

Sweet potatoes—Harvest the roots when they are large enough for use before the first frost. Avoid bruising or scratching the potatoes during handling. Ideal storage conditions are at a temperature of 55 degrees Fahrenheit and a relative humidity of 85 percent. The basement or crawl space of a house may suffice.

Swiss chard—Harvest by breaking off the developed outer leaves 1 inch above the soil. New leaves will grow, providing a continuous harvest.

Tomatoes—Harvest the fruits at the most appealing stage of ripeness, when they are bright red. The flavor is best at room temperature, but ripe fruit may be held in the refrigerator at 45 to 50 degrees Fahrenheit for seven to ten days.

Turnips—Harvest the roots when they are 2 to 3 inches in diameter but before heavy fall frosts occur. The tops may be used as salad greens when the leaves are 3 to 5 inches long. Turnips can be eaten almost any way potatoes can be—mashed, roasted, or even fried. The greens can be eaten raw, steamed, or boiled.

Watermelons—Harvest when the watermelon produces a dull thud rather than a sharp, metallic sound when thumped—this means the fruit is ripe. Other ripeness indicators are a deep yellow rather than a white color where the melon touches the ground, brown tendrils on the stem near the fruit, and a rough, slightly ridged feel to the skin surface.

Community Gardens

A community garden is considered any piece of land that many people garden. These gardens can be located in urban areas, out in the country, or even in the suburbs, and they grow everything from flowers to vegetables to herbs—anything that the community members want to produce. Community gardens help promote the growing of fresh food, have a positive impact on neighborhoods by cleaning up vacant lots, educate youth about gardening and working together as a community, and bring whole communities together in one common goal. Whether the garden is just one plot or many individual gardens in a specified area, a community garden is a wonderful way to reach out to fellow neighbors while growing fresh foods wherever you live.

10 Steps to Beginning Your Own Community Garden

The following steps are adapted from the American Community Garden Association.

1. Organize a meeting for all those interested. Determine whether a garden is really needed and wanted, what kind it should be (vegetable, flower, or a combination, and organic or not), whom it will involve, and who benefits. Invite neighbors, tenants, community organizations, gardening and horticultural societies, and building superintendents (if it is at an apartment building)—in other words, anyone who is likely to be interested.

2. Establish a planning committee. This group can be made up of people who feel committed to the creation of the garden and who have the time to devote to it, at least in the initial stages. Choose well-organized people as garden coordinators and form committees to tackle specific tasks such as funding and partnerships, youth activities, construction, and communication.

3. Compile all of your resources. Do a community asset assessment. What skills and resources already exist in the community that can aid in the garden's creation? Contact local municipal planners about possible sites, as well as horticultural societies and other local sources of information and assistance. Look within your community for people with experience in landscaping and gardening.

4. Look for a sponsor. Some gardens "self-support" through membership dues, but for many, a sponsor is essential for donations of tools, seeds, or money. Churches, schools, private businesses, or parks and recreation departments are all possible sponsors.

5. Choose a site for your garden. Consider the amount of daily sunshine (vegetables need at least six hours a day), availability of water, and soil testing for possible pollutants. Find out who owns the land. Can the gardeners get a lease agreement for at least three years? Will public liability insurance be necessary?

6. Prepare and develop the chosen site. In most cases, the land will need consider-

- Seeds, seedlings, and organic material, such as compost, manure, or peat moss
- Long-handled shovels, hoes, rakes, garden spades, and three pronged hand cultivators
- Scissors, knives, and containers (baskets, bowls, or cardboard boxes)

Decide What to Plant:
- Think about what vegetables to grow and decide on the number of plants needed. Involve the children in this part of the planning process.

Design the Site:
- Draw a picture of the garden and plan out what plants will grow in which rows. Figure how far apart the rows should be. Find out how wide the plants will get, add the two widths together, and divide by 2. That's how far apart the rows should be. This will make it easier on planting day.

Once the Garden is Designed—Then What?

Test the Soil:
- If the soil has not been tested, conduct a soil test. Call a local Cooperative Extension Service office, listed under county or state government, for the name and location of a laboratory to do testing for lead content and soil pH or see page 14 to do your own soil testing. A soil test tells two things: (a) lead level of the soil; and (b) whether the soil is *acid* (sour), *alkaline* (sweet), or *neutral* (neither sour nor sweet). Lead is a poison and if it gets into the plants, it will get into the food. Plants will not grow well in soil that is either too acid or too alkaline.

Get the Tools:
- Long-handled shovels, gardening spades, spading forks, hoes, and rakes are all excellent tools for beginning a garden. To care for the garden, use hand tools such as three-pronged hand cultivators, hose and nozzle, and/or watering cans. If the group doesn't have its own garden tools, find someone who has what is needed and ask to borrow the tools. Or check yard sales to buy used tools.

Prepare the Soil:
- Once the soil is dry enough, dig it and loosen it. Involve children and youth in the process of preparing the soil.
- Remove grass and weeds (roots and all). Take the time to do this well.
- Dig the soil as deep as the blade of the spade and turn the soil. Or, find someone to till the soil with a rototiller.
- If the soil test showed the soil to be too acid, add limestone (lime); if the soil is too alkaline, add ground agricultural sulfur. Sprinkle either lime or sulfur evenly over the garden soil.
- Add organic material such as compost, manure, or peat moss. This helps feed the plants and enriches the soil. Spread evenly on top of the turned soil in a layer no deeper than 3 inches.
- Blend everything using a spading fork, until the soil is so soft that planting can be done with the hands.
- Rake the soil until the soil is smooth and level, with no hills or holes. This will allow the water to seep down to the roots.

GET READY TO PLANT

Children will enjoy accompanying adults on a trip to purchase seeds or seedlings (also called transplants) for the vegetables you intend to plant. Some plants do better if you start with seedlings rather than seeds. Seedlings are the fastest way to grow plants, and the easiest.

- To identify what you have planted, buy or make stakes, and write the names of the plants on the stakes with a waterproof marker. Save the stakes for later use.
- Here is another place that children and youth can help. If seeds are planted: make a shallow straight line (furrow) in the soil with a finger.
- Put the seeds in the furrow about half an inch apart or as suggested on the seed packaging.
- When the seeds are in the furrow, squeeze the furrow closed with thumb and finger.
- Water the soil right after the seeds are planted—water the plants so that the water comes out like a shower.
- Place the marker stakes in the soil at one end of the row to identify what has been planted.
- If another kind of plant is planted, check the distance between the rows following your design.
- If seedlings are planted in each row, mark the spot where the plants should go, by poking a hole in the soil using a finger or the end of a pole. Do the entire row at one time.
- Set each plant in the soil so that it isn't too high or too low but just above the root ball. Cover the root ball with soil and press the soil gently so there are no empty spaces near the roots.
- Feed the seedlings with a mixture of fertilizer and water. Water each plant once, let the water soak in, and water a second time. Depending on what plants are grown, this feeding may need to be done every two to three weeks.

Tending the Garden:

Visit the garden daily—check if the garden needs watering, weeding, feeding, and thinning. Make sure to bring the proper tools. Take the children and youth to the garden and have them help care for the plants. Two hours is plenty of time for them to work in the garden at any one time. Children and youth can do weeding, thinning, and harvesting of crops.

board in the garden, and have regular community celebrations. Community gardens are all about creating and strengthening communities.

Tools Needed to Create and Maintain a Community Garden

When purchasing tools for your community garden, buy high-quality tools. These will last longer and are an investment that will benefit your garden and those working in it for years to come. Every community garden should be equipped with these ten essential tools:

Fork: You can't dig and divide perennials without a heavy-duty fork.

Gloves: Leather gloves hold up best. If you have roses, get a pair that is thick enough to resist thorn pricks.

Hand fork or cultivator: A hand fork helps cultivate soil, chop up clumps, and work fertilizer and compost into the soil. A hand fork is necessary for cultivating in closely planted beds.

Hand pruners: There are different types and sizes of pruners depending on the type and size of the job. Hand pruners are for cutting small diameters, up to the thickness of your little finger.

able preparation for planting. Organize volunteer work crews to clean it, gather materials, and decide on the design and plot arrangement.

7. Organize the layout of the garden. Members must decide how many plots are available and how they will be assigned. Allow space for storing tools and making compost, and allot room for pathways between each plot. Plant flowers or shrubs around the garden's edges to promote goodwill with non-gardening neighbors, pedestrians, and municipal authorities.

8. Plan a garden just for kids. Consider creating a special garden for the children of the community—including them is essential. Children are not as interested in the size of the harvest but rather in the process of gardening. A separate area set aside for them allows them to explore the garden at their own speed and can be a valuable learning tool.

9. Draft rules and put them in writing. The gardeners themselves devise the best ground rules. We are more willing to comply with rules that we have had a hand in creating. Ground rules help gardeners to know what is expected of them. Think of it as a code of behavior. Some examples of issues that are best dealt with by agreed-upon rules are: What kinds of dues will members pay? How will the money be used? How are plots assigned? Will gardeners share tools, meet regularly, and handle basic maintenance?

10. Keep members involved with one another. Good communication ensures a strong community garden with active participation by all. Some ways to do this are: form a telephone tree, create an e-mail list, install a rainproof bulletin

Hoe: A long-handled hoe is key to keeping weeds out of your garden.

Hose: This is the fastest way to transport lots of water to your garden plants. Consider using drip irrigation hoses or tape to apply a steady stream of water to your plants.

Shovels and spades: There are several different types and shapes of shovels and spades, each with its own purpose. There are also different types of handles for either—a D shape, a T shape, or none at all. A shovel is a requisite tool for planting large perennials, shrubs, and trees; breaking ground; and moving soil, leaves, and just about anything else. The sharper the blade, the better.

Trowel: A well-made trowel is your most important tool. From container gardening to large beds, a trowel will help you get your plants into the soil.

Watering can: A watering can creates a fine, even stream of water that delivers with a gentleness that won't wash seedlings or sprouting seeds out of their soil.

Wheelbarrow: Wheelbarrows come in all different sizes (and prices). They are indispensable for hauling soil, compost, plants, mulch, hoses, tools, and everything else you'll need to make your garden a success.

Things to Consider

There are many things that need to be considered when you create your own community garden. Remember that community gardens take a lot of work and foresight in order for them to be successful and beneficial for all those involved.

The Organization of the Garden

Will your garden establish rules and conditions for membership—such as residence, dues, and agreement with any drafted rules and regulations? It is important to know who will be able to use the garden and who will not. Furthermore, deciding how the garden will be parceled out is another key topic to discuss before beginning your community garden. Will the plots be assigned by family size or need? Will some plots be bigger to accommodate larger families? Will your garden incorporate children's plots as well?

Insurance

It is becoming increasingly difficult to obtain leases from landowners without liability insurance. Garden insurance is a new thing for many insurance carriers, and their underwriters are reluctant to cover community gardens. It helps if you know what you want out of your community garden before you start talking to insurance agents. Two tips: Work with an agent from a firm that deals with many different carriers (so you can get the best policy for your needs); and you will probably have better success with someone local who has already done this type of policy or who works with social service agencies in the area. Shop around until you find a policy that fits the needs of your community garden and its users.

Set Up a Garden Association

Many garden groups are organized very informally and operate successfully. Leaders "rise to the occasion" to propose ideas and carry out tasks. However, as the workload expands, many groups choose a more formal structure for their organization.

A structured program is a conscious, planned effort to create a system so that each person can participate fully and the group can perform effectively. It's vital that the leadership be responsive to the members and their needs.

If your group is new, have several planning meetings to discuss your program and organization. Try out suggestions raised at these meetings, and after a few months of operation, you'll be in a better position to develop bylaws or organizational guidelines. A community garden project should be kept as simple as possible.

Creating Bylaws

Bylaws are rules that govern the internal affairs of an organization. Check out bylaws from other community garden organization when creating your own. The bylaws of your community garden can be as simple as rules and regulations. This is usually plenty adequate for small community garden affairs.

Bylaws cover these topics:

- The full name of the organization and address
- The organizing members and their addresses

- The purpose, goal, and philosophy of the organization
- Membership eligibility and dues
- Timeline for regular meetings of the committee
- How the bylaws can be rescinded or amended
- Maintenance and cleanup of the community garden
- A hold harmless clause: "We the undersigned members of the [name] garden group hereby agree to hold harmless [name landowner] from and against any damage, loss, liability, claim, demand, suit, cost, and expense directly or indirectly resulting from, arising out of or in connection with the use of the [name] garden by the garden group, its successors, assigns, employees, agents, and invitees."

Sample Guidelines and Rules for Garden Members

Here are some sample guidelines community garden members may need to follow:

- I will pay a fee of $___ to help cover garden expenses.
- I will have something planted in the garden by [date] and keep it planted all summer long.
- If I must abandon my plot for any reason, I will notify the garden leadership.
- I will keep weeds at a minimum and maintain the areas immediately surrounding my plot.
- If my plot becomes unkempt, I understand I will be given one week's notice to clean it up. At that time, it will be reassigned or tilled in.
- I will keep trash and litter out of the plot, as well as adjacent pathways and fences.
- I will participate in the fall cleanup of the garden.
- I will plant tall crops where they will not shade neighboring plots.
- I will pick only my own crops unless given permission by another plot user.
- I will not use fertilizers, insecticides, or weed repellents that will in any way affect other plots.
- I understand that neither the garden group nor owners of the land are responsible for my actions. I therefore agree to hold harmless the garden group and owners of the land for any liability, damage, loss, or claim that occurs in connection with use of the garden by me or any of my guests.

Preventing Vandalism of Your Community Garden

Vandalism is a common fear among community gardeners. Try to deter vandalism by following these simple, preventative methods:

- Make a sign for the garden. Let people know that the garden is a community project.
- Put up fences around your garden. Fences can be of almost any material. They serve as much to mark possession of a property as to prevent entry.
- Invite everyone in the neighborhood to participate in the garden project from the very beginning. If you exclude people, they may become potential vandals.
- Plant raspberries, roses, or other thorny plants along the fence as a barrier to anyone trying to climb the fence.
- Make friends with neighbors whose windows overlook the garden. Trade them flowers and vegetables for a protective eye.
- Harvest all ripe fruit and vegetables on a daily basis to prevent the temptation of outsiders to harvest your crops.
- Plant a "vandal's garden" at the entrance. Mark it with a sign: "If you must take food, please take it from here."

People Problems and Solutions

Angry neighbors and bad gardeners pose problems for a community garden. Neighbors may complain to municipal governments about messy, unkempt gardens or rowdy behavior; most gardens cannot afford poor relations with neighbors, local politicians, or potential sponsors. Therefore, choose bylaws carefully so you have procedures to follow when members fail to keep their plots clean and up to code. A well-organized garden with strong leadership and committed members can overcome almost any obstacle.

School Gardens

A school garden provides children with an ideal outdoor classroom. Within a single visit to a garden, a student can record plant growth, study decomposition while turning a compost pile, and learn more about plants, nature, and the outdoors in general. Gardens also provide students with opportunities to make healthier

---

10 BENEFITS OF CREATING A COMMUNITY GARDEN

1. Improves the quality of life for people using the garden.
2. Provides a pathway for neighborhood and community development and promotes intergenerational and cross-cultural connections.
3. Stimulates social interaction and reduces crime.
4. Encourages self-reliance.
5. Beautifies neighborhoods and preserves green space.
6. Reduces family food budgets while providing nutritious foods for families in the community.
7. Conserves resources.
8. Creates an opportunity for recreation, exercise, therapy, and education.
9. Creates income opportunities and economic development.
10. Reduces city heat from streets and parking lots.

---

food choices, learn about nutrient cycles, and develop a deeper appreciation for the environment, community, and each other. While school gardens are typically used for science classes, they can also teach children about the history of their community (what there town was like hundreds of years ago and what people did to farm food), and be incorporated into math curriculum and other school subjects.

How to Start a School Garden

School gardens do not need to start on a grand scale. In fact, individual classrooms can grow their very own container gardens just by planting seeds in small pots, watering them daily, placing them in a sunny corner of the room, and watching the seedlings grow. However, if there is space for a larger, outdoor garden, this is the ideal place to teach children about working as a community, about plants and vegetables, and responsibility. A school garden should eventually become a permanent addition to the school and be maintained year-round.

When starting a school garden, it is important to find someone to coordinate the garden program. This is the perfect way to get parents involved in the school's garden. Establish a volunteer garden committee and assign certain parents particular tasks in the planning, upkeep (even during the summer months when school is generally not in session), and funding for the school's garden.

It is important, so the garden is not neglected, to plan particular classroom activities and lessons that will incorporate the garden and its plants. Assigning students various jobs that relate to the garden will be a wonderful way of introducing them to gardening as well as responsibility and community.

After all the initial planning is done, it is time to choose a spot for the school garden. A place in the lawn that receives plenty of sunlight and that will be close enough to the building for easy access is ideal. It should also be near an outdoor spigot so the plants can be easily watered. If there is enough space, it might be beneficial to have a garden shed, where gardening gloves, tools, buckets, hoses, and other items can be stored for use in the garden. Once you have chosen your spot, it's time to start digging and planting!

Both new and established gardens benefit from the use of compost and mulch. Many schools purchase compost when they initially establish their garden, and then they start making their own compost—which is a wonderful scientific lesson for students as well. You can use grass clippings, yard trimmings, rotten vegetables, and even food scraps from the cafeteria or students' lunches to build and maintain your compost pile. While some schools choose to make compost piles in the garden, others compost with worm boxes right in the classroom!

Depending on funding and the needs and desires of the school, these gardens can become quite elaborate, with fences, ponds, trellises, trees and shrubs, and other structures. However, all a school garden truly needs is a little bit of dirt and a few plants (preferably an assortment of vegetables, fruit, and wildflowers) that students can study and even eat. Whether the school garden is established only for one class or grade level, or if it is going to be available to everyone at the school, is a factor that will determine the types of plants and the size of the overall garden.

Whether big or small, complex or simple, school gardens provide a wonderful, enriching learning experience for children and their parents alike.

Farmers' Markets

Farmers' markets are an integral part of the urban–farm linkage and have continued to rise in popularity, mostly due to growing consumer interest in obtaining fresh products directly from the farm. Farmers' markets allow consumers to have access to locally grown, farm-fresh produce, enable farmers the opportunity to develop a personal relationship with their customers, and cultivate consumer loyalty with the farmers who grow the produce. Direct marketing of farm products through farmers' markets continues to be an important sales outlet for agricultural producers nationwide. Today, there are more than 4,600 farmers' markets operating throughout the nation.

Who Benefits from Farmers' Markets?

- Small farm operators: Those with less than $250,000 in annual receipts who work and manage their own operations meet this definition (94 percent of all farms).

- Farmers and consumers: Farmers have direct access to markets to supplement farm income. Consumers have access to locally grown, farm-fresh produce and the opportunity to personally interact with the farmer who grows the produce.
- The community: Many urban communities—where fresh, nutritious foods are scarce—gain easy access to quality food. Farmers' markets also help to promote nutrition education, wholesome eating habits, and better food preparation, as well as boosting the community's economy.

Getting Involved in a Farmers' Market

A farmers' market is a great place for new gardeners to learn what sorts of produce customers want, and it also promotes wonderful community relationships between the growers and the buyers. If you have a well-established garden, and know a few other people who also have fruits, vegetables, and even garden flowers to spare, you may want to consider organizing and implementing your own local farmers' market (if your town or city already has an established farmers' market, you may want to go visit it one day and ask the farmers how you could join, or contact your local Cooperative Extension Service

for more information). Joining existing farmers' markets may require that you pay an annual fee, and your produce may also be subject to inspection and other rules established by the market's organization or the local government.

Establishing a farmers' market is not simply setting up a stand in front of your home and selling your vegetables—though you can certainly do this if you prefer. A farmers' market must have a small group of people who are all looking to sell their produce and garden harvests. It is important, before planning even begins, to hold a meeting and discuss the feasibility of your venture. Is there other local competition that might impede on your market's success? Are there enough people and enough produce to make a farmers' market profitable and sustainable? What kind of monetary cost will be incurred by establishing a farmers' market? It is also a good idea to think about how you can sponsor your market—such as though members, nonprofit organizations, or the chamber of commerce. If at all possible, it is best to try to get your entire neighborhood and local government involved and promote the idea of fresh, home-grown fruits and vegetables that will be available to the community through the establishment of a farmers' market.

Once you've decided that you want to go ahead with your plan, you should establish rules for your market. Such rules and regulations should determine if there will be a board of directors, who will be responsible for the overall management of the market, who can become a member, where the market will be located (preferably in a place with ample parking, good visibility, and cover in case of bad weather), and when and for how long the market will be open to the public. It is also a good idea to discuss how the produce should be priced and make sure you are following all local and state regulations.

Ideally, it is beneficial for your market's vendors and the community to gather support and involvement from local farmers in your area. That way, your garden vegetable and fruit stand will be supplemented with other locally grown produce and crops from farms, which will draw more people to your market.

Once your farmers' market is up and running, it is important to maintain a good rapport with the community. Be friendly when customers come to your stand, and make sure that you are offering quality products for them to purchase. Price your vegetables and fruits fairly and make your displays pleasing to the eye, luring customers to your stall. Make sure your produce is clearly marked with the name and price, so your customers will have no doubt as to what they are buying. Make your stall as personal as possible, and always, always interact with the customer. In this way, you'll begin to build relationships with your community members and hopefully continue to draw in business for yourself and the other vendors in the farmers' market.

If you are looking to find a local farmers' market near you (either to try to join or just to visit), here is a link to a Web site that offers extensive listings for farmers' markets by state: http://apps.ams.usda.gov/farmersmarkets.

Attracting Birds, Butterflies, and Bees to Your Garden

A wonderful part of having a garden is the wildlife it attracts. The types of trees, shrubs, vines, plants, and flowers you choose for your garden and yard affect the types of wildlife that will visit. Whether you are looking to attract birds, butterflies, or bees to your garden, here are some specific types of plants that will bring these creatures to your yard.

Plant Species for Birds

Following is a list of trees, shrubs, and vines that will attract various birds to your yard and garden. Be sure to check with your local nursery to find out which plants are most suitable for your area.

Trees:

- American beech
- American holly
- Balsam fir
- Crabapple
- Flowering dogwood
- Oak

Shrubs:

- Common juniper
- Hollies
- Sumacs
- Viburnums

Vines:

- Strawberry
- Trumpet honeysuckle
- Virginia creeper
- Wild grape

Flowers and Nectar Plants for Hummingbirds, Butterflies, and Bees

If you are looking at attract hummingbirds, butterflies, and bees to your garden, consider planting these nectar-producing shrubs and flowers. Again, check with your local nursery to make sure these plants are suitable for your geographic area. Some common nectar plants are:

- Aster
- Azalea
- Butterfly bush
- Clover and other legumes
- Columbine
- Coneflower
- Honeysuckle
- Lupine
- Milkweeds
- Zinnia

Part 2

The Country Kitchen

"The kitchen is a country in which there are always discoveries to be made."

—Grimod de la Reyniere, 1758–1838

Baking Bread	96
Baking Cakes, Pies, Cookies, and Other Desserts	110
Eating Well	124
Food Co-ops: Grocery Shopping in Community	127
Maple Sugaring	130
Edible Wild Plants and Mushrooms	132
Making Sausage	140
Curing Virginia wHam	141
Make Your Own Foods	145
Sharing Your Bounty	151

Baking Bread

Bread has been a dining staple for thousands of years. The art of breadmaking has evolved over time, but the basic principles remain unchanged. Bread is made from flour of wheat or other grains, with the addition of water, salt, and a fermenting ingredient (such as yeast or another leavening agent). After you've baked a few loaves, you'll start to get a feel for what the dough should look and feel like. Then you can start experimenting with different flours, or additions of fruits, nuts, seeds, herbs, and more.

Quick Breads

Muffins, banana bread, zucchini bread, and many other sweet breads are often leavened with agents other than yeast, such as baking soda or baking powder. These breads are easy to make and require far less preparation time than yeast breads. They're also very versatile; once you master the basic recipe you can add almost any fruit, nut, or flavoring to make a uniquely delicious treat.

Basic Quick Bread Recipe

This basic recipe will make two loaves or twelve large muffins. Fold in 1 to 2 cups of mashed fruit, whole berries, nuts, or chocolate chips before pouring the batter into the pans.

- 3½ cups flour (use at least 2 cups of a gluten-rich flour)
- 2 teaspoons baking powder
- 1 teaspoon baking soda
- 1 teaspoon salt
- 1 to 2 teaspoons spices or herbs, if desired
- 1¼ cups sugar
- ¾ cup butter, oil, or fruit puree
- 3 eggs
- ¾ cup milk

1. In a large mixing bowl combine all dry ingredients except sugar.
2. In a separate bowl, beat together sugar and butter, oil, or fruit puree. Add eggs and beat until light and fluffy.
3. Add butter and sugar mixture and milk alternately to the dry ingredients, stirring just until combined. Fold in additional fruit, nuts, or flavors of your choice.
4. For bread, pour into a greased bread pan and bake at 350°F for 1 hour. For muffins, fill muffin cups ⅔ full and bake at 350°F for 20 to 25 minutes.

Cinnamon Bread

- 2 eggs
- ½ cup butter
- 1 cup sugar
- ½ cup milk
- 1¼ cups flour
- 2½ teaspoons baking powder
- 1 teaspoon cinnamon
- 1 teaspoon butter, melted
- 2 tablespoons sugar and 2 tablespoons cinnamon, mixed together

1. Beat together the eggs, butter, and sugar until fluffy.
2. In a separate bowl, combine the dry ingredients. Add the dry mixture and

the milk to the butter mixture and mix until combined.

3. Bake in a greased bread pan at 300°F for almost an hour. When done pour melted butter over top and sprinkle with cinnamon and sugar mixture.

One-Hour Brown Bread

- 1 cup cornmeal
- 1 cup white flour
- ½ teaspoon salt
- 1 teaspoon baking soda
- 1 cup water, boiling
- 1 egg
- ½ cup molasses
- ½ cup sugar

1. Combine cornmeal, flour, and salt.
2. Add the baking soda to boiling water and stir. Add to dry ingredients.
3. Beat together egg, molasses, and sugar and add to dry ingredients. Mix until combined. Pour batter into an empty coffee can with a cover (or cover with foil).
4. Place a cake rack in the bottom of a dutch oven or large pot. Place the covered can on the rack and pour boiling water into the pot until it reaches half way up the can. Cover the pot, turn the unit on very low, and steam for one hour.

Cranberry Coffee Cake

- 2 tablespoons butter
- ¼ cup firmly packed brown sugar
- 1 cup cooked or canned cranberry sauce

- ¼ cup pecans, chopped
- 1 tablespoon grated orange rind
- 1½ cups sifted flour
- 2 teaspoons, double acting baking powder
- ¼ cup sugar
- ⅓ cup shortening
- 1 egg, beaten
- ½ cup milk

1. Melt butter in 9-inch ring mold. Spread brown sugar over bottom of pan.
2. Combine cranberry sauce, pecans, and orange rind. Spread over brown sugar in bottom of pan.
3. Sift together flour, baking powder, and sugar.
4. Cut in shortening until dough resembles coarse meal. Combine egg with milk. Add all at once, mixing only to dampen flour. Turn into pan.
5. Bake at 400°F for 25 to 30 minutes. Cool 5 minutes and invert onto plate. Serve warm.

Date-Orange Bread

- 2 tablespoons butter or margarine, melted
- ¾ cup orange juice
- 2 tablespoons grated orange rind
- ½ cup finely cut dates
- 1 cup sugar
- 1 egg, slightly beaten
- ½ cup coarsely chopped pecan
- 2 cups sifted all-purpose flour
- ½ teaspoon baking soda
- 1 teaspoon baking powder
- ½ teaspoon salt

1. Combine first seven ingredients.
2. Mix and sift remaining ingredients; stir into wet mixture. Mix well, but quickly, being careful not to overbeat.
3. Turn into greased loaf pan. Bake in moderate oven, 350°F, for 50 minutes or until done. Remove from pan and let cool right side up, on a wire rack.

Pineapple Nut Bread

- 2¼ cups sifted flour
- ¾ cup sugar
- 1½ teaspoons salt
- 3 teaspoons baking powder
- ½ teaspoon baking soda
- 1 cup prepared bran cereal
- ¾ cup walnuts, chopped
- 1½ cups crushed pineapple, undrained
- 1 egg, beaten
- 3 tablespoons shortening, melted

1. Sift flour, sugar, salt, baking powder, and soda together.
2. Mix together remaining ingredients and combine with dry mixture.
3. Bake in greased loaf pan at 350°F for 1 ¼ hrs. This bread keeps moist a week or ten days, and slices best when a day or more old.

Date Muffins

- 1¾ cups sifted enriched flour
- 2 tablespoons sugar
- 2¼ teaspoons baking powder
- ¾ teaspoon salt
- ½ to ¾ cup coarsely cut pitted dates
- 1 egg, well-beaten
- ¾ cup milk
- ⅓ cup melted shortening or salad oil

1. Sift dry ingredients into mixing bowl and stir in dates. Make a well in center.
2. Combine egg, milk, and salad oil; add all at once to dry ingredients. Stir quickly only till dry ingredients are moistened.
3. Drop batter by tablespoons into greased muffin pans. Fill two-thirds full. Bake in oven, at 400°F, for about 25 minutes. Makes one dozen.

Yeast Bread

Once you've made a loaf of homemade yeast bread, you'll never want to go back to buying packaged bread from the grocery store. Homemade bread tastes and smells heavenly and the baking process itself can be very rewarding. Store homemade bread in a paper or resealable plastic bag and eat within a day or two for best results. Bread that begins to get stale can be cubed and made into stuffing or croutons.

Before you start baking, it's helpful to understand the various components that make up bread.

Baking Bread

Wheat

Wheat is the most common flour used in bread making, as it contains gluten in the right proportion to make bread rise. Gluten, the protein of wheat, is a gray, tough, elastic substance, insoluble in water. It holds the gas developed in bread dough by fermentation, which otherwise would escape. Though there are many ways to make gluten-free bread, flour that naturally contains gluten will rise more easily than gluten-free grains. In general, combining smaller amounts of other flours (rye, corn, oat, etc.) with a larger proportion of wheat flour will yield the best results.

A grain of wheat consists of (1) an outer covering, or husk, which is always removed before milling; (2) bran, a hard shell that contains minerals and is high in fiber; (3) the germ, which contains the fat and protein content and is the part that can be planted and cultivated to grow more wheat; and (4) the endosperm, which is the wheat plant's own food source and is mostly starch and protein. Whole wheat contains all of these components except for the husk. White flour is only the endosperm.

Yeast

Yeast is a microscopic fungus that consists of spores, or germs. These spores grow by budding and division, multiply very rapidly under favorable conditions, and produce fermentation. Fermentation is the process by which, under influence of air, warmth, and a fermenting ingredient, sugar (or dextrose, starch converted into sugar) is changed into alcohol and carbon dioxide.

Baking Bread

FLOUR	DESCRIPTION
All-purpose	A blend of high- and low-gluten wheat. Slightly less protein than bread flour. Best for cookies and cakes.
Amaranth	Gluten-free. Made from seeds of amaranth plants. Very high in fiber and iron.
Arrowroot	Gluten-free. Made from the ground-up root. Clear when cooked, which makes it perfect for thickening soups or sauces.
Barley	Ground barley grain. Very low in gluten. Use as a thickener in soups or stews or mixed with other flours in baked goods.
Bran	Made from the hard outer layer of wheat berries. Very high in protein, fiber, vitamins, and minerals.
Bread	Made from hard, high-protein wheat with small amounts of malted barley flour and vitamin C or potassium bromate. It has a high gluten content, which helps bread to rise. Excellent for bread, but not as good for use in cookies or cakes.
Chickpea	Gluten-free. Made from ground chickpeas. Used frequently in Indian, Middle Eastern, and some French Provençal cooking.
Buckwheat	Gluten-free. Highly nutritious with a slightly nutty flavor.
Oat	Gluten-free, though those with Celiac Disease or severe gluten sensitivities may not be able to tolerate. Made from ground oats. High in fiber.
Quinoa	Gluten-free. Made from ground quinoa, a grain native to the Andes in South America. Slightly yellow or ivory-colored with a mild nutty flavor. Very high in protein.
Rye	Milled from rye berries and rye grass. High in fiber and low in gluten. Light rye has had more of the bran removed through the milling process than dark rye. Slightly sour flavor.
Semolina	Finely ground endosperm of durum wheat. Very high in gluten. Often used in pasta.
Soy	Gluten-free. Made from ground soybeans. High in protein and fiber.
Spelt	Similar to wheat, but with a higher protein and nutrient content. Contains gluten but is often easier to digest than wheat. Slightly nutty flavor.
Tapioca	Gluten-free. Made from the cassava plant. Starchy and slightly sweet. Generally used for thickening soups or puddings, but can also be used along with other flours in baked goods.
Teff	Gluten-free. Higher protein content than wheat and full of fiber, iron, calcium, and thiamin.
Whole wheat	Includes the bran, germ, and endosperm of the wheat berry. Far more nutritious than white flour, but has a shorter shelf life.

THE JUNIOR HOMESTEADER

The Grain Game

Play a "grain game" to learn about grain-based foods. Taste different foods made from grains and learn where grains are grown in the United States.

Materials Needed
- Food samples
- Paper cups, napkins, or paper towels
- Container of water

Goals
- To identify different foods made from grains.
- To expand the variety of foods eaten by tasting different kinds of foods made from grains.

- To identify where different grains are grown in the United States.

Key Concepts
- There are a wide variety of foods made from grains.
- Eating foods made from different grains adds a diversity of tastes to meals.

Preparation
Purchase foods that are made from the grains that will be discussed in this lesson. Examples of foods that might be purchased are: corn tortilla, rye bread, pumpernickel bread, oatmeal muffins or oatmeal cookies, and rice cakes. Additional breads you might consider including, if they are available, are:

scones—a British sweet biscuit
chapatis—a flat bread eaten in India and in East Africa

pita bread—a flat bread also known as "pocket bread"

lavash—a paper-thin Russian bread used for wrapping food

matzoh—a flat, cracker-like bread

corn bread—a bread made from cornmeal

Cut foods into bite-sized pieces.

Background

A grain is a single seed of a cereal grass. Some of the cereal grains grown in the United States are wheat, corn, rye, rice, barley, and oats. More foods are made with wheat than any other cereal grain.

Each grain tastes differently and adds delicious taste, nutrition, and variety to meals. Grain-based foods provide complex carbohydrates, which are an important source of energy for the body.

Grains also provide vitamins that help keep the body strong and healthy such as B vitamins, minerals such as iron, and dietary fiber, which keeps the digestive systems healthy. Grain products belong in the Breads, Cereals, Rice, and Pasta Group of the Food Guide Pyramid. The following are examples of grain-based foods categorized by their main grain ingredient.

Oats

oatmeal	ready-to-eat oat cereal
oatmeal cookie	granola
oatmeal muffin	muesli

Rye

rye bread	pumpernickel bread
rye flatbread	rye crackers

Rice

wild rice	sticky rice
white rice	rice noodles
basmati rice	rice cake
texmati rice	rice pudding
jasmine rice	rice cereal (infant)
brown rice	rice balls
Spanish rice	popped wild rice
ready-to-eat rice cereal	cream-of-rice cereal
risotto	

Wheat

white bread	pita bread
wheat bread	matzoh
noodles	pancake
spaghetti	crepe
biscuit	cream-of-wheat cereal
fry bread	wheat flakes
flour tortilla	popover
wonton wrapper	couscous
cracker	tabbouleh
waffle	cake
graham cracker	
scone	

Corn

corn bread	grits
corn tortilla	corn flakes
popcorn	cornmeal mush
hominy	hushpuppy

Baking Bread

Setup and Introduction

1. Tell children that they will be playing a game about grains.
2. Explain that before they begin the Grain Game they will first need to understand that a grain is a seed from a cereal grass. Some cereal grains grown in the United States for food are wheat, corn, rice, oat, and rye (barley and millet are also cereal grains but are used less often in the United States).

3. Divide the group into two teams. The group leader should moderate the game. First, explain to children how the game is played:

- The group leader will call out a type of grain (e.g., wheat, corn, rice, oats, rye).
- One team will begin the game by calling out a food made from that grain.
- The other team will respond by calling out a different food made from that grain.

- Each time a correct food is called out, that team gets 1 point. When a team calls out an incorrect food, that team will not get a point but the point will go to the other team.
- The teams will alternate calling out a different food until no more can be named.
- When no more can be named, the group leader calls out the name of a new grain and the game continues.
- Play rounds of this game until wheat, corn, rice, oats, and rye are covered.

Game Closure

Suggested discussion questions following the game:

What foods made from grains do you eat now?
What new foods made from grains did you learn about today?
Which of the foods named today would you want to taste?

Tasting Activity

1. Tell the kids that now they are going to taste some foods made from different grains.
2. Have everyone wash their hands.
3. Have a volunteer distribute cups of water and paper plates or napkins. The water is for sipping between food samples.
4. Children should take a sample from each food plate, taste the food, and try to figure out which grain it's made with. Have volunteers name each food and its grain ingredient.
5. Suggested discussion questions:

Which foods did you like best? Why?
Which were new to you?

Suggested discussion starters:

Name the states where grains are grown.
Where are most grains grown? Why?
How do the grains grown in the different states get to other parts of the country to be made into foods?

MILLING YOUR OWN GRAINS

You can grind grains into flour at home using a mortar and pestle, a coffee or spice mill, manual or electric food grinders, a blender, or a food processor. Grains with a shell (quinoa, wheat berries, etc.) should be rinsed and dried before milling to remove the layer of resin from the outer shell that can impart a bitter taste to your flour. Rinse the grains thoroughly in a colander or mesh strainer, then spread them on a paper or cloth towel to absorb the extra moisture. Transfer to a baking sheet and allow to air dry completely (to speed this process you can put them in a very low oven for a few minutes). When the grains are dry, they're ready to be ground.

Dry yeast is most commonly used for baking. Most grocery stores sell regular active dry and instant yeast. Instant yeast is more finely ground and thus absorbs the moisture faster, speeding up the leavening process and making the bread rise more rapidly.

Active dry yeast should be proofed before using. Mix one packet of active dry yeast with ¼ cup warm water and 1 teaspoon sugar. Stir until yeast dissolves. Allow it to sit for 5 minutes, or until it becomes foamy.

Making Bread

Making bread is a fairly simple process, though it does require a chunk of time. Keep in mind, though, that you can be doing other things while the bread is rising or baking. The process is fairly straightforward and only varies slightly by kind of bread.

Mix together the flour, sugar, salt, and any other dry ingredients. Form a well in the center and add the dissolved yeast and any other wet ingredients. Mix all the ingredients together.

Gather the dough into a ball and place it on a lightly floured surface. Flour your hands to keep the dough from sticking to your fingers. Knead the dough by folding it toward you and then pushing it away with the palms of your hands. Continue kneading for five to ten minutes, or until the dough is soft and elastic.

Place the dough in a lightly greased pan, cover with a dish towel, and allow to rise in a warm place until it doubles in size.

Punch the dough down to expel the air and place it in a greased and lightly floured baking pan. Cover and let rise a second time until it doubles in size.

Bake the bread in a preheated oven according to the recipe. Bread is done when it is golden brown and sounds hollow when you tap the top.

Remove bread from the pan by loosening the sides with a knife or spatula and tipping the pan upside down onto a wire rack.

Biscuits

Any bread recipe can be made into biscuits instead of one large loaf. To shape bread dough into biscuits, pull or cut off pieces, making them all as close to uniform in size as possible. Flour palms of hands slightly and shape each piece individually. Using the thumb and first two fingers of one hand, and holding it in the palm of the other hand, move the dough round and round, folding the dough towards the center. When smooth, turn it over and roll between palms of hands. Place in greased pans nearly together, brushed between with a little melted butter, which will allow biscuits to separate after baking.

GLUTEN-FREE BREAD

Making good gluten-free bread isn't always easy, but there are several things you can do to improve your chances of success:

- Choose flours that are high in protein, such as sorghum, amaranth, millet, teff, oatmeal, and buckwheat
- Use all room temperature ingredients. Yeast thrives in warm environments.
- Add a couple teaspoons of xantham gum to your dry ingredients.
- Add eggs and dry milk powder to your bread. These will add texture and help the bread to rise.
- Crush a vitamin C tablet and add it to your dry ingredients. The acidity will help the yeast do its job.
- Substitiute carbonated water or gluten-free beer for other liquids in the recipe.
- If you're following a traditional bread recipe, add extra liquid (water, carbonated water, milk, fruit juice, or olive oil) to get a soft and sticky consistency. The batter should be a little too sticky to knead. For this reason, bread machines are great for making gluten-free bread.

Multigrain Bread

- ¼ cup yellow cornmeal
- ¼ cup packed brown sugar
- 1 teaspoon salt
- 2 tablespoons vegetable oil
- 1 cup boiling water
- 1 package active dry yeast
- ¼ cup warm (105 to 115ºF) water
- ¼ cup whole wheat flour
- ¼ cup rye flour
- 2¼–2¾ cups all-purpose flour

1. Mix cornmeal, brown sugar, salt, and oil with boiling water; cool to lukewarm (105 to 115ºF).
2. Dissolve yeast in ¼ cup warm water; stir into cornmeal mixture. Add whole wheat and rye flours and mix well. Stir in enough all-purpose flour to make dough stiff enough to knead.
3. Turn dough onto lightly floured surface. Knead until smooth and elastic, about 5 to 10 minutes.
4. Place dough in lightly oiled bowl, turning to oil top. Cover with clean towel; let rise in warm place until double, about 1 hour.
5. Punch dough down; turn onto clean surface. Cover with clean towel; let rest 10 minutes. Shape dough and place in greased 9 x 5 inch pan. Cover with clean towel; let rise until almost double, about 1 hour.
6. Preheat oven to 375ºF. Bake 35 to 45 minutes or until bread sounds hollow when tapped. Cover with aluminum foil during baking if bread is browning too quickly. Remove bread from pan and cool on wire rack.

Oatmeal Bread

- 1 cup rolled oats
- 1 teaspoon salt
- 1½ cups boiling water
- 1 package active dry yeast
- ¼ cup warm water (105 to 115ºF)
- ¼ cup light molasses
- 1½ tablespoons vegetable oil
- 2 cups whole wheat flour
- 2–2½ cups all-purpose flour

1. Combine rolled oats and salt in a large mixing bowl. Stir in boiling water; cool to lukewarm (105 to 115ºF).
2. Dissolve yeast in ¼ cup warm water in small bowl.
3. Add yeast water, molasses, and oil to cooled oatmeal mixture. Stir in whole wheat flour and 1 cup all-purpose flour. Add additional all-purpose flour to make a dough stiff enough to knead.
4. Knead dough on lightly floured surface until smooth and elastic, about 5 minutes.
5. Place dough in lightly oiled bowl, turning to oil top. Cover with clean towel; let rise in warm place until double, about 1 hour.
6. Punch dough down; turn onto clean surface. Shape dough and place in greased 9 x 5 inch pan. Cover with clean towel; let rise in a warm place until almost double, about 1 hour.
7. Preheat oven to 375ºF. Bake 50 minutes or until bread sounds hollow when tapped. Cover with aluminum foil during baking if bread is browning too quickly. Remove bread from pan and cool on wire rack.

Raised Buns (Brioche)

- ½ cup milk
- ⅓ cup butter
- ¼ cup sugar
- ¾ teaspoon salt
- 1 package yeast (active, dry or compressed)
- ¼ cup lukewarm water
- 3 eggs, well-beaten
- 2½ cups enriched flour
- ¼ teaspoon lemon rind, or ⅛ teaspoon crushed cardamom seeds

1. Save 2 tablespoons egg to brush brioches before baking. Scald milk; stir in butter, sugar, and salt; cool to lukewarm. Sprinkle or crumble yeast into water in large bowl; stir to dissolve. Add lukewarm milk mixture and beaten eggs; mix well.

2. Sift flour; add 1 ½ cups of it to mixture and beat by hand 8 minutes or with electric mixer at medium speed for 3 minutes. Add remaining sifted flour and lemon rind; beat to smooth, heavy batter. Cover with towel; let rise in warm place 2 hours or until doubled. Stir down; cover tightly; chill at least 5 hours or overnight.

3. Stir down again. Mixture is soft now. Grease hands slightly, place dough on lightly floured board, and knead a few times. With sharp knife, cut off small pieces of dough. Roll with hands to about ½ inch in diameter, and about 10 inches long. Coil loosely in circle, winding around toward center. Top with small ball of dough. Cover with a towel and let rise ½ hour or until doubled. Brush with egg yolk beaten with 1 teaspoon water. Bake in oven 400°F for 12 to 15 minutes. For an extra touch, top with thin frosting while still warm. Makes 12 to 18 buns.

THE JUNIOR HOMESTEADER

Family Taste Celebration

Many families use special foods for family gatherings and celebrations big and small. Take a few minutes to think about the special foods you prepare.

Encourage your child to explore his or her food heritage. Often, foods that were prepared as everyday foods many years ago have become today's celebration foods. For example, great grandmom's crumb cake, which she made every week, you may make once a year as part of a holiday breakfast. Or the homemade ravioli that was made weekly has been replaced by the store-bought variety. Suggest that your child talk to older family members about the foods they or their grandparents ate when they were younger.

Here are a few questions to get them started:

What countries did our relatives come from? _____

_____

What recipes or foods did you eat when you were younger that can be traced back to these countries?

_____

_____

Do you have any recipes that have been handed down from generation to generation?

_____

_____

Discuss with your child what was discovered about his or her family food heritage. Discuss the family recipes and make a shopping list for one of these recipes. Have your child track down the ingredients when you go to the grocery store. Together with your child, prepare the recipe and enjoy a celebration of family history.

THE JUNIOR HOMESTEADER

Bread in a Bag

Materials needed:

- A heavy-duty zipper-lock freezer bag (1 gallon size)
- Measuring cup
- Measuring spoons
- Cookie sheet
- Pastry towel or cloth
- 13 x 9-inch baking pan
- 8½ x 4½-inch glass loaf pan

Ingredients:

- 2 cups all-purpose flour, divided
- 1 package rapid rise yeast
- 3 tablespoons sugar
- 3 tablespoons nonfat dry milk
- 1 teaspoon salt
- 1 cup hot water (125°F)
- 3 tablespoons vegetable oil
- 1 cup whole-wheat flour

1. Combine 1 cup all-purpose flour, yeast, sugar, dry milk, and salt in a freezer bag. Squeeze upper part of the bag to force out air and then seal the bag.
2. Shake and work the bag with fingers to blend the ingredients.
3. Add hot water and oil to the dry ingredients in the bag. Reseal the bag and mix by working with fingers.
4. Add whole-wheat flour. Reseal the bag and mix ingredients thoroughly.
5. Gradually add remaining cup of all-purpose flour to the bag. Reseal and work with fingers until the dough becomes stiff and pulls away from sides of the bag.
6. Take dough out of the bag, and place on floured surface.
7. Knead dough 2 to 4 minutes, until smooth and elastic.
8. Cover dough with a moist cloth or pastry towel; let dough stand for 10 minutes.
9. Roll dough to 12 x 7-inch rectangle. Roll up from narrow end. Pinch edges and ends to seal.
10. Place dough in a greased glass loaf pan; cover with a moist cloth or pastry towel.
11. Place baking pan on the counter; half fill with boiling water.
12. Place cookie sheet over the baking pan and place loaf pan on top of the cookie sheet; let dough rise 20 minutes or until dough doubles in size.
13. Preheat oven, 375°F, while dough is rising (about 15 minutes).
14. Place loaf pan in oven and bake at 375°F for 25 minutes or until baked through.

Baking Cakes, Pies, Cookies, and Other Desserts

There is something uniquely satisfying and comforting about baking. To start with ingredients as plain as flour, eggs, and sugar and end up with a rich cake, soft and gooey cookies, or a heaping pie can seem like a small miracle. Try your hand at some of these classic desserts, or use them as inspiration for creating your own recipes.

Cakes

Yellow Layer Cake

- ½ cup butter
- 1½ cups sugar
- 4 eggs, separated
- 2 cups flour
- 2 teaspoons baking powder
- ½ teaspoon baking soda
- ¾ cup sweet milk
- 1 teaspoon vanilla

1. Cream butter thoroughly, add sugar gradually, and cream together until light and fluffy. Add eggs yolks and beat well.

2. Sift flour, baking powder, and baking soda together and add alternately with milk, beating after each addition until smooth. Add vanilla and fold in stiffly beaten egg whites.
3. Bake in two greased layer pans in 350°F oven about 25 minutes.

Chocolate Cake

- 4 ounces unsweetened baking chocolate
- 1½ cups sugar, divided
- 1½ cups milk, divided
- 2 teaspoons vanilla
- ½ cup butter
- 2 cups flour
- 1 teaspoon baking soda
- ¼ teaspoon salt
- 2 eggs, beaten

TIP

Buttermilk can be made by combining 1 cup milk with 1 teaspoon vinegar. Stir and let sit for 3 to 5 minutes.

1. Melt chocolate in a saucepan over low heat. Add ½ cup sugar, 1 cup milk, and vanilla.
2. In a mixing bowl, cream butter and 1 cup sugar. Sift together flour, baking soda, and salt.
3. Add eggs and remaining ½ cup milk alternately with dry ingredients to the chocolate mixture.
4. Bake in two greased layer pans at 350°F about 25 minutes.

Banana Layer Cake

- 2½ cups cake flour
- 1 teaspoon baking soda
- ½ teaspoon baking powder
- ¾ cup butter or shortening
- 1½ cups sugar
- 2 eggs
- ¼ cup buttermilk

- 1 teaspoon vanilla
- 1¼ cups ripe bananas (mashed)

1. Sift flour with soda and baking powder.
2. Cream butter and sugar together until light and fluffy. Add the whole eggs, beating well after the addition of each.
3. Add buttermilk and vanilla to bananas.
4. Add alternately the flour and banana mixture to butter mixture, beginning and ending with the dry ingredients. Beat until smooth and well blended.
5. Divide evenly into two 9-inch greased cake pans (lightly floured). Bake 25 to 30 minutes at 350°F.

Cherry-nut Cake

- 2 cups flour
- 1 cup sugar
- 2 teaspoons baking powder
- ¼ teaspoon salt
- 2 eggs
- ½ cup milk
- ½ cup butter
- 1 teaspoon vanilla
- 1 cup maraschino cherries, chopped
- ½ cup walnuts, chopped

1. Sift together flour, sugar, baking powder, and salt. In a separate bowl, beat together eggs, milk, butter, and vanilla.
2. Add dry ingredients to wet ingredients and mix until combined. Add cherries and walnuts and stir.
3. Bake in two greased and floured cake pans at 350° for 20 to 30 minutes. Frost as desired.

Applesauce Cake

- 1⅔ cups flour
- 1 teaspoon baking soda
- ½ teaspoon salt
- ½ teaspoon cloves
- 1 teaspoon cinnamon
- 1 teaspoon nutmeg
- 1 cup brown sugar
- ½ cup butter
- 1 cup unsweeetened applesauce
- 1 cup raisins or nuts, chopped

1. Sift together flour, baking soda, salt, and spices.
2. In a separate bowl, beat together sugar and butter until light and fluffy. Add applesauce and continue to mix until combined.
3. Add dry ingredients to wet ingredients and mix until combined. Add raisins or nuts (if desired) and stir until combined.
4. Bake in a greased loaf pan for 1 hour at 350°F. Allow to sit in oven for 10 minutes more with heat off.

Fruit Cake

- 2½ cups sifted flour
- 1 teaspoon baking powder
- 1 pound candied pineapple
- ¼ pound citron
- 1 pound raisins
- 1 pound candied cherries
- ½ cup butter
- ¾ cup sugar
- 3 eggs, beaten
- ½ teaspoon baking soda dissolved in 1 teaspoon hot water
- ¾ cup buttermilk
- 1 pound coconut
- ½ cup blanched almonds

Baking Cakes, Pies, Cookies, and Other Desserts

1. Sift together flour and baking powder.
2. Cut the fruit into small pieces. If citron and pineapple are too heavily coated with sugar, either wash or scrape the sugar off. Sprinkle with some of flour mixture and work in.
3. Cream butter and sugar, add eggs, and beat until light and fluffy. Add other ingredients and mix with a spoon, adding fruit and nuts last.
4. Use two bread tins well greased. Set them on shelf halfway up in oven and bake at 275 to 300°F for 1 ½ to 2 hours. The cake should be light brown on top when done. For smaller cakes use small pans. Four or five small bread tins (7 ½ x 3 ½ inches) work nicely, and make nice Christmas gifts. If desired, a small amount of brandy added to the batter gives it an excellent flavor.

Christmas Cupcakes

- 1½ cups seedless raisins
- ½ cup nuts of your choice
- 1 tablespoon grated orange rind
- ½ cup butter
- 1 cup sugar
- 2 eggs, slightly beaten
- 2 cups sifted flour
- ½ teaspoon salt
- ½ teaspoon vanilla
- 1 teaspoon baking soda
- 1 cup buttermilk

Glaze

- ¼ cup sugar
- 3 tablespoons orange juice

1. Chop or grind raisins and nuts together. Add orange rind.
2. Cream butter and sugar together until light and fluffy; add eggs.
3. Mix flour and salt. Add vanilla and baking soda to buttermilk.
4. Add flour and liquid to butter mixture alternately, stirring well after each addition. Fold in raisin mixture.
5. Fill 3-inch greased and floured muffin pans two-thirds full. Bake in moderate oven at 350°F for 40 to 50 minutes. Meanwhile, make the glaze. Mix sugar with orange juice, and let stand 30 minutes.
6. Cool cupcakes for 5 minutes. before removing from pans. Dip top of each warm cupcake in glaze. Makes about 18.

Gingerbread

- 2 eggs
- ¾ cup sugar
- ½ cup butter, melted
- 1 cup molasses
- 2½ cups sifted flour
- 1 teaspoon salt
- 1 teaspoon cinnamon
- 1 teaspoon ginger
- ½ teaspoon cloves
- 2 teaspoons baking soda
- 1 cup boiling water

1. Beat eggs well and gradually add sugar. Add melted butter and molasses and mix well.
2. Combine flour, salt, cinnamon, ginger, cloves, and baking soda. Add flour mixture to egg mixture. Mix well. The batter will be stiff.
3. Add 1 cup boiling water and stir until smooth.
4. Bake in greased and floured 9 x 12 ½-inch pan 350°F about 40 minutes.

Spice Cake

- ⅓ cup butter
- 1 cup brown sugar
- 2 eggs (save 1 white for meringue)
- 1¾ cups flour
- ¾ teaspoon baking soda
- 1 teaspoon salt
- ½ teaspoon cloves

CAKE TROUBLES

If the cake doesn't rise enough, the problem may be:
- You didn't use enough leavening. Use the type called for, and measure exact amount.
- The cake was baked in too long a pan.
- The oven was too hot for proper rising.

If the cake falls, the cause may be:
- Too much shortening.
- Too much leavening.
- Too much liquid.
- Too much sugar.
- The cake baked too long or at too low a temperature.
- You tried to remove the cake from the pan before it cooled.

If the top of the cake gets crusty, the cause may be:
- The oven was too hot.
- The cake baked too long.

- ¼ teaspoon ginger
- 1 teaspoon cinnamon
- ¼ teaspoon nutmeg
- ¾ cup sour milk

1. Beat together butter, sugar, salt, and eggs.
2. Sift together flour, baking soda, salt, and spices. Add flour mixture alternately with milk into first mixture.
3. Pour into shallow 8 ½-inch square pan that has been greased and lightly dusted with flour. Bake at 350°F for 40 minutes.

Boiled Butterscotch Frosting

- 2 cups brown sugar
- 1 cup granulated sugar
- 1 cup sour cream or milk
- 1 tablespoon butter
- 1 teaspoon vanilla

1. Combine first three ingredients in a saucepan and boil until the mixture forms a soft ball when a small amount is dropped in a glass of water.
2. Add butter and vanilla, remove from heat, and beat until creamy.

Sea Foam Icing

- 1½ cups brown sugar
- ⅓ cup water
- 2 egg whites, unbeaten
- ⅛ teaspoon cream of tartar
- 1 teaspoon vanilla

1. Boil sugar and water for 3 minutes. Cool slightly.
2. Put egg whites and cream of tartar into mixing bowl and beat on high. Immediately add the hot syrup. Beat for 5 minutes and add vanilla.

Pies

Easy Pie Crust

This recipe will make enough dough for one double-crust or two single-crust pies.

- 2½ cups all-purpose flour
- 1 cup salted butter
- 4 to 8 tablespoons ice water

1. Combine flour and butter in a food processor and pulse until mixture resembles coarse crumbs.
2. Add ice water, 1 tablespoon at a time, pulsing between each addition. As soon as the dough clings together, form into two balls.
3. Roll dough to ¼-inch thickness on a floured cutting board. Dust dough with flour as needed to keep it from sticking to the rolling pin. If not using the dough immediately, cover the balls of dough with plastic wrap and refrigerate until ready to use.

Lemon Meringue Pie

- 1¼ cups cold water
- 4 level tablespoons cornstarch
- 1 cup sugar + 6 tablespoons sugar, divided
- 1 tablespoon flour

- 3 eggs, separated
- 1 grated lemon rind
- 3 to 4 tablespoons lemon juice
- 1 tablespoon butter
- pinch salt
- 1 9-inch baked pie shell (see Easy Pie Crust)

1. Preheat oven to 350° F.
2. Mix cornstarch with cold water. In a saucepan, whisk together 1 cup sugar, flour, and cornstarch mixture. Cook the mixture on top of a double boiler until thick.
3. In a bowl, whisk 3 egg yolks. Gradually combine the mixture with the egg yolks. Add grated rind of lemon, lemon juice, butter, and salt. Continue to cook the mixture, stirring constantly until thick.

4. Remove from heat and pour filling into baked pie shell.
5. In a bowl, whip 3 egg whites while gradually adding 6 tablespoons sugar. Spread meringue on top of filling evenly. Bake pie at 325°F about 15 minutes or until meringue is golden brown.
6. Remove from oven and allow the pie to cool before serving.

Open Blueberry Pie (No Bake)

- 2 cups blueberries (Canned berries may be used. Drain and measure juice. Use juice instead of water.)
- 1 cup sugar
- 1 cup water
- 3 tablespoons tapioca
- pinch salt
- Cream, whipped

Cornflake crust:

- ¼ pound butter
- 4 or more cups of crushed cornflakes
1. Melt butter in a bowl. Add 4 or more cups of crushed cornflakes to bowl. Stir gently and press mixture into pie plate.
2. Chill crust until cold and set.

PREBAKING A PIE SHELL

Many of these recipes call for a baked pie shell. To prebake, or "blind bake" a pie crust, roll it out and place it in the pie plate. Line the crust with parchment paper and sprinkle dried beans or rice on top. This will keep the crust from getting puffy. Bake at 425°F for about 20 minutes. Remove beans or rice and parchment paper and the crust will be ready to fill.

thick. Remove from heat; add orange juice. Beat egg whites until stiff, beating in remainder of sugar gradually. Fold in peach mixture and pack lightly into baked crust. Bake in a moderate oven at 325°F until firm, about 30 minutes. Chill until serving.

Chocolate Chiffon Pie

- 2 1-ounce squares unsweetened chocolate
- ½ cup boiling water
- 1 tablespoon (1 envelope) unflavored gelatin
- ¼ cup cold water
- 4 eggs, separated, whites stiffly beaten
- 1 cup sugar, separated
- ¼ teaspoon salt
- 1 teaspoon vanilla
- ½ cup sugar
- 1 9-inch baked pie shell
- Cream, whipped

1. Melt chocolate in boiling water. Soften gelatin in cold water and add to chocolate; stir until gelatin dissolves.
2. Add egg yolks, beaten lightly with ½ cup sugar. Add salt and vanilla.
3. Beat remaining ½ cup sugar into egg whites. Fold into chocolate mixture; pour into pie shell. Top with whipped cream.

Pumpkin Chiffon Honey Pie

- 1 tablespoon unflavored gelatin
- ¼ cup cold water
- 3 eggs, separated

6. Pour into baked pie shell. Refrigerate until set.

Peach Chiffon Pie

- ¼ cup flour
- 1 cup sugar, divided
- 1¼ cups canned peaches, crushed
- ½ cup orange juice
- 3 egg whites
- 1 9-inch baked pie shell

1. Mix flour with ½ cup sugar. Add crushed peaches. Cook in double boiler until

3. Put blueberries, sugar, water, tapioca, and salt into a saucepan and cook, stirring constantly, until thickened.
4. Cool mixture and pour it into the chilled crust. Chill pie until ready to serve.
5. Cover pie with whipped cream before serving.

Coconut Bavarian Pie

- 1 envelope unflavored gelatin
- ¼ cup cold water
- 1 cup milks
- 3 eggs, separated
- ½ cup granulated sugar, divided
- ¼ teaspoon salt
- 1 cup heavy cream, whipped
- 1 teaspoon vanilla
- ½ cup shredded coconut
- 1 9-inch baked pie shell (see Easy Pie Crust)

1. Soften gelatin in cold water.
2. Scald milk in double boiler.
3. Combine egg yolks and ¼ cup sugar; stir in milk. Return to double boiler; cook over hot (not boiling) water, until custard coats spoon.
4. Stir in gelatin until dissolved. Refrigerate until it is as thick and syrupy as unbeaten egg white.
5. Beat egg whites with salt till quite stiff. Fold in custard mixture, then whipped cream, vanilla, and coconut.

- ¾ cup honey
- 1½ cups canned pumpkin
- ½ cup milk
- ½ teaspoon salt
- 1 teaspoon cinnamon
- 3 tablespoons sugar
- 1 9-inch baked pie shell

1. Soak gelatin in cold water for 5 minutes.
2. Beat egg yolks and combine with honey, pumpkin, milk, salt, and cinnamon in top of double boiler. Cook until thick, stirring constantly, about 8 to 10 minutes. Add softened gelatin and stir until dissolved.
3. Beat egg whites until frothy; add sugar gradually, and continue beating until whites stand in soft peaks. Fold meringue into pumpkin mixture. Fill pie shell. Chill several hours.

Apple Crumb Pie

- 6 cups peeled, sliced apples
- 1 9-inch pastry shell, unbaked
- ⅓ cup sugar
- 1 teaspoon cinnamon

Crumb topping:

- ⅓ cup sugar
- ½ cup flour
- 4 tablespoons butter (chilled)

Place apple slices in the pie shell. Combine cinnamon and sugar and pour evenly over the apples. Combine crumb topping ingredients, rubbing together with your fingers until crumbly. Sprinkle topping evenly over the pie. Bake at 450°F for 10 minutes; reduce heat to 375°F and bake 30 minutes or until apples are tender.

A BAKER'S DOZEN BAKING TIPS

- Keep an apple in your brown sugar. Store the pair in a tightly closed container and the sugar will stay soft and moist.
- Roll out your pastry dough on waxed paper. Use a sponge to dampen the work surface before placing the paper on it. Dust the paper with flour and draw a circle in the flour slightly larger your pie pan. Once the pastry is rolled to the size of your circle, pick up the paper and pastry, flip it upside down into your pan, and then pull the paper away from the crust.
- Run a measuring cup under hot water for a few seconds before using it to measure honey, oil, nut butters, etc.
- Never grease the cake pan when making sponge, angel food, or chiffon cakes. For other sorts of cakes, greasing the pan will help the cakes to rise evenly and slide out of the pan without falling part.

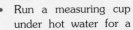

- To tell if a cake is done, stick a sharp knife into the center and pull it out. If the knife is dry, the cake is done.
- To easily peel peaches or plums for use in pies, dip them in boiling water for a few seconds. The peels will slide right off.
- Leftover pie dough can be slathered with butter and jam, rolled into a log, and then cut in slices and baked until golden.
- To prevent the bottom pie crust from getting soggy from berries or other juicy fruits, brush the crust with an egg white. You can also brush the top crust with egg white to create a golden glaze.
- Pie crusts roll out best when the dough is kept cold. To this end, you can substitute a bottle filled with ice water for the rolling pin.
- To make cookies soft and chewy, pull them out of the oven while the centers are still slightly gooey. Allow to cool on the pan for a few minutes, and then move to a cooling rack.
- Most cookie dough can be stored in the freezer for up to three months. Thaw in the refrigerator before using.
- To make egg whites stiffen more quickly, add a pinch of salt before beating.
- When separating eggs, if a bit of egg yolk slips into the whites, moisten a cloth with cold water and touch it to the yolk. The yolk will stick to the cloth.

Cookies

Ginger Cookies

- ¾ cup butter
- 1 cup white sugar
- 4 tablespoons molasses
- 1 egg
- 2 cups flour
- 2 teaspoons baking soda
- ⅛ teaspoon salt
- 1 teaspoon cinnamon
- ½ teaspoon cloves
- 1 teaspoon ginger

1. Beat together butter and sugar until light and fluffy. Add molasses and egg and continue to beat until thoroughly combined.
2. In a separate mixing bowl, combine dry ingredients. Add dry ingredients to wet ingredients and mix until well combined.
3. Chill dough for at least an hour, then roll into small balls. Roll in sugar. Place on cookie sheet 2 to 3 inches apart. Do not flatten them out. Bake at 350°F for 10 minutes.

Chocolate Brownies

- ½ cup butter
- 2 blocks (2 ounces) baking chocolate, melted
- 1 cup sugar
- 3 eggs, well-beaten
- ¾ cup flour
- ½ teaspoon baking powder
- ½ teaspoon salt
- 1 teaspoon vanilla
- 1 cup walnuts, chopped

1. Cream butter, add melted chocolate, and beat together.
2. In a separate bowl, combine sugar and eggs and add to chocolate mixture.
3. Sift flour, baking powder, and salt together and add to wet ingredients, beating thoroughly. Add vanilla and nuts.
4. Pour into a greased, shallow square pan. Bake in oven at 350°F for 20 to 25 minutes. Cut into squares while warm.

Date Bars

- 3 eggs
- 2 cups dates, chopped
- 1 cup nuts
- 1 teaspoon vanilla
- ¾ cup sugar
- 1 cup flour
- 2 teaspoons baking powder

1. Beat eggs until frothy. Add dates, nuts, sugar, and vanilla.
2. Sift flour and baking powder and add to first mixture.

3. Bake in shallow pan at 325°F for 45 minutes. Cut into squares.

Chocolate Pinwheel Cookies

- ½ cup butter
- ¾ cup sugar
- 1 teaspoon vanilla
- 1 egg
- 1¼ cups sifted flour
- ¼ teaspoon baking powder
- ¼ teaspoon salt
- 1 square unsweetened chocolate (melted)

1. Cream butter and sugar until light and fluffy. Add vanilla and egg; beat until light.
2. Add sifted dry ingredients. Halve dough and add chocolate to one half. Chill several hours.
3. Roll white dough on waxed paper to form a 9 x 12-inch rectangle. Roll out chocolate dough to same size on waxed paper.
4. Invert chocolate dough onto white dough, pull off paper, and press gently with rolling pin. Roll tightly, wrap in paper, and chill thoroughly. Slice roll into ⅛-inch-thick cookies. Bake 10 minutes at 350°F.

Raisin Penny-pinchers

- ¾ cup soft butter
- 1 cup brown sugar
- 2 eggs, unbeaten
- ½ cup undiluted evap. milk
- 2 cups sifted flour
- ½ teaspoon salt
- 2 cups rolled oats
- ½ cup nuts, chopped

- 1 cup seedless raisins
- 1 teaspoon grated lemon rind

1. Cream butter and sugar until light and fluffy. Add eggs and blend well.
2. Add milk, flour, and salt and stir until blended.
3. Add remaining ingredients and mix thoroughly.
4. Drop by level tablespoons onto a well-greased cookie sheet. Bake for 8 to 10 minutes at 375°F. Makes 60 cookies.

German Christmas Cookies

- 1 cup butter or margarine
- 2 cups sugar
- 3 eggs
- 1 teaspoon vanilla
- 6 tablespoons milk
- About 6 cups flour, enough to make stiff dough
- 1 tablespoon anise seed

1. Cream together butter and sugar till very smooth. Add eggs (well beaten). Work in vanilla and milk and sifted flour alternately. Add anise seeds and mix. Dough must be stiff to roll easily. Roll very thin, and cut with fancy cutters. Bake at 350°F till delicate brown (about 10 minutes).

Apple Cider Doughnuts

Cider doughnuts are a fall treat often found at apple orchards or farm stands. They're delicious with a cup of coffee or hot chocolate, and kids will have fun helping to cut out the dough and rolling the finished doughnuts in sugar.

- ½ cup granulated sugar
- ½ cup brown sugar
- 4 tablespoons butter, softened
- 2 large eggs
- 1 cup apple cider, boiled down until it is reduced to about ⅓ cup
- ½ cup buttermilk
- 1 teaspoon vanilla
- 2 ½ cups all-purpose flour
- 1 cup whole wheat flour
- 2 teaspoons baking powder
- 1 teaspoon baking soda
- ½ teaspoon cinnamon
- ½ teaspoon salt
- ⅛ teaspoon ground nutmeg
- ⅛ teaspoon cloves
- 1 medium apple, peeled and finely chopped
- canola oil for frying
- extra cinnamon and sugar or powdered sugar for dusting

1. Beat the sugar and butter until light and fluffy. Add the eggs and continue beating until smooth. Then add the reduced cider, buttermilk, and vanilla.
2. In a separate bowl, combine remaining dry ingredients. Add the wet ingredients to the dry and stir just until combined. Fold in chopped apple.
3. Cover the dough and allow to chill for about an hour.
4. Divide the dough in half, knead on a lightly floured surface for about a minute, and then roll out to a ½-inch thickness, using more flour if the dough is sticky. Use a doughnut cutter to cut out the shapes, adding any dough scraps to the second piece of dough and repeating.
5. Pour oil into a large pot until it is about 3 inches deep. Heat oil to 350°F. Drop several doughnuts in and allow to cook for 2 minutes on each side, using a slotted spoon or tongs to flip the doughnuts. When doughnuts are golden brown, transfer to a cooling rack covered with paper towel.
6. Once doughnuts are cool, roll lightly in cinnamon and sugar or powdered sugar.

Swedish Heirloom Cookies

- 1 cup butter
- 1 cup confectioner's sugar
- ½ teaspoon salt
- 1¼ cups almonds, ground
- 2 cups sifted flour
- 1 tablespoon water
- 1 tablespoon vanilla

1. Cream butter. Gradually add confectioner's sugar and salt, creaming well. Add ground almonds, blend in sifted flour gradually, and mix thoroughly. Add water and vanilla, mixing thoroughly with fork. Shape into balls or crescents using 1 level tablespoon of dough for each cookie. Place on ungreased baking sheet. Flatten slightly. Bake at 350°F for 10 minutes. Roll in confectioner's sugar while still warm.

Brazil Nut Shortbreads

- 1 cup butter
- ½ cup sugar
- 2 cups flour
- 1 cup Brazil nuts, sliced chunks, plus additional for decorating top of cookies

1. Cream butter and sugar well. Add flour and nuts and mix well. Chill dough for at least an hour and then shape cookies with hands into balls the size of a small walnut. Flatten and press a chunk of nut into the center of each cookie. Bake on an ungreased cookie sheet 15 to 20 minutes at 300°F. Makes 4 to 5 dozen.

Granola

Granola is as delicious and satisfying as it is healthy. Make this recipe your own by adding your favorite dried fruits, nuts, and seeds. Carob or chocolate chips can also be added for a more decadent treat.

- 3 cups oats
- ⅓ cup wheat germ
- 2 teaspoons cinnamon
- 1 teaspoon salt
- ¼ cup brown sugar
- ¼ cup vegetable oil
- ½ cup honey
- ¼ cup molasses
- 1 teaspoon vanilla
- ⅔ cup nuts or seeds
- ⅔ cup dried fruit

1. Preheat oven to 325ºF and line a baking sheet with parchement paper.
2. Combine the oats, wheat germ, cinnamon, salt, and brown sugar in a medium bowl.
3. In a separate bowl, combine the honey, molasses, vegetable oil, and vanilla. Pour over the oat mixture and stir until oats are thoroughly coated.
4. Spread mixture on cookie sheet and bake for 10 minutes. Remove pan from oven, stir the granola, add nuts, and bake for another 10 minutes or until granola begins to brown.
5. Remove from oven, stir in fruit, and store in an airtight container.

Eating Well

Why Eat Organically Grown Food?

Organically grown produce is becoming more and more readily available, regardless of where you live. If you grow your own fruits and vegetables or have ready access to a farmers' market, eating organically may be cheaper than purchasing commercially grown produce at the supermarket. Though organic foods bought at a grocery store may be 10 to 50 percent more expensive than their traditionally grown companions, the benefits are often worth the cost. Here are just a few of many reasons to eat organically grown produce:

- **Improved taste.** Tests comparing various gardening methods have shown that fruit grown organically has a higher natural sugar content and firmer flesh, and is less apt to bruise easily. Do your own taste test and you'll easily tell the difference!

- **Fewer health risks.** Pesticides have been linked to cancer and other diseases.
- **Help support smaller farms.** Most organic farms are small, family-owned endeavors. By purchasing organic produce, you'll be helping them survive and thrive.
- **Help the environment.** According to the EPA (Environmental Protection Agency), agriculture is responsible for 70 percent of the pollution in the United States' streams and rivers. Organic farmers don't use the synthetic pesticides and fertilizers that cause this pollution.
- **Better nutrition.** Higher levels of lycopene, polyphenols, and flavonols have been found in organically produced fruits and vegetables. Phytonutrients (many of which are antioxidants involved in the plant's own defense system) may be higher in organic produce because crops rely more on their own defenses in the absence of regular applications of chemical pesticides.

Why Eat Locally Grown Food?

Even if you grow the majority of your own produce, you may want to supplement your menu with food from other sources. When doing so, there are lots of great reasons to choose foods grown near where you live. Here are a few:

- **Support the local economy.** According to a study by the New Economics Foundation in London, a dollar spent locally generates twice as much income for the local economy.
- **Fresher food.** Produce you buy at the supermarket has likely been in transit for several days or even weeks, and during that time it's been declining in flavor and nutrition. Produce you buy from a farmers' market or local farm stand was likely picked the same day.
- **Help the environment.** By eating food grown locally, you're cutting down on the number of miles it had to travel to get to you, thus lowering fuel emissions.
- **Stay attuned to the seasons.** Eating locally means that you may not get asparagus in October or sweet

potatoes in April, but those foods will not be at their best quality in those months anyway. You'll get more nutritious, better tasting food if you eat it at its growing peak. You'll also feel more connected to the natural seasonal rhythms.

- **Encourage variety.** By supporting the local farmers, you give them the opportunity to try less common or heirloom varieties that wouldn't travel as well, produce as high a yield, or have the shelf life of most supermarket varieties.

Food Co-ops: Grocery Shopping in Community

What Is a Food Co-op?

Food co-ops are non-profit, democratic, and member-owned businesses that provide low-cost organic or natural foods to members and, in some cases, non-members.

Since food co-ops are established and operated by members, each member has a voice regarding what types of foods will be sold, maintenance issues, and management of the stores. Food co-ops are democratically run, so each member has one vote in any type of election. Members generally elect a board of directors to oversee the everyday running of the co-op and to hire staff.

What Are the Different Types of Food Co-ops?

Typically, there are two types of food co-ops: the co-op grocery store and the buying club. Each is owned and run by members but they do vary in their structure and number of members.

Co-op Grocery Store

Co-op grocery stores are basically regular grocery stores that are member-owned and -operated and provide low cost, healthy foods to members and often to the public as well. There are around 500 co-op grocery stores in the United States alone.

Buying Club

A buying club consists of a small group of people (friends, neighbors, families, or colleagues) who get together and buy food in bulk from a co-op distributor (a co-op warehouse or natural foods distributor) or from local farms. By ordering in bulk, the members are able to save money on grocery items. The members of the buying club share the responsibilities of collecting money from the other members, placing orders with the distributor, picking up the orders from the drop-off site, and distributing the food to the individual members or families.

How Do You Become a Food Co-op Member?

In order to become a member of a food co-op, you must pay a small initial fee and then typically invest a certain amount of money into the co-op to purchase a share. Sometimes members can accumulate more shares (by paying an annual fee, for example). Members can also help run the co-op by volunteering their time. Members reap the benefits of their membership by having access to discounted prices on food products. However, if you decide not to become a member, some food co-ops still allow non-members to shop at their stores without the membership discount.

How Do You Become a Buying Club Member?

If you are looking to join a buying club in your area, it is best to contact your regional co-op distributor and ask them for information on local buying clubs. Check out the Web sites of local distributors to see if they have links to buying clubs near you. Or ask friends and neighbors if they're aware of any buying clubs that are active in your area.

How Do You Start Your Own Food Co-op?

Here are some steps to follow in order to establish your own food co-op:

1. Invite potential members to meet and discuss the start-up of a food co-op. Identify how a food co-op may help the finances of the members.
2. Hold a meeting in which potential members vote to continue the process of forming a co-op and then select a committee for this purpose.
3. Determine how often the co-op will be used by the potential members.
4. Discuss the results of any surveys at another meeting and then vote to see if the plans should proceed.
5. Do a needs analysis (determine what the members will need in order to establish a food co-op).
6. Hold a meeting to discuss the outcome of the needs analysis and vote (anonymously) on whether or not to proceed with the co-op.
7. Develop a business plan for the co-op and decide the financial contribution needed to start the co-op.
8. At another meeting, have members vote on the business plan and if members want to continue, decide on whether or not to keep the committee members.

9. Prepare all legal documents and incorporate.

10. Hold a meeting for all potential members to review and accept the bylaws (terms of operation, responsibilities of members and board of directors). Hold an election for the board of directors.

11. At the first board of directors meeting, elect officers and assign them certain responsibilities in carrying out the business plan.

12. Hold a membership drive—try to recruit new members to the food co-op.

13. Pool monetary resources and create a loan application package.

14. Employ a manager for the co-op store.

15. Find a building or storefront to house the co-op.

16. Start your business!

Examples of bylaws for a food co-op:

- Establish membership requirements.
- Formulate the rights and responsibilities of all members and the board of directors.
- Stipulate the grounds for member expulsion.
- Establish rules for calling and implementing membership meetings.
- Establish how members will vote.
- Provide election procedures for board members and officers.
- Specify the number of board members and officers and how long their terms in office will be and what sort of compensation they will be awarded.
- Establish what time and where meetings will be held.
- Specify the co-op's fiscal year dates.
- Provide information on the distribution of net earnings.
- Include any other rules of management for the co-op.

How Do You Start Your Own Buying Club?

To start your own buying club, you will need to collect a group of people (preferably more than five households). If no one in your new group has any experience with organizing a buying club, it may be beneficial for you to temporarily join a buying club to see how it works. Once you are confident in your understanding of a buying club, it's time to begin!

1. Find a co-op distributor's (wholesaler's) pricing guide to share with the others in your buying club, so you all have an understanding of the products available and the savings from which you'll benefit. If you'll be buying from local farms, discuss pricing and bulk discounts with the farmers.

2. Have a meeting and invite all those who are interested in joining your buying club. Emphasize that a buying club requires its members to share in all responsibilities—from placing orders to picking up deliveries to collecting the money—and that they will all reap the benefits of obtaining great organic and natural foods at wholesale prices.

3. Establish an organizational committee. Discuss areas such as coordination, price guide distribution, orderings, potential delivery location, what supplies will be needed, bookkeeping, and how to orient new members.

4. Draw up any membership requirements you think necessary.

5. Brainstorm possible delivery sites, such as churches, firehouses, or other public buildings. Your optimal site should be able to accommodate a large truck and have long hours of operation. Make sure you will have enough space at the site to go through the products and distribute them accordingly.

6. Develop a name for your buying club and fill out a membership application with the co-op distributor of your choice. You should receive some sort of confirmation, complete with order deadlines, date of delivery, and a simple orientation to the buying club.

7. Start enjoying your healthy foods for lower prices!

Maple Sugaring

The production of maple syrup and maple sugar is purely an American industry, Canada being the only country outside of the United States where they are made. The earliest explorers in this country found the Native Americans making sugar from maple trees, and in some sections producing it in quantity for trade. The settlers began to make maple products as well and to attempt to improve their manufacture. For many years, maple sugar was the only sugar used, and despite refinements, beyond the tapping and boiling, the general process remains the same as at that time.

All the maples have sweet sap, but only from a few of the species has sugar been made in worthwhile quantities. The first place is held by the sugar maple and a variety of it—the black maple. These can be found in the Northeastern region of the United States, as well as the northern Midwest. Other varieties, including the red maple, the silver maple, and the Oregon maple, can be tapped, but will produce smaller quantities of syrup. It takes approximately 40 gallons of sap to make 1 gallon of syrup, so it is generally not worthwhile to tap the less productive tree varieties.

Tapping

The quantity of sap that a tree yields stands in direct relation to the size of its crown. It is good to make it a rule to tap only one place on a tree; by doing so the life of the tree is prolonged. Large trees might be tapped in two and sometimes three places without injury, but not in two places so near together that the sap from the two is collected in one bucket. Each hole should heal over in as quickly as one season.

Before tapping, the side of the tree should be brushed with a stiff broom to remove all loose bark and dirt and a spot selected where the bark looks healthy, some distance from the scar of a previous tapping. Care should also be taken to tap where a bucket attached to the spout inserted in the hole will hang level and be partly supported by the tree itself. The distance from the ground should be about waist high, convenient for the sap collector. In general it is best to tap on the side of the tree where other trees do not shade the spot. The main requisite in tapping a tree is a good sharp bit with which a clean-cut hole can be made. A rough, feathered hole soon becomes foul, stopping the flow. After the tapping, all shavings should be removed to make the hole clean. The bark should never be cut away before boring the hole, as this shortens the life of the tree.

General practice concerning the size of the hole seems to indicate that three-eighths to half an inch is the best dia-meter; then, if the season is long and a warm spell interrupts the flow, the holes can be reamed out to one-half to five-eighths of an inch, and thereby secure an increased run. A thirteen thirty-seconds of an inch bit is often used. The bit should be especially sharp and should bring the shavings to the surface. Its direction is slightly upward into the tree. The slant allows the hole to drain readily.

The depth of the hole should be regulated by the size of the tree, as only the layers next to the bark are alive and contain enough sap to flow freely. Toward the interior the flow diminishes. With the ordinary tree a hole less than 1½ to 2 inches deep is best. In small trees make a short incision just through the sapwood. In any case boring should be stopped when dark colored shavings appear, as this shows dead wood and that the sapwood has been passed through. It is good policy to tap early in the season in order to obtain the earlier runs, which are generally the sweetest. "Sugar weather" begins sometime between mid-February to mid-March, when the days are becoming warm, the temperature going above 32° F, and the nights are still frosty.

The spout, or spile, is the tube through which the sap flows into the bucket. It is usually of metal, but hollow reeds are sometimes used. The best are perfectly cylindrical and of an even taper, making them easy to insert and to remove without interfering with the wood tissue. The perfect spout should be strong enough to support the bucket of sap safely, and for obvious reasons should bring the whole weight on the bark of the tree and not on the inner tissue or sapwood. A spout should have a hook or stop on which the bucket is to hang, unless the bucket may hang on the spout itself, and it is best to have a spout with a small hole, because one with a large hole allows the bore to dry out faster when there are strong winds. Buckets are typically of galvanized metal (free from corrosion or rust, covered, and fitting well to the tree). Sometimes old plastic one-gallon milk cartons are used, since their narrow neck prevents debris from entering. Make sure to clean any container thoroughly before use.

The sap should be collected each day and not be allowed to accumulate. It is also necessary to keep the buckets and containers clean, and they should be washed in warm water after each run. So long as it is cold, you may store the sap outdoors for up to three days in any large metal or plastic container. When pouring the sap into its collecting device, stretch a flannel cloth over the top of the tanks and pour the sap through this to remove any twigs, leaves, or pieces of dirt.

Syrup and Sugar Making

Once you have enough sap to start making syrup, you may start to boil it down. Use any outdoor method, from bonfire to coal-burning range, camp stove to commercial evaporator, but avoid boiling sap inside, as it results in a sticky residue on your walls.

Use two pans, one to evaporate excess moisture from the sap and concentrate it into syrup, and one as a finishing pan, in which you will finish boiling it. The evaporator should have a large bottom surface area, and the sap should not be deeper than 1 to 1½ inches in the pan at any time. The size of the pan depends on how much sugaring you intend to do—it is best if it can hold at least one gallon. Put in an inch or two of sap, boil, and add more, a little at a time, so as not to stop boiling or materially change the density of the boiling liquid; then, when this charge is concentrated, or has reached approximately 6 degrees above the temperature at which water boils (use a candy thermometer to monitor it), the syrup should be drawn off. Care must be exercised not to allow the

remaining syrup in the pan to he burned. While evaporating, use a kitchen strainer to skim off the froth, and keep a spoon for stirring on hand.

To finish the syrup, pour it through a piece of felt cloth, or two pieces of thick flannel, into the finishing pot. Once the syrup has reached 7 degrees above the boiling point of water, it is ready for storage.

Glass containers are the best method for keeping syrup, although airtight plastic and metal containers will also work. When carefully canned, syrup will keep from one season to another without souring or bursting the jar. It is best to store syrup immediately after finishing, while still hot, and then keep at an even, cold temperature. Temperatures around freezing, however, should not be used, as this may crystallize the syrup.

"Sugaring-off" applies to the further treatment of the maple syrup by which it is made into a solid product. The ordinary iron pot of the kitchen is filled nearly half full with the syrup and this concentrated over the fire. Use a candy thermometer to determine the proper point of stopping the boiling. In the first runs of sap the boiling should be carried up to 26 to 28 degrees above the boiling point of water at that elevation to make a medium hard sugar. With later runs the finishing temperature should be 28 to 38 degrees above the boiling point. After the thick syrup has reached the proper boiling point, it should be taken from the fire and stirred until somewhat cooled. This gives it a uniform grain and color in the mold. As in syrup making one should "sugar-off" a charge before adding any more syrup. The hard-

ness of the sugar produced is to a large extent controlled by its moisture content. High temperatures are required to evaporate more of the water, but note that for softer sugars you should use slightly lower temperatures.

Like brown sugar, maple sugar does not keep well in a moist atmosphere. It tends to absorb water, molds rather quickly, and if finished at too low a temperature the sugar is soft and the liquid portion drains out. Therefore sugar which is to be stored should always be boiled to a high temperature. It can be wrapped in paper, but should not be put in covered containers unless these are absolutely sealed. It is best to store the sugar in a warm room of even temperature. If the cakes are sealed without access to air, a cold place can be used, but make sure they are kept dry.

Edible Wild Plants and Mushrooms

Wild Vegetables, Fruits, and Nuts

Agave

Description: Agave plants have large clusters of thick leaves that grow around one stalk. They grow close to the ground and only flower once before dying.

Location: Agave like dry, open areas and are found in the deserts of the American West.

Edible Parts and Preparing: Only agave flowers and buds are edible. Boil these before consuming. The juice can be collected from the flower stalk for drinking.

Other Uses: Most agave plants have thick needles on the tips of their leaves that can be used for sewing.

Asparagus

Description: When first growing, asparagus looks like a collection of green fingers. Once mature, the plant has fernlike foliage and red berries (which are toxic if eaten). The flowers are small and green and several species have sharp, thornlike projections.

Location: It can be found growing wild in fields and along fences. Asparagus is found in temperate areas in the United States.

Edible Parts and Preparing: It is best to eat the young stems, before any leaves grow. Steam or boil them for ten to fifteen minutes before consuming. The roots are a good source of starch, but don't eat any part of the plant raw, as it could cause nausea or diarrhea.

Beech

Description: Beech trees are large forest trees. They have smooth, light gray bark, very dark leaves, and clusters of prickly seedpods.

Location: Beech trees prefer to grow in moist, forested areas. These trees are found in the Temperate Zone in the eastern United States.

Edible Parts and Preparing: Eat mature beechnuts by breaking the thin shells with your fingers and removing the sweet, white kernel found inside. These nuts can also be used as a substitute for coffee by roasting them until the kernel turns hard and golden brown. Mash up the kernel and boil or steep in hot water.

Cattail

Description: These plants are grasslike and have leaves shaped like straps. The male flowers grow above the female flowers, have abundant, bright yellow pollen, and die off quickly. The female flowers become the brown cattails.

Location: Cattails like to grow in full-sun areas near lakes, streams, rivers, and brackish water. They can be found all over the country.

Edible Parts and Preparing: The tender, young shoots can be eaten either raw or cooked. The rhizome (rootstalk) can be pounded and made into flour. When the cattail is immature, the female flower can be harvested, boiled, and eaten like corn on the cob.

Other Uses: The cottony seeds of the cattail plant are great for stuffing pillows. Burning dried cattails helps repel insects.

Blackberry and Raspberry

Description: These plants have prickly stems that grow upright and then arch back toward the ground. They have alternating leaves and grow red or black fruit.

Location: Blackberry and raspberry plants prefer to grow in wide, sunny areas near woods, lakes, and roads. They grow in temperate areas.

Edible Parts and Preparing: Both the fruits and peeled young shoots can be eaten. The leaves can be used to make tea.

Burdock

Description: Burdock has wavy-edged, arrow-shaped leaves. Its flowers grow in burrlike clusters and are purple or pink. The roots are large and fleshy.

Location: This plant prefers to grow in open waste areas during the spring and summer. It can be found in the Temperate Zone in the north.

Edible Parts and Preparing: The tender leaves growing on the stalks can be eaten raw or cooked. The roots can be boiled or baked.

Chicory

Description: This is quite a tall plant, with clusters of leaves at the base of the stem and very few leaves on the stem itself. The flowers are sky blue in color and open only on sunny days. It produces a milky juice.

Location: Chicory grows in fields, waste areas, and alongside roads. It grows primarily as a weed all throughout the country.

Edible Parts and Preparing: The entire plant is edible. The young leaves can be eaten in a salad. The leaves and roots may also be boiled as you would regular vegetables. Roast the roots until they are dark brown, mash them up, and use them as a substitute for coffee.

Cranberry

Description: The cranberry plant has tiny, alternating leaves. Its stems crawl along the ground and it produces red berry fruits.

Location: Cranberries only grow in open, sunny, wet areas. They thrive in the colder areas in the northern states.

Edible Parts and Preparing: The berries can be eaten raw, though they are best when cooked in a small amount of water, adding a little bit of sugar if desired.

Dandelion

Description: These plants have jagged leaves and grow close to the ground. They have bright yellow flowers.

Location: Dandelions grow in almost any open, sunny space in the United States.

Edible Parts and Preparing: All parts of this plant are edible. The leaves can be eaten raw or cooked and the roots boiled. Roasted and ground roots can make a good substitute for coffee.

Other Uses: The white juice in the flower stem can be used as glue.

Elderberry

Description: This shrub has many stems containing opposite, compound leaves. Its flower is white, fragrant, and grows in large clusters. Its fruits are berry-shaped and are typically dark blue or black.

Location: Found in open, wet areas near rivers, ditches, and lakes, the elderberry grows mainly in the eastern states.

Edible Parts and Preparing: The flowers can be soaked in water for eight hours and then the liquid can be drunk. The fruit is also edible but don't eat any other parts of the plant—they are poisonous.

Hazelnut

Description: The nuts grow on bushes in very bristly husks.

Location: Hazelnut grows in dense thickets near streambeds and in open areas and can be found all over the United States.

Edible Parts and Preparing: In the autumn, the hazelnut ripens and can be cracked open and the kernel eaten. Eating dried nuts is also tasty.

Juniper

Description: Also known as cedar, this shrub has very small, scaly leaves that are densely crowded on the branches. Berrylike cones on the plant are usually blue and are covered with a whitish wax.

forests. It can be found throughout the United States.

Edible Parts and Preparing: The young shoots and leaves are edible. To eat, boil the plant for ten to fifteen minutes.

Oak

Description: These trees have alternating leaves and acorns. Red oaks have bristly leaves and smooth bark on the upper part of the tree and their acorns need two years to reach maturity. White oaks have leaves with no bristles and rough bark on the upper part of the tree. Their acorns only take one year to mature.

Location: Found in various locations and habitats throughout the country.

Edible Parts and Preparing: All parts of the tree are edible, but most are very bitter. Shell the acorns and soak them in water for one or two days to remove their tannic acid. Boil the acorns to eat or grind them into flour for baking.

Palmetto Palm

Description: This is a tall tree with no branches and has a continual leaf base on the trunk. The leaves are large, simple, and lobed and it has dark blue or black fruits that contain a hard seed.

Location: This tree is found throughout the southeastern coast.

Edible Parts and Preparing: The palmetto palm fruit can be eaten raw. The seeds can also be ground into flour, and the heart of the palm is a nutritious source of food, but the top of the tree must be cut down in order to reach it.

Persimmon

Description: The persimmon tree has alternating, elliptical leaves that are dark green in color, and inconspicuous flowers. It has orange fruits that are very sticky and contain many seeds.

Location: Growing on the margins of forests, it resides in the eastern part of the country.

Location: They grow in open, dry, sunny places throughout the country.

Edible Parts and Preparing: Both berries and twigs are edible. The berries can be consumed raw or the seeds may be roasted to make a substitute for coffee. Dried and crushed berries are good to season meat. Twigs can be made into tea.

Lotus

Description: This plant has large, yellow flowers and leaves that float on or above the surface of the water. The lotus fruit has a distinct, flattened shape and possesses around twenty hard seeds.

Location: Found on fresh water in quiet areas, the lotus plant is native to North America.

Edible Parts and Preparing: All parts of the lotus plant are edible, raw or cooked. Bake or boil the fleshy parts that grow underwater and boil young leaves. The seeds are quite nutritious and can be eaten raw or they can be ground into flour.

Marsh Marigold

Description: Marsh marigold has round, dark green leaves and a short stem. It also has bright yellow flowers.

Location: The plant can be found in bogs and lakes in the northeastern states.

Edible Parts and Preparing: All parts can be boiled and eaten. Do not consume any portion raw.

Mulberry

Description: The mulberry tree has alternate, lobed leaves with rough surfaces and blue or black seeded fruits.

Location: These trees are found in forested areas and near roadsides in temperate and tropical regions of the United States.

Edible Parts and Preparing: The fruit can be consumed either raw or cooked and it can also be dried. Make sure the fruit is ripe or it can cause hallucinations and extreme nausea.

Nettle

Description: Nettle plants grow several feet high and have small flowers. The stems, leafstalks, and undersides of the leaves all contain fine, hairlike bristles that cause a stinging sensation on the skin.

Location: This plant grows in moist areas near streams or on the edges of

Edible Parts and Preparing: The leaves provide a good source of vitamin C and can be dried and soaked in hot water to make tea. The fruit can be consumed either baked or raw and the seeds may be eaten once roasted.

Pine

Description: Pine trees have needlelike leaves that are grouped into bundles of one to five needles. They have a very pungent, distinguishing odor.

Location: Pines grow best in sunny, open areas and are found all over the United States.

Edible Parts and Preparing: The seeds are completely edible and can be consumed either raw or cooked. Also, the young male cones can be boiled or baked and eaten. Peel the bark off of thin twigs and chew the juicy inner bark. The needles can be dried and brewed to make tea that's high in vitamin C.

Other Uses: Pine tree resin can be used to waterproof items. Collect the resin from the tree, put it in a container, heat it, and use it as glue or, when cool, rub it on items to waterproof them.

Plantain

Description: The broad-leafed plantain grows close to the ground and the flowers are situated on a spike that rises from the middle of the leaf cluster. The narrow-leaf species has leaves covered with hairs that form a rosette. The flowers are very small.

Location: Plantains grow in lawns and along the side of the road in the northern Temperate Zone.

Edible Parts and Preparing: Young, tender leaves can be eaten raw and older leaves should be cooked before consumption. The seeds may also be eaten either raw or roasted. Tea can also be made by boiling 1 ounce of the plant leaves in a few cups of water.

Pokeweed

Description: A rather tall plant, pokeweed has elliptical leaves and produces many large clusters of purple fruits in the late spring.

Location: Pokeweed grows in open and sunny areas in fields and along roadsides in the eastern United States.

Edible Parts and Preparing: If cooked, the young leaves and stems are edible. Be sure to boil them twice and discard the water from the first boiling. The fruit is also edible if cooked. Never eat any part of this plant raw, as it is poisonous.

Prickly Pear Cactus

Description: This plant has flat, pad-like green stems and round, furry dots that contain sharp-pointed hairs.

Location: Found in arid regions and in dry, sandy areas in wetter regions, it can be found throughout the United States.

Edible Parts and Preparing: All parts of this plant are edible. To eat the fruit, peel it or crush it to make a juice. The seeds can be roasted and ground into flour.

Reindeer Moss

Description: This is a low plant that does not flower. However, it does produce bright red structures used for reproduction.

Location: It grows in dry, open areas in much of the country.

Edible Parts and Preparing: While having a crunchy, brittle texture, the whole plant can be eaten. To remove some of the bitterness, soak it in water and then dry and crush it, adding it to milk or other foods.

Sassafras

Description: This shrub has different leaves—some have one lobe, others two lobes, and others have none at all. The flowers are small and yellow and appear in the early spring. The plant has dark blue fruit.

Location: Sassafras grows near roads and forests in sunny, open areas. It is common throughout the eastern states.

Edible Parts and Preparing: The young twigs and leaves can be eaten either fresh or dried—add them to soups. Dig out the underground portion of the shrub, peel off the bark, and dry it. Boil it in water to make tea.

Other Uses: Shredding the tender twigs will make a handy toothbrush.

Spatterdock

Description: The leaves of this plant are quite long and have a triangular notch at the base. Spatterdock has yellow flowers that become bottle-shaped fruits, which are green when ripe.

Location: Spatterdock is found in fresh, shallow water throughout the country.

Edible Parts and Preparing: All parts of the plant are edible and the fruits have brown seeds that can be roasted and ground into flour. The rootstock can be dug out of the mud, peeled, and boiled.

Strawberry

Description: This is a small plant with a three-leaved pattern. Small white flowers appear in the springtime and the fruit is red and very fleshy.

Location: These plants prefer sunny, open spaces, are commonly planted, and appear in the northern Temperate Zone.

Edible Parts and Preparing: The fruit can be eaten raw, cooked, or dried. The plant leaves may also be eaten or dried to make tea.

Thistle

Description: This plant may grow very high and has long-pointed, prickly leaves.

Location: Thistle grows in woods and fields all over the country.

Edible Parts and Preparing: Peel the stalks, cut them into smaller sections, and boil them to consume. The root may be eaten raw or cooked.

Walnut

Description: Walnuts grow on large trees and have divided leaves. The walnut has a thick outer husk that needs to be removed before getting to the hard, inner shell.

Location: The black walnut tree is common in the eastern states.

Edible Parts and Preparing: Nut kernels become ripe in the fall and the meat can be obtained by cracking the shell.

Water Lily

Description: With large, triangular leaves that float on water, these plants have fragrant flowers that are white or red. They also have thick rhizomes that grow in the mud.

Location: Water lilies are found in many temperate areas.

Edible Parts and Preparing: The flowers, seeds, and rhizomes can be eaten either raw or cooked. Peel the corky rind off of the rhizome and eat it raw or slice it thinly, dry it, and grind into flour. The seeds can also be made into flour after drying, parching, and grinding.

Wild Grapevine

Description: This vine will climb on tendrils, and most of these plants produce

deeply lobed leaves. The grapes grow in pyramidal bunches and are black-blue, amber, or white when ripe.

VIOLETS

Violets can be candied and used to decorate cakes, cookies, or pastries. Pick the flowers with a tiny bit of stem, wash, and allow to dry thoroughly on a paper towel or a rack. Heat ½ cup water, 1 cup sugar, and ¼ teaspoon almond extract in a saucepan. Use tweezers to carefully dip each flower in the hot liquid. Set on wax paper and dust with sugar until every flower is thoroughly coated. If desired, snip off remaining stems with small scissors. Allow flowers to dry for a few hours in a warm, dry place.

Location: Climbing over other vegetation on the edges of forested areas, they can be found in the eastern and southwestern parts of the United States.

Edible Parts and Preparing: Only the ripe grape and the leaves can be eaten.

Wild Onion and Garlic

Description: These are recognized by their distinctive odors.

Location: They are found in open areas that get lots of sun throughout temperate areas.

Edible Parts and Preparing: The bulbs and young leaves are edible and can be consumed either raw or cooked.

Wild Rose

Description: This shrub has alternating leaves and sharp prickles. It has red, pink, or yellow flowers and fruit (rose hip) that remains on the shrub all year.

Location: These shrubs occur in dry fields throughout the country.

Edible Parts and Preparing: The flowers and buds are edible raw or boiled. Boil fresh, young leaves to make tea. The rose hips can be eaten once the flowers fall and they can be crushed once dried to make flour.

Edible Wild Mushrooms

A walk through the woods will likely reveal several varieties of mushrooms, and chances are that some are the types that are edible. However, because some mushrooms are very poisonous, it is important never to try a mushroom of which you are unsure. Never eat any mushroom that you cannot positively identify as edible. Also, never eat mushrooms that appear wilted, damaged, or rotten.

Here are some common edible mushroom that you can easily identify and enjoy.

Chanterelles

These trumpet-shaped mushrooms have wavy edges and interconnected blunt-ridged gills under the caps. They are varied shades of yellow and have a fruity fragrance. They grow in summer and fall on the ground of hardwood forests. Because chanterelles tend to be tough, they are best when slowly sautéed or added to stews or soups.

Notes: Beware of Jack O'Lantern mushrooms, which look and smell similarly to chanterelles. Jack O'Lanterns have sharp knifelike gills instead of the blunt gills of chanterelles, and generally grow in large clusters at the base of trees or on decaying wood.

Coral Fungi

These fungi are aptly named for their bunches of upward-facing branching stems, which look strikingly like coral. They are whitish, tan, yellowish, or sometimes pinkish or purple. They may reach 8 inches in height. They grow in the summer and fall in shady, wooded areas.

Notes: Avoid coral fungi that are bitter, have soft, gelatinous bases, or turn brown when you poke or squeeze them. These may have a laxative effect, though are not life-threatening.

Morels

Morels are sometimes called sponge, pine cone, or honeycomb mushrooms because of the pattern of pits and ridges that appears on the caps. They can be anywhere from 2 to 12 inches tall. They may be yellow, brown, or black and grow in spring and early summer in wooded areas and on river bottoms. To cook, cut in half to check for insects, wash, and sauté, bake, or stew.

Notes: False morels can be poisonous and appear similar to morels because of their brainlike irregularly shaped caps. However, they can be distinguished from true morels because false morel caps bulge inward instead of outward. The caps have lobes, folds, flaps, or wrinkles, but not pits and ridges like a true morel.

Puffballs

These round or pear-shaped mushrooms are often mistaken for golf balls or eggs. They are always whitish, tan, or gray and sometimes have a thick stem. Young puffballs tend to be white and older ones yellow or brown. Fully matured puffballs have dark spores scattered over the caps. Puffballs are generally found in late summer and fall on lawns, in the woods, or on old tree stumps. To eat, peel off the outer skin and eat raw or batter-fried.

Notes: Slice each puffball open before eating to be sure it is completely white inside. If there is any yellow, brown, or black, or if there is a developing mushroom inside with a stalk, gills, and cap, do not eat! Amanitas, which are very poisonous, can appear similar to puffballs when they are young. Do not eat if the mushroom gives off an unpleasant odor.

⯯ **Chanterelles**

Shaggy Mane Mushrooms

This mushroom got its name from its cap, which is a white cylinder with shaggy, upturned, brownish scales. As the mushroom matures, the bottom outside circumference of the cap becomes black. Shaggy manes are generally 4 to 6 inches tall and grow in all the warm seasons in fields and on lawns.

Shaggy manes are tastiest eaten when young, but they're easiest to identify once the bottoms of the caps begin to turn black. They are delicious sautéed in butter or olive oil and lightly seasoned with salt, garlic, or nutmeg.

⤜ Coral mushroom.

⤒ Puffball mushrooms.

⤜ Shaggy mane mushroom.

Morels. ⤜

Making Sausage

Important Considerations in Sausage Making

Temperatures

Meat products are extremely perishable and must be maintained under refrigeration (40°F or below). When you have finished processing a product, return it to the refrigerator. After the product has been formulated, smoke and cook the product to the required temperature and then return the product to refrigeration. Sausage that hasn't been properly refrigerated will spoil and can make you sick.

Sanitation

There is no substitute for keeping the tables, utensils, and ingredients clean and free from dirt and contamination. Use plenty of hot water and soap before and after processing sausages. Always keep your hands clean.

Taking Notes

Just as you keep a copy of a good recipe, you should keep notes on the formulation and processing procedures of your favorite smoked and cooked sausage. Ingredients, times, temperatures, and end results should be noted. This will help to make a better sausage the next time.

Fat Content

Different sausages have different amounts of fat. Avoid making the formula too lean as the sausage will be dry and hard. Fresh pork sausage contains 30 to 45 percent fat. Smoked or roasted sausage contains 20 to 30 percent fat. Formulate the fat content just as you would the other ingredients in a sausage.

Nitrates and Nitrites

These curing ingredients are required to achieve the characteristic flavor, color, and stability of cured meat. Nitrate and nitrite are converted to nitric oxide by microorganisms and combine with the meat pigment myoglobin to give the cured meat color. However, more importantly, nitrite provides protection against the growth of botulism-producing organisms, acts to show rancidity, and stabilizes the flavor of the cured meat. Potassium nitrate (saltpeter) was the salt historically used for curing. However, we now know that saltpeter is much stronger than necessary for curing meat.

Much controversy has surrounded the use of nitrite in recent years—for good reason. Too many nitrites can cause all sorts of health problems, including cancer. However, the amount used in sausage making is not enough to cause concern and is far less of a worry than the botulism that can result if it is not used. A commercial premixed cure can be used when nitrate or nitrite is called for in the recipe. The premixes have been diluted with salt so that the small quantities of nitrites and nitrates can more easily be weighed. This reduces the possibility of serious error in handling pure nitrate or nitrite. Several premixes are available. Many local grocery stores stock Morton® Tender Quick® Product and other brands of premix cure.

Because the amount of premixed cure will vary depending on what brand is used, it is important to follow the directions on the package. The recipes below are only for fresh sausage, which does not require a cure. Fresh sausage is delicious, simple to make, and can be frozen if you do not want to cook and serve it right away.

Storage

The length of time a sausage can be stored depends on the type of sausage. Fresh sausage is highly perishable and will only last seven to ten days. However, it may be frozen for four to six months if wrapped in moisture-vapor proof wrap (freezer paper). Smoked sausages may last from two to four weeks under refrigeration.

Fresh Pork Sausage

Fresh pork sausage is a mixture of pork meats, salt, and spices which have been ground or chopped with no added water or extenders. Fat content usually ranges from 35 to 50 percent depending upon individual preference. For a spicier sausage, add a teaspoon of crushed red pepper or ½ teaspoon ground red pepper.

Ingredients

- 1½ lbs. 60% lean ground pork
- 1 teaspoon sugar
- 1 teaspoon salt
- 3 tablespoons fresh minced parsley
- 1 teaspoon sage

Directions

Mix spices with ground meat. Form into nine small patties and cook in a cast iron skillet until brown on both sides and cooked in the middle. Or stuff in natural casings (pork rounds) or collagen casings.

All Beef Sausage

This recipe can be halved or quartered for smaller portions. You can substitute 85% lean ground beef for the beef chucks and plates in this recipe if you don't want to pull out the meat grinder. You can also mix up a large batch of the spices ahead of time to have on hand and use a portion every time you make sausage.

Ingredients

- 7½ lbs. fresh boneless beef chucks (85% lean)
- 2½ lbs. fresh boneless beef plates (50% fat)
- ½ cup salt
- 2 tablespoons sugar
- 3 tablespoons ground white pepper
- 1 teaspoons mashed peeled garlic
- 1½ tablespoons paprika
- 2½ teaspoons ground coriander

1. Grind beef chucks through ³/₈-inch plate and beef plates through ¼-inch plate. Combine meat ingredients adding salt, sugar, and seasoning and mix well for five minutes.
2. Transfer to stuffer and stuff in cellulose or fibrous casings three to four inches in diameter or large beef casing. Refrigerate or freeze immediately.

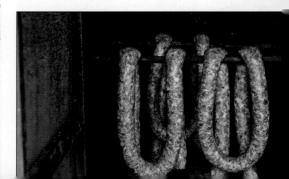

Polish Sausage (Kielbasa)

Polish sausage is made of coarsely ground lean pork with some added beef. The basic spices for this well-known sausage are garlic and marjoram. The recipe can be halved or quartered for smaller portions.

Ingredients

- 8 lbs. pork shoulder or lean trim (75% lean)
- 2 lbs. beef trimmings (80% lean)
- 4.8 oz. ice or water
- 7¼ tablespoons salt
- 4 tablespoons sugar
- 2½ tablespoons white pepper
- 2¼ teaspoons mustard seed
- 4 teaspoons marjoram
- 1½ teaspoons garlic powder
- 2½ teaspoons nutmeg
- 1⅓ cups nonfat dry milk

1. Grind beef and pork through ¼-inch plate. Add spices and water, mix thoroughly. Grind through ³⁄₁₆-inch plate.
2. Stuff into natural hog casings. Refrigerate or freeze immediately.

Venison or Game Sausage

Venison is high quality, delicious, and nutritious meat. Care should be used in handling venison just as you would any other meat. Most of the flavor in a meat product is in the fat; therefore, in making a breakfast type sausage using game meat, pork fat is used. An average deer will yield 50 to 60 pounds of venison. This recipe is perfect if you want to save half your venison for steaks or other use and use half for sausage. If you have a smaller quantity of venison, just divide the recipe accordingly.

Ingredients

- 25 lbs. lean venison or trimmings
- 25 lbs. fat pork (jowls or fresh bellies)
- 2 cups salt
- ⅔ cup black pepper
- ⅓ cup ground ginger
- ½ cup rubbed sage
- ¼ cup crushed red pepper (optional)
- ¼ cup ground red pepper (optional)

1. Cut lean venison and pork into small pieces, add spices, and mix. Grind twice through ⅛-inch or ³⁄₁₆-inch plate.
2. Sausage may be stuffed or pattied. Refrigerate or freeze immediately.

Cooked Bratwurst

Bratwurst is a typical fresh German style sausage. It is a mild sausage, often served with sauerkraut.

Ingredients

- 10 lbs. pork trim (70% lean)
- 3 lbs. ice or water
- 6½ tablespoons salt
- 1½ tablespoons ground white pepper
- 1¾ teaspoons sugar
- ½ tablespoon mace
- 1 tablespoon onion powder

1. Grind pork through ¼-inch plate and mix with salt, water, and spices. Stuff in natural hog casings or 32-mm collagen casings.
2. Steam or water cook at 170°F to an internal temperature of 155°F. Store in the refrigerator or freezer.

Curing Virginia Ham

Meat has been preserved for centuries by drying, salting, and curing. Virginia ham was one of the first agricultural products exported from North America. Today, Virginia ham is still loved around the world for its distinctive savory taste. To cure ham at home, a few simple rules should be followed.

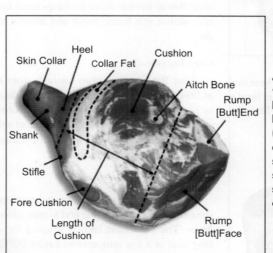

Heel
Skin Collar
Collar Fat
Cushion
Aitch Bone
Rump [Butt]End
Shank
Stifle
Fore Cushion
Length of Cushion
Rump [Butt]Face

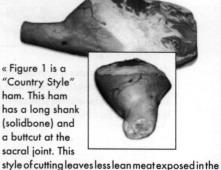

« Figure 1 is a "Country Style" ham. This ham has a long shank (solidbone) and a buttcut at the sacral joint. This style of cutting leaves less lean meat exposed in the shank and butt areas, which reduces the possibility of spoilage.

Figure 2 is a "Regular" cut ham. This style of cut is satisfactory for curing and aging hams under conditions of controlled temperature and humidity. The shank is cut short, exposing an open bone with marrow and lean tissue around the bone. The butt is cut between the second and third Sacral Vertebrae which results in a larger lean cut butt face than on "Country Style" hams. »

Curing Virginia Ham

Start with a Good Ham

To get a high quality cured ham, you have to start with a high quality fresh ham. Choose ham from young, healthy, fast-growing hogs with a desirable lean-to-fat ratio. If you are not butchering your own animal, fresh hams can be purchased from grocery stores or a local butcher. Hams for curing should have a long thick cushion (Figure 1), a deep and wide butt face, minimal seam, and external fat as seen on the collar (Figure 1) and alongside the butt face (Figure 3A), and weigh less than 24 pounds. Heavier hams are normally fatter and are more likely to spoil before the cure adjuncts penetrate to prevent deterioration.

Keep the Hams Properly Chilled

Fresh hams should be kept chilled below 40°F before being cured and then kept between 36° and 40°F during curing.

Cure Application

Hams can be cured in just salt, or you can add a dry sugar cure for a richer, sweeter flavor. For each 25 lbs. of fresh meat, use:

- 2 pounds salt
- ½ pound sugar
- ½ ounce coarse kosher salt

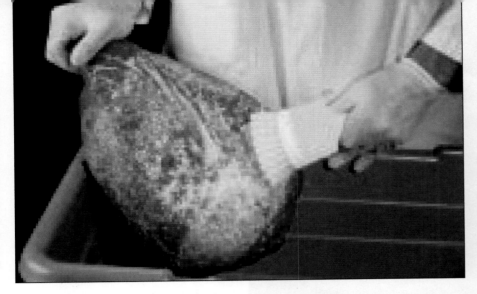

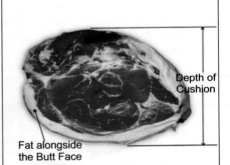

Figure 3A illustrates a high-quality ham that has a firm, bright-colored lean with at least a small amount of marbling (specks of fat in the lean) in the butt face.

Depth of Cushion

Fat alongside the Butt Face

Figure 3B reveals the type of ham to avoid. Its muscles are soft, usually pale in color and lack marbling. They also "weep" excessively and will shrink more during curing. The open seams between the muscles allow bacterial and insect invasion.

Mix these ingredients thoroughly and divide into two equal parts. Apply the first half on day one and the second portion on day seven of the curing period. Rub the curing mixture into all lean surfaces (Figure 1) of the ham. Cover the skin and fat (little will be absorbed through these surfaces).

Virginia style hams should be cured seven days per inch of cushion depth (Figure 3A), or one and a half days per pound of ham. Keep accurate records of placing hams in cure and mark on your calendar when its time to remove hams from cure. During the curing period, keep hams at a temperature of 36 to 40°F.

After Curing—Soak and Wash

When the curing period has passed, the hams should be placed in a tub of clean, cold water for one hour. This will dissolve most of the surface curing mix and make the meat receptive to smoke. After soaking, scrub the ham with a stiff bristle brush and allow it to dry.

Cure Equalization

After the ham has been washed, store it at 50-60°F for approximately fourteen days to permit the cure adjuncts to be distributed evenly throughout the ham. The product will shrink approximately 8–10 percent during cure application and equalization.

Final Steps

In Southeastern Virginia, most hams are smoked to accelerate drying and to give added flavor. The Smithfield ham is smoked for a long time at a low temperature (under 90°F). Wood from hardwood species of trees (trees that shed their leaves in the fall) should be used to produce the smoke. Hickory is the most popular, but apple, plum, peach, oak, maple, beech, ash, or cherry may be used. Pine, cedar, spruce, and other "needle leaf" trees are *not*

to be used for smoking meat since they give off resin which has a bitter taste and odor.

The fire should be a "cool" smoldering type that produces dense smoke. The temperature of the smokehouse should be kept below 90°F. Hams should be hung in a smokehouse* so that they do not touch each other. Smoke until they become chestnut brown in color, which may take one to three days.

In Southwest Virginia, hams are not traditionally smoked after curing, washing, and equalizing. Instead, the final step is to rub hams with the following mixture.

For 25 lbs. of ham:

- ½ pound black pepper
- 1 cup molasses
- ¼ pound of brown sugar
- ¼ ounce coarse kosher salt
- ¼ ounce of cayenne pepper

Then bag the hams as shown.

Age the Hams for 45 to 180 Days

As with wine or cheese, it is during the aging process that rich flavor develops. Age hams for 45 to 180 days at 75 to 95°F and a relative humidity of 55 to 65 percent. Use an exhaust fan controlled by a humidistat to limit mold growth and prevent excessive drying. Air circulation is needed, particularly during the first seven to ten days of aging, to dry the ham surface. Approximately 8 to 12 percent of the initial weight is lost.

Pests

As unpleasant as it is to think about, it's important to know that pests are drawn to cured meat and can ruin all your hard work if you don't take certain precautions. The insects attracted to cured meat are the cheese skipper, larder beetle, and red-legged ham beetle. Mites, which are not officially insects, also may infest cured meats.

1. **Cheese Skipper**

This insect gets its name from the jumping habit of the larvae which bore through cheese and cured meats. Meat infested with this insect quickly rots and becomes slimy. Adult flies are two-winged and are one-third the size of houseflies. They lay their eggs on meat and cheese and multiply rapidly.

2. **Larder Beetle**

This insect is dark brown and has a yellowish band across its back. The adult is about ⅓ inch long. Its larvae feed on or immediately beneath the cured meat surface, but do not rot the meat. The larvae are fuzzy, brownish, and about ⅓ inch long at maturity.

3. **Red-Legged Ham Beetle** The larvae are purplish and about ⅓ inch long. They bore through the meat and cause it to dry rot. Adults are about ¼ inch long, brilliant greenish blue with red legs and are red at the bases of their antennae. They feed on the meat surface.

4. **Mites** Mites are whitish and about 1/32 inch long at maturity. Affected parts of meat infested with mites appear powdery.

To help prevent insect and mite infestation, start the curing and aging during cold weather when these insects are inactive. Proper cleaning of the aging and storage areas is essential since the cheese skipper feeds and breeds on grease and tiny scraps of meat lodged in cracks. Cracks should be sealed with putty or plastic wood after cleaning.

Screens should be installed to prevent entrance of flies, ants, and other insects that carry mites.

Double entry doors are recommended to reduce infestation of insects.

If any product becomes infested after precautions have been taken, it should be removed from the storeroom and the infested area should be trimmed. The trim should be deep enough to remove larvae that have penetrated along the bone and through the fat. The uninfested portion is safe to eat, but should be prepared and consumed promptly. The exposed lean of the trimmed areas should be protected by greasing it with salad oil or melted fat to delay molding or drying.

Protect the hams by placing a barrier between the meat and the insects. Heavy brown grocery bags with no rips or tears in them are ideal to use for this purpose. Place the ham in a bag and fold and tie the top. Then, place the bagged ham in a second bag, fold and tie. The hams wrapped by this method can be hung in a dry, cool, protected room to age. This room should be clean, tight, and well ventilated.

Preparing the Ham

The traditional four-step method of preparing Virginia ham is to:

1. Wash ham with a stiff bristled brush, removing as much of the salt as possible.
2. Place the ham in a large container, cover with cold water, and allow it to stand 10–12 hours or overnight.
3. Lift the ham from the water and place it in a deep kettle with the skin side up and cover with fresh, cold water.
4. Cover the kettle, heat to a boil, but reduce heat as soon as the water boils. Simmer 20 to 25 minutes per pound until done.

Another method of cooking is to soak, scrub, and place the ham in a covered roaster, fat side up. Then, pour 2 inches of water into the roaster and place it in a 325°F oven. Cook approximately 20 to 25 minutes per pound. Baste frequently. Cook to an internal temperature of 155°F as indicated by a meat thermometer placed in the thickest position of the ham cushion. If you do not have a thermometer, test for doneness by moving the flat (pelvic) bone. It should move easily when ham is done. Lift ham from kettle. Remove skin. Sprinkle with brown sugar and/or bread crumbs and brown lightly in a 375°F oven, or use one of the suggested glazes.

Orange Glaze: Mix 1 cup brown sugar, juice, and grated rind of one orange, spread over fat surface. Bake until lightly browned in a 375°F oven. Garnish with orange slices.

Mustard Glaze: Mix ¼ cup brown sugar, 2 teaspoons prepared mustard, 2 tablespoons vinegar, and 1 tablespoon water. Spread over fat surface and bake as directed above.

Spice Glaze: Use 1 cup brown sugar and 1 cup juice from spiced peaches or crab apples. Bake as directed above. Garnish with the whole pickled fruit.

Cooking Ham Slices

Baking: Place thick slice in covered casserole and bake in 325°F oven. Brown sugar and cloves, fruit juice, or mustard-seasoned milk may be used over the ham during baking. Uncover the last 15 to 20 minutes for browning.

Broiling: Score fat edges and lay on broiler rack. Place 4 inches from broiler and broil for specified time, turning only once.

Curing Virginia Ham

Frying: Trim the skin off the ham slices. Cut the outer edge of fat in several places to prevent it from curling during cooking. Place a small amount of fat in a moderately hot skillet. When it has melted, add ham slices. Cook ham slowly, turning often. Allow about 10 minutes total cooking time for thin slices. Remove ham from pan and add a small amount of water to raise the drippings for red-eye gravy. To decrease the salty taste, fry ham with a small amount of water in the skillet.

How To Carve a Ham

The most delightful flavor of Virginia ham can be enjoyed from thin slices. Thus, a very sharp knife, preferably long and narrow, is needed. With the ham on a platter, dressed side up, make a cut perpendicular to the bone about 6 inches in from the end of the hock. Then follow the steps in figure 8.

How To Debone Cooked Ham

The ham is easier to slice when the bones are removed while the ham is warm.

- Place skinned ham fat side down on three or four strips of firm white cloth 3 inches wide and long enough to reach around ham and tie. Do not tie until bone is removed.

- Remove flat aitch bone (pelvic) by scalping around it.
- Take sharp knife, and, beginning at hock end, cut to bone the length of ham. Follow bones with point of knife as you cut.
- Loosen meat from bones. Remove bones.
- Tie cloth strips together, pulling ham together as you tie.
- Chill in the refrigerator overnight. Slice very thin, or have the ham sliced by machine.

Make Your Own Foods

Make Your Own Butter

Making butter the old-fashioned way is incredibly simple and very gratifying. It's a great project to do with kids, too. All you need are a jar, a marble, some fresh cream, and about 20 minutes.

1. Start with about twice as much heavy whipping cream as you'll want butter. Pour it into the jar, drop in the marble, close the lid tightly, and start shaking.
2. Check the consistency of the cream every three to four minutes. The liquid will turn into whipped cream, and then eventually you'll see little clumps of butter forming in the jar. Keep shaking for another few minutes and then begin to strain out the liquid into another jar. This is buttermilk, which is great for use in making pancakes, waffles, biscuits, and muffins.
3. The butter is now ready, but it will store better if you wash and work it. Add ½ cup of ice cold water and continue to shake for two or three minutes. Strain out the water and repeat. When the strained water is clear, mash the butter to extract the last of the water, and strain.
4. Scoop the butter into a ramekin, mold, or wax paper.

If desired, add salt or chopped fresh herbs to your butter just before storing or serving. Butter can also be made in a food processor or blender to speed up the processing time.

Make Your Own Yogurt

Yogurt is basically fermented milk. You can make it by adding the active cultures *Streptococcus thermophilus* and *Lactobacillus bulgaricus* to heated milk, which will produce lactic acid, creating yogurt's tart flavor and thick consistency. Yogurt is simple to make and is delicious on its own, as a dessert, in baked goods, or in place of sour cream.

Yogurt is thought to have originated many centuries ago among the nomadic tribes of Eastern Europe and Western Asia. Milk stored in animal skins would acidify and coagulate. The acid helped preserve the milk from further spoilage and from the growth of pathogens (disease-causing microorganisms).

Ingredients
Makes 4 to 5 cups of yogurt

- **1 quart milk** (cream, whole, low-fat, or skim)—In general the higher the milk fat level in the yogurt, the creamier and smoother it will taste. **Note:** If you use home-produced milk it *must* be pasteurized before preparing yogurt. See box at the top of page for tips on pasteurizing milk.
- **Nonfat dry milk powder**—Use ⅓ cup powder when using whole or low-fat milk, or use ⅔ cup powder when using skim milk. The higher the milk solids, the firmer the yogurt will be. For even more firmness add gelatin (directions below).

- **Commercial, unflavored, cultured yogurt**—Use ¼ cup. Be sure the product label indicates that it contains a live culture. Also note the content of the culture. *L. bulgaricus* and *S. thermophilus* are required in yogurt, but some manufacturers may in addition add *L. acidophilus* or *B. bifidum*. The latter two are used for slight variations in flavor, but more commonly for health reasons attributed to these organisms. All culture variations will make a successful yogurt.
- **2 to 4 tablespoons sugar or honey (optional)**
- **1 teaspoon unflavored gelatin (optional)**—For a thick, firm yogurt, swell 1 teaspoon gelatin in a little milk for 5 minutes. Add this to the milk and nonfat dry milk mixture before cooking.

Supplies
- **Double boiler or regular saucepan**—1 to 2 quarts in capacity larger than the volume of yogurt you wish to make.
- **Cooking or jelly thermometer**—A thermometer that can clip to the side of the saucepan and remain in the milk works best. Accurate temperatures are critical for successful processing.
- **Mixing spoon**
- **Yogurt containers**—cups with lids or canning jars with lids.
- **Incubator**—a yogurt-maker, oven, heating pad, or warm spot in your kitchen. To use your oven, place yogurt containers into deep pans of

110°F water. Water should come at least halfway up the containers. Set oven temperature at lowest point to maintain water temperature at 110°F. Monitor temperature throughout incubation, making adjustments as necessary.

Processing

1. **Combine ingredients and heat.** Heating the milk is necessary in order to change the milk proteins so that they set together rather than form curds and whey. Do not substitute this heating step for pasteurization. Place cold, pasteurized milk in top of a double boiler and stir in nonfat dry milk powder. Adding nonfat dry milk to heated milk will cause some milk proteins to coagulate and form strings. Add sugar or honey if a sweeter, less tart yogurt is desired. Heat milk to 200°F, stirring gently and hold for 10 minutes for thinner yogurt, or hold 20 minutes for thicker yogurt. Do not boil. Be careful and stir constantly to avoid scorching if not using a double boiler.

2. **Cool and inoculate.** Place the top of the double boiler in cold water to cool milk rapidly to 112 to 115°F. Remove 1 cup of the warm milk and blend it with the yogurt starter culture. Add this to the rest of the warm milk. The temperature of the mixture should now be 110 to 112°F.

HOW TO PASTEURIZE RAW MILK

If you are using fresh milk that hasn't been processed, you can pasteurize it yourself. Heat water in the bottom section of a double boiler and pour milk into the top section. Cover the milk and heat to 165°F while stirring constantly for uniform heating. Cool immediately by setting the top section of the double boiler in ice water or cold running water. Store milk in the refrigerator in clean containers until ready for making yogurt.

3. **Incubate.** Pour immediately into clean, warm con-tainers; cover and place in prepared incubator. Close the incubator and incubate about 4 to 7 hours at 110°F, ± 5°F. Yogurt should set firm when the proper acid level is achieved (pH 4.6). Incubating yogurt for several hours past the time after the yogurt has set will produce more acidity. This will result in a more tart or acidic flavor and eventually cause the whey to separate.

4. **Refrigerate.** Rapid cooling stops the development of acid. Yogurt will keep for about ten to twenty-one days if held in the refrigerator at 40°F or lower.

Yogurt Types

Set yogurt: A solid set where the yogurt firms in a container and is not disturbed.

Stirred yogurt: Yogurt made in a large container then spooned or otherwise dispensed into secondary serving containers. The consistency of the "set" is broken and the texture is less firm than set yogurt. This is the most popular form of commercial yogurt.

Drinking yogurt: Stirred yogurt into which additional milk and flavors are mixed. Add fruit or fruit syrups to taste. Mix in milk to achieve the desired thickness. The shelf life of this product is four to ten days, since the pH is raised by the addition of fresh milk. Some whey separation will occur and is natural. Commercial products recommend a thorough shaking before consumption.

Fruit yogurt: Fruit, fruit syrups, or pie filling can be added to the yogurt. Place them on top, on bottom, or stir them into the yogurt.

Troubleshooting

- If milk forms some clumps or strings during the heating step, some milk proteins may have jelled. Take the solids out with a slotted spoon or, in difficult cases, after cooking pour the milk mixture through a clean colander or cheesecloth before inoculation.

- When yogurt fails to coagulate (set) properly, it's because the pH is not low enough. Milk proteins will coagulate when the pH has dropped to 4.6. This is done by the culture growing and producing acids. Adding culture to very hot milk (+115°F) can kill bacteria. Use a thermometer to carefully control temperature.

- If yogurt takes too long to make, it may be because the temperature is off. Too hot or too cold of an incubation temperature can slow down culture growth. Use a thermometer to carefully control temperature.

- If yogurt just isn't working, it may be because the starter culture was of poor quality. Use a fresh, recently purchased culture from the grocery store each time you make yogurt.

- If yogurt tastes or smells bad, it's likely because the starter culture is contaminated. Obtain new culture for the next batch.

- Yogurt has over-set or incubated too long. Refrigerate yogurt immediately after a firm coagulum has formed.

- If yogurt tastes a little odd, it could be due to overheating or boiling of the milk. Use a thermometer to carefully control temperature.

- When whey collects on the surface of the yogurt, it's called syneresis. Some syneresis is natural. Excessive separation of whey, however, can be caused by incubating yogurt too long or by agitating the yogurt while it is setting.

Storing Your Yogurt

- Always pasteurize milk or use commercially pasteurized milk to make yogurt.

- Discard batches that fail to set properly, especially those due to culture errors.

- Yogurt generally has a ten- to twenty-one-day shelf life when made and stored properly in the refrigerator below 40°F.

- Always use clean and sanitized equipment and containers to ensure a long shelf life for your yogurt. Clean equipment and containers in hot water with detergent, then rinse well. Allow to air dry.

Make Your Own Cheese

There are endless varieties of cheese you can make, but they all fall into two main categories: soft and hard. Soft cheeses (like cream cheese) are easier to make because they don't require a cheese press. The curds in hard cheeses (like cheddar) are pressed together to form a solid block or wheel, which requires more time and effort, but hard cheeses will keep longer than soft cheeses, and generally have a much stronger flavor.

Cheese is basically curdled milk and is made by adding an enzyme (typically rennet) to milk, allowing curds to form, heating the mixture, straining out the whey, and finally pressing the curds together. Cheeses such as *queso fresco* or *queso blanco* (traditionally eaten in Latin American countries) and *paneer* (traditionally eaten in India), are made with an acid such as vinegar or lemon juice instead of bacterial cultures or rennet.

You can use any kind of milk to make cheese, including cow's milk, goat's milk, sheep's milk, and even buffalo's milk (used for traditional mozzarella). For the richest flavor, try to get raw milk from a local farmer. If you don't know of one near you, visit www.realmilk.com/where.html for a listing of raw milk suppliers in your state.

You can use homogenized milk, but it will produce weaker curds and a milder flavor. If your milk is pasteurized, you'll need to "ripen" it by heating it in a double boiler until it reaches 86°F and then adding 1 cup of unpasteurized, preservative-free cultured buttermilk per gallon of milk and letting it stand 30 minutes to three hours (the longer you leave it, the sharper the flavor will be). If you cannot find unpasteurized buttermilk, diluting 1/8 teaspoon calcium chloride (available from online cheesemaker suppliers) in 1/4 cup of water and adding it to your milk will create a similar effect.

Rennet (also called rennin or chymosin) is sold online at cheesemaking sites in tablet or liquid form. You may also be able to find Junket rennet tablets near the pudding and gelatin in your grocery store. One teaspoon of liquid rennet is the equivalent of one rennet tablet, which is enough to turn 5 gallons of milk into cheese (estimate four drops of liquid rennet per gallon of milk). Microbial rennet is a vegetarian alternative that is available for purchase online.

Preparation

It's important to keep your hands clean and all equipment sterile when making cheese.

1. Wash hands and all equipment with soapy detergent before and after use.
2. Rinse all equipment with clean water, removing all soapy residue.
3. Boil all cheesemaking equipment between uses.
4. For best-quality cheese, use new cheesecloth each time you make cheese. (Sterilize cheesecloth by first washing, then boiling.)
5. Squeaky clean is clean. If you can feel a residue on the equipment, it is not clean.

Yogurt Cheese

This soft cheese has a flavor similar to sour cream and a texture like cream cheese. A pint of yogurt will yield approximately 1/4 pound of cheese. The yogurt cheese has a shelf life of approximately seven to fourteen days when wrapped and placed in the refrigerator and kept at less than 40°F. Add a little salt and pepper and chopped fresh herbs for variety.

1. Line a large strainer or colander with cheesecloth.
2. Place the lined strainer over a bowl and pour in plain, whole-milk yogurt. Do not use yogurt made with the addition of gelatin, as gelatin will inhibit whey separation.
3. Let yogurt drain overnight, covered with plastic wrap. Empty the whey from the bowl.
4. Fill a strong plastic storage bag with some water, seal, and place over the cheese to weigh it down. Let the cheese stand another eight hours and then enjoy!

Queso Blanco

Queso blanco is a white, semi-hard cheese made without culture or rennet. It is eaten fresh and may be flavored with peppers, herbs, and spices. It is considered a "frying cheese," meaning it does not melt and may be deep-fried or grilled. *Queso blanco* is best eaten fresh, so try this small recipe the first time you make it. If it disappears quickly, next time double or triple the recipe. This recipe will yield about ½ cup of cheese.

- 2 cups milk
- 4 teaspoons white vinegar
- Salt
- Minced jalapeño, black pepper, chives, or other herbs to taste

1. Heat milk to 176°F for 20 minutes.
2. Add vinegar slowly to the hot milk until the whey is semi-clear and the curd particles begin to form stretchy clumps.

Stir for 5 to 10 minutes. When it's ready you should be able to stretch a piece of curd about ⅓ inch before it breaks.
3. Allow to cool, and strain off the whey by filtering through a cheesecloth-lined colander or a cloth bag.
4. Work in salt and spices to taste.
5. Press the curd in a mold or simply leave in a ball.

Credit: Timothy W. Lawrence

« This press works for both making cider and for pressing cheese. The apple press part is a scissors jack (found at an auto parts store) mounted to one of the top timbers. The opposite end is a grinder, made up of two oak rollers. Stainless steel screws serve as teeth to mash the apples, which are then strained in the mesh-lined bucket. You can then press the apples on the opposite end.

The cheese press is a wooden arm mounted across the two top timbers. Another arm goes straight down into the cheese mold. A water-filled jug is hung from the end. The pressure is varied by adding or subtracting water from the jug.

6. *Queso blanco* may keep for several weeks if stored in a refrigerator, but is best eaten fresh.

Ricotta Cheese

Making ricotta is very similar to making *queso blanco*, though it takes a bit longer. Start the cheese in the morning for use at dinner, or make a day ahead. Use it in lasagna, in desserts, or all on its own.

- 1 gallon milk
- ¼ teaspoon salt
- ⅓ cup plus 1 teaspoon white vinegar

1. Pour milk into a large pot, add salt, and heat slowly while stirring until the milk reaches 180°F.
2. Remove from heat and add vinegar. Stir for 1 minute as curds begin to form.
3. Cover and allow to sit undisturbed for 2 hours.
4. Pour mixture into a colander lined with cheesecloth, and allow to drain for two or more hours.
5. Store in a sealed container for up to a week.

Mozzarella

This mild cheese will make your home-made pizza especially delicious. Or slice it and eat with fresh tomatoes and basil from the garden. Fresh cheese can be stored in salt water but must be eaten within two days.

- 1 gallon 2% milk
- ¼ cup fresh, plain yogurt (see recipe on page 150)
- One tablet rennet or 1 teaspoon liquid rennet dissolved in ½ cup tap water
- Brine: use 2 pounds of salt per gallon of water

1. Heat milk to 90°F and add yogurt. Stir slowly for 15 minutes while keeping the temperature constant.
2. Add rennet mixture and stir for 3 to 5 minutes.
3. Cover, remove from heat, and allow to stand until coagulated, about 30 minutes.
4. Cut curd into ½-inch cubes. Allow to stand for 15 minutes with occasional stirring.
5. Return to heat and slowly increase temperature to 118°F over a period of 45 minutes. Hold this temperature for an additional 15 minutes.
6. Drain off the whey by transferring the mixture to a cheesecloth-lined colander. Use a spoon to press the liquid out of the curds. Transfer the mat of curd to a flat pan that can be kept warm in a low oven. Do not cut mat, but turn it over every 15 minutes for a 2-hour period. Mat should be tight when finished.
7. Cut the mat into long strips 1 to 2 inches wide and place in hot water (180°F). Using wooden spoons, tumble and stretch it under water until it becomes elastic, about 15 minutes.
8. Remove curd from hot water and shape it by hand into a ball or a loaf, kneading in the salt. Place cheese in cold water (40°F) for approximately 1 hour.
9. Store in a solution of 2 teaspoons salt to 1 cup water.

Cheddar Cheese

Cheddar is a New England and Wisconsin favorite. The longer you age it, the sharper the flavor will be. Try a slice with a wedge of homemade apple pie.

Ingredients
- 1 gallon milk
- ¼ cup buttermilk
- 1 tablet rennet, or 1 teaspoon liquid rennet
- 1½ teaspoons salt

MAKE YOUR OWN SIMPLE CHEESE PRESS

1. Remove both ends of a large coffee can or thoroughly cleaned paint can, saving one end. Use an awl or a hammer and long nail to pierce the sides in several places, piercing from the inside out.
2. Place the can on a cooling rack inside a larger basin. Leave the bottom of the can in place.
3. Use a saw to cut a ¾-inch-thick circle of wood to create a "cheese follower." It should be small enough in diameter to fit easily in the can.
4. Place cheese curds in the can, and top with the cheese follower. Place several bricks wrapped in cloth or foil on top of the cheese follower to weigh down curds.
5. Once the cheese is fully pressed, remove the bricks and bottom of the can. Use the cheese follower to push the cheese out of the can.

Credit: Timothy W. Lawrence

HOMEMADE CRACKERS

Crackers are very easy to make and can be varied endlessly by adding seasonings of your choice. Try sprinkling a coarse sea salt and dried oregano or cinnamon and sugar over the crackers just before baking. Serve with homemade cheese.

- 1½ cups all-purpose flour
- 1½ cups whole wheat flour
- 1 teaspoon salt
- 1 cup warm water
- ⅓ cup olive oil
- Herbs, spices, or coarse sea salt as desired

1. Stir together the dry ingredients in a mixing bowl. Add the water and olive oil and knead until dough is elastic and not too sticky (about 5 minutes in an electric mixer with a dough attachment or 10 minutes by hand).
2. Allow dough to rest at room temperature for about half an hour. Preheat oven to 450°F.
3. Flour a clean, dry surface and roll dough to about ⅛ inch thick. Cut into squares and place on a cookie sheet. Sprinkle with desired topping. Bake for about 5 minutes or until crackers are golden brown.

Directions
1. Combine milk and buttermilk and allow the mixture to ripen overnight.
2. The next day, heat milk to 90°F in a double boiler and add rennet.
3. After about 45 minutes, cut curds into small cubes and let sit 15 minutes.
4. Heat very slowly to 100°F and cook for about an hour or until a cooled piece of curd will keep its shape when squeezed.
5. Drain curds and rinse out the double boiler.
6. Place a rack lined with cheesecloth inside the double boiler and spread the curds on the cloth. Cover and reheat at about 98° F for 30 to 40 minutes. The curds will become one solid mass.
7. Remove the curds, cut them into 1-inch wide strips, and return them to the pan. Turn the strips every 15 to 20 minutes for one hour.
8. Cut the strips into cubes and mix in salt
9. Let the curds stand for 10 minutes, place them in cheesecloth, and press in a cheese press with 15 pounds for 10 minutes, then with 30 pounds for an hour.
10. Remove the cheese from the press, unwrap it, dip in warm water, and fill in any cracks.
11. Wrap again in cheesecloth and press with 40 pounds for twenty-four hours.
12. Remove from the press and let the cheese dry about five days in a cool, well-ventilated area, turning the cheese twice a day and wiping it with a clean cloth. When a hard skin has formed, rub with oil or seal with wax. You can eat

the cheese after six weeks, but for the strongest flavor, allow cheese to age for six months or more.

Make Your Own Ice Cream

Supplies
- 1-pound coffee can
- 3-pound coffee can
- Duct tape
- Ice
- 1 cup salt

Ingredients
- 2 cups half and half
- ½ cup sugar
- 1 teaspoon vanilla

Directions
1. Mix all the ingredients in the 1-pound coffee can. Cover the lid with duct tape to ensure it is tightly sealed.
2. Place the smaller can inside the larger can and fill the space between the two with ice and salt.
3. Cover the large can and seal with duct tape. Roll the can back and forth for 15 minutes. To reduce noise, place a towel on your working surface, or work on a rug.
4. Dump out ice and water. Stir contents of small can. Store ice cream in a glass or plastic container (if you leave it in the can it may take on a metallic flavor).

If desired, add cocoa powder, coffee granules, crushed peppermint sticks or other candy, or fruit.

Brew Your Own Root Beer

The primary ingredients that give root beer its distinctive flavor are sassafras, vanilla,

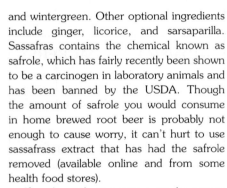

and wintergreen. Other optional ingredients include ginger, licorice, and sarsaparilla. Sassafras contains the chemical known as safrole, which has fairly recently been shown to be a carcinogen in laboratory animals and has been banned by the USDA. Though the amount of safrole you would consume in home brewed root beer is probably not enough to cause worry, it can't hurt to use sassafrass extract that has had the safrole removed (available online and from some health food stores).

If you're gathering your roots, be sure to rinse them thoroughly and cut thicker ones in half. When using leaves, be sure to use only the leaves and not the stems, roots, or flowers. Leaves should be rinsed thoroughly and then dried.

Root Beer
- ¼ ounce wintergreen leaves
- ¼ ounce sarsaparilla root
- ½ ounce sassafras root bark or 8 to 10 drops sassafras extract
- 1 1-inch piece ginger root, unpeeled and thinly sliced
- ½ ounce burdock root
- ¼ ounce licorice root
- ⅛ teaspoon active dry yeast
- 1 cup molasses
- 1 cup sugar
- 1 teaspoon vanilla extract
- 1 gallon water

1. Place roots and and leaves in 2 quarts water in a large pot and bring to a boil. Remove from heat and allow to steep for 2 hours.
2. Strain through a cheesecloth-lined strainer into a clean pot. Discard roots

and leaves. Add remaining 2 quarts of water to infused liquid.
3. Add yeast, molasses, vanilla, and sugar to liquid, cover, and set aside to ferment for about 20 minutes.
4. Use a funnel to fill four 1-liter plastic soda bottles. Leave at least 2 inches of space at the top of each bottle. Screw on lids and store at room temperature for 12 hours.
5. Chill root beer in the refrigerator for at least two days, or until it reaches the desired fizziness. After four or five days the root beer will become slightly alcoholic. Remove caps slowly to allow gas to escape gradually.

≽ The leaves of the Sassafras tree are very distinctive.

Sharing Your Bounty

Plant a Row for the Hungry (PAR) Program

One in ten households in the United States experiences hunger or the risk of hunger every year. This is an astounding number of adults and children who are not receiving the proper foods, nutrients, and sustenance to maintain a healthy lifestyle. What can you do as a gardener to help provide fresh produce to the hungry in your community? Churches, schools, and other local organizations try to provide food for those who are hungry by giving to local food banks and shelters and by establishing soup kitchens and other programs. Often these food drives can only accept nonperishable items, but, recognizing the need for fresh fruits and vegetables in a healthy diet, many organizations are beginning to take produce on a conditional basis. If they can use the produce in their prepared meals or distribute them to needy families before the food begins to perish, they will accept it.

Started in 1995 in Anchorage, Alaska, by Jeff Lowenfels as a public service program of the Garden Writers Association (GWA) and Foundation, PAR is a way for those who garden and grow their own vegetables and fruit to help combat hunger and poverty in their local communities. PAR encourages local gardeners to plant one extra row of produce in their gardens and then to donate that harvest to neighbors or others in the community who struggle to feed themselves and their families. Donating excess vegetables and fruit to your local food bank, soup kitchen, shelter, or other food agency to help feed the hungry in your local area is a wonderful way to share your love of gardening and to be an active and important member of your community.

Anyone in the United States or Canada can participate in the PAR program, and members grow and donate over one million tons of food a year to help fight hunger in their local communities. Some people establish a community garden that solely grows food to give to local pantries and soup kitchens. Others simply grow additional crops and give individually to those in need.

If you want to become involved in the PAR program, it is best to call or e-mail the GWA in order to receive more information about PAR programs in your area. It is also beneficial to call or visit local food agencies and see what crops they need most, or what types of fruits and vegetables they would like to provide for their customers. Most agencies seek out food that can be shelved for a few days if necessary and also foods that are high in nutritional value. They also accept fresh herbs that they can then use when making foods, such as soups. Flowers are also acceptable donations.

When thinking about growing excess vegetables and fruit in your garden for donation, it is good to select easily grown crops that will help to maximize your harvest and will encourage you to be enthusiastic about growing and tending the plants. Some easily grown and highly desirable plants are: beans, cucumbers, peas, radishes, summer squash, and tomatoes. Then, once your produce is ready for harvesting, gather up the excess and take it to a selected PAR drop-off site or, if acceptable, directly to the food agency.

Sharing Your Bounty

There are some guidelines for growing fruits and vegetables that are acceptable for donation. Always be sure to contact your local food banks and soup kitchens for information on what they need in terms of fresh produce. Be sure to space your plantings apart so that your growing season and harvesting season are extended over a longer period of time. That way, you'll be able to donate longer. Choose to grow produce that lasts well and stays fresher longer. Pick your ripe produce promptly and be sure to clean it of any dirt (but don't wash it). And, of course, don't ever give away overripe or spoiled produce.

How to Start Your Own PAR Program

What if your local community does not yet have a PAR program? It is easy to establish your own program by contacting PAR (either by e-mail, by visiting the GWA website, or by calling toll free; see contact information below). This way, you will be able to gain access to information on creating a successful PAR program. Next, you should try to recruit volunteers (neighbors, community gardeners, garden clubs, garden centers, and nurseries) to be a part of the program. It is important to establish a local coordinator who can be a "go-to" person for any questions volunteers may have about the program, growing tips, and drop-off sites.

In addition to gathering volunteers, it is essential to find a food distribution agency partner who wants to collect the donated produce. This may be a food bank, food pantry, soup kitchen, or local shelter. It is also wise to find someone to market and publicize your local program. Getting the word out about the program will help gather support and more gardeners who want to help. There should also be someone appointed to help coordinate and oversee the drop-off sites and to take food to the designated area if need be.

For a PAR program to be a success, it is a good idea to reach out to your local extension services; community, church, and school gardens; businesses; and local food agencies to see if a specific PAR garden can be established. It is also wise to ask farmers to donate their unsold produce from farmers' markets and trucks in the area. All of these various elements will help you to create a successful fresh food donation program in your area and will ensure that those who are hungry in your community will not be lacking in fresh produce in their diets.

For more information on Plant a Row for the Hungry, please call, e-mail, or visit the Garden Writers Association at:

1-877-GWAA-PAR

par@gwaa.org

www.gwaa.org

To find a local PAR contact in your area, please call the Foodchain at: 1-800-845-3008.

Part 3

Canning and Preserving

"It is not always granted to the sower to live to see the harvest. All work that is worth anything is done in faith."

—Albert Schweitzer

"Striving for success without hard work is like trying to harvest where you haven't planted."

—David Bly

Canning	156
Drying and Freezing	222

Canning

Canning began in France, at the turn of the nineteenth century, when Napoleon Bonaparte was desperate for a way to keep his troops well fed while on the march. In 1800 he decided to hold a contest, offering 12,000 francs to anyone who could devise a suitable method of food preservation. Nicolas François Appert, a French confectioner, rose to the challenge, considering that if wine could be preserved in bottles, perhaps food could be as well. He experimented until he was able to prove that heating food to boiling after it had been sealed in airtight glass bottles prevented the food from deteriorating. Interestingly, this all took place about 100 years before Louis Pasteur found that heat could destroy bacteria. Nearly ten years after the contest began, Napoleon personally presented Nicolas with the cash reward.

Canning practices have evolved over the last two centuries, but the principles remain the same. In fact, the way we can foods today is basically the same way our grandparents and great grandparents preserved their harvests for the winter months.

On the next few pages you will find descriptions of proper canning methods, with details on how canning works and why it is both safe and economical. Much of the information here is from the USDA, which has done extensive research on home canning and preserving. If you are new to home canning, read this section carefully as it will help to ensure success with the recipes that follow.

Whether you are a seasoned home canner or this is your first foray into food

preservation, it is important to follow directions carefully. With some recipes it is okay to experiment with varied proportions or added ingredients, and with others it is important to stick to what's written. In many instances it is noted whether or not creative liberty is a good idea for a

particular recipe, but if you are not sure, play it safe—otherwise you may end up with a jam that is too runny, a vegetable that is mushy, or a product that is spoiled. Take time to read the directions and prepare your foods and equipment adequately and you will find that home canning is safe, economical, tremendously satisfying, and a great deal of fun!

The Benefits of Canning

Canning is fun, economical, and a good way to preserve your precious produce. As more and more farmers' markets make their way into urban centers, city dwellers are also discovering how rewarding it is to make seasonal treats last all year round. Besides the value of your labor, canning home-grown or locally grown food may save you half the cost of buying commercially canned food. Freezing food may be simpler, but most people have limited freezer space, whereas cans of food can be stored almost anywhere. And what makes a nicer, more thoughtful gift than a jar of homemade jam, tailored to match the recipient's favorite fruits and flavors?

The advantages of home canning are lost when you start with poor quality foods, when jars fail to seal properly, when food spoils, and when flavors, texture, color, and nutrients deteriorate during prolonged storage. The tips that follow explain many of these problems and recommend ways to minimize them.

How Canning Preserves Foods

The high percentage of water in most fresh foods makes them very perishable. They spoil or lose their quality for several reasons:

- Growth of undesirable microorganisms—bacteria, molds, and yeasts
- Activity of food enzymes
- Reactions with oxygen
- Moisture loss

Microorganisms live and multiply quickly on the surfaces of fresh food and on the inside of bruised, insect-damaged, and diseased food. Oxygen and enzymes are present throughout fresh food tissues.

Proper canning practices include:

- Carefully selecting and washing fresh food
- Peeling some fresh foods
- Hot packing many foods
- Adding acids (lemon juice, citric acid, or vinegar) to some foods

The nutritional value of home canning is an added benefit. Many vegetables begin to lose their vitamins as soon as they are harvested. Nearly half the vitamins may be lost within a few days unless the fresh produce is kept cool or preserved. Within one to two weeks, even refrigerated produce loses half or more of certain vitamins. The heating process during canning destroys from one-third to one-half of vitamins A and C, thiamin, and riboflavin. Once canned, foods may lose from 5 to 20 percent of these sensitive vitamins each year. The amounts of other vitamins, however, are only slightly lower in canned compared with fresh food. If vegetables are handled properly and canned promptly after harvest, they can be more nutritious than fresh produce sold in local stores.

hold food in boiling water or with a fitted rack to steam foods. Useful for loosening skins on fruits to be peeled or for heating foods to be hot packed.

Boiling-water canner—A large, standard-sized, lidded kettle with jar rack designed for heat-processing seven quarts or eight to nine pints in boiling water.

Botulism—An illness caused by eating a toxin produced by growth of *Clostridium botulinum* bacteria in moist, low-acid food containing less than 2 percent oxygen and stored between 40 and 120°F. Proper heat processing destroys this bacterium in canned food. Freezer temperatures inhibit its growth in frozen food. Low moisture controls its growth in dried food. High oxygen controls its growth in fresh foods.

Canning—A method of preserving food that employs heat processing in airtight, vacuum-sealed containers so that food can be safely stored at normal home temperatures.

Canning salt—Also called pickling salt. It is regular table salt without the anti-caking or iodine additives.

Citric acid—A form of acid that can be added to canned foods. It increases the acidity of low-acid foods and may improve their flavor.

Cold pack—Canning procedure in which jars are filled with raw food. "Raw pack" is the preferred term for describing this practice. "Cold pack" is often used incorrectly to refer to foods that are open-kettle canned or jars that are heat-processed in boiling water.

Enzymes—Proteins in food that accelerate many flavor, color, texture, and nutritional changes, especially when food is cut,

- Using acceptable jars and self-sealing lids
- Processing jars in a boiling-water or pressure canner for the correct amount of time

Collectively, these practices remove oxygen; destroy enzymes; prevent the growth of undesirable bacteria, yeasts, and molds; and help form a high vacuum in jars. High vacuums form tight seals, which keep liquid in and air and microorganisms out.

marmalades; and all fruits except figs. Acid foods may be processed in boiling water.

Ascorbic acid—The chemical name for vitamin C; commonly used to prevent browning of peeled, light-colored fruits and vegetables.

Blancher—A 6- to 8-quart lidded pot designed with a fitted, perforated basket to

---

TIP

A large stockpot with a lid can be used in place of a boiling-water canner for high-acid foods like tomatoes, pickles, apples, peaches, and jams. Simply place a rack inside the pot so that the jars do not rest directly on the bottom of the pot.

Canning Glossary

Acid foods—Foods that contain enough acid to result in a pH of 4.6 or lower. Includes most tomatoes; fermented and pickled vegetables; relishes; jams, jellies, and

sliced, crushed, bruised, or exposed to air. Proper blanching or hot-packing practices destroy enzymes and improve food quality.

Exhausting—Removing air from within and around food and from jars and canners. Exhausting or venting of pressure canners is necessary to prevent botulism in low-acid canned foods.

Headspace—The unfilled space above food or liquid in jars that allows for food expansion as jars are heated and for forming vacuums as jars cool.

Heat processing—Treatment of jars with sufficient heat to enable storing food at normal home temperatures.

Hermetic seal—An absolutely airtight container seal that prevents reentry of air or microorganisms into packaged foods.

Hot pack—Heating of raw food in boiling water or steam and filling it hot into jars.

Low-acid foods—Foods that contain very little acid and have a pH above 4.6. The acidity in these foods is insufficient to prevent the growth of botulism bacteria. Vegetables, some varieties of tomatoes, figs, all meats, fish, seafood, and some dairy products are low-acid foods. To control all risks of botulism, jars of these foods must be either heat processed in a pressure canner or acidified to a pH of 4.6 or lower before being processed in boiling water.

Microorganisms—Independent organisms of microscopic size, including bacteria, yeast, and mold. In a suitable environment, they grow rapidly and may divide or reproduce every ten to thirty minutes. Therefore, they reach high populations very quickly. Microorganisms are sometimes intentionally added to ferment foods, make antibiotics, and for other reasons. Undesirable microorganisms cause disease and food spoilage.

Mold—A fungus-type microorganism whose growth on food is usually visible and colorful. Molds may grow on many foods, including acid foods like jams and jellies and canned fruits. Recommended heat processing and sealing practices prevent their growth on these foods.

Mycotoxins—Toxins produced by the growth of some molds on foods.

Open-kettle canning—A non-recommended canning method. Food is heat-processed in a covered kettle, filled while hot into sterile jars, and then sealed. Foods canned this way have low vacuums or too much air, which permits rapid loss of quality in foods. Also, these foods often spoil because they become recontaminated while the jars are being filled.

Pasteurization—Heating food to temperatures high enough to destroy disease-causing microorganisms.

bacteria exist either as spores or as vegetative cells. The spores, which are comparable to plant seeds, can survive harmlessly in soil and water for many years. When ideal conditions exist for growth, the spores produce vegetative cells, which multiply rapidly and may produce a deadly toxin within three to four days in an environment consisting of:

- A moist, low-acid food
- A temperature between 40 and 120°F, and
- Less than 2 percent oxygen.

Botulinum spores are on most fresh food surfaces. Because they grow only in the absence of air, they are harmless on fresh foods. Most bacteria, yeasts, and molds are difficult to remove from food surfaces. Washing fresh food reduces their numbers only slightly. Peeling root crops, underground stem crops, and tomatoes reduces their numbers greatly. Blanching also helps, but the vital controls are the method of canning and use of the recommended research-based processing times. These processing times ensure destruction of the largest expected number of heat-resistant microorganisms in home-canned foods.

Properly sterilized canned food will be free of spoilage if lids seal and jars are stored below 95°F. Storing jars at 50 to 70°F enhances retention of quality.

pH—A measure of acidity or alkalinity. Values range from 0 to 14. A food is neutral when its pH is 7.0. Lower values are increasingly more acidic; higher values are increasingly more alkaline.

PSIG—Pounds per square inch of pressure as measured by a gauge.

Pressure canner—A specifically designed metal kettle with a lockable lid used for heat processing low-acid food. These canners have jar racks, one or more safety devices, systems for exhausting air, and a way to measure or control pressure. Canners with 20- to 21-quart capacity are common. The minimum size of canner that should be used has a 16-quart capacity and can hold seven one-quart jars. Use of pressure saucepans with a capacity of less than 16 quarts is not recommended.

Raw pack—The practice of filling jars with raw, unheated food. Acceptable for canning low-acid foods, but allows more rapid quality losses in acid foods that are heat-processed in boiling water. Also called "cold pack."

Style of pack—Form of canned food, such as whole, sliced, piece, juice, or sauce. The term may also be used to specify whether food is filled raw or hot into jars.

Vacuum—A state of negative pressure that reflects how thoroughly air is removed from within a jar of processed food; the higher the vacuum, the less air left in the jar.

Proper Canning Practices

Growth of the bacterium *Clostridium botulinum* in canned food may cause botulism—a deadly form of food poisoning. These

fruits, pickles, sauerkraut, jams, jellies, marmalade, and fruit butters.

Although tomatoes usually are considered an acid food, some are now known to have pH values slightly above 4.6. Figs also have pH values slightly above 4.6. Therefore, if they are to be canned as acid foods, these products must be acidified to a pH of 4.6 or lower with lemon juice or citric acid. Properly acidified tomatoes and figs are acid foods and can be safely processed in a boiling-water canner.

Botulinum spores are very hard to destroy at boiling-water temperatures; the higher the canner temperature, the more easily they are destroyed. Therefore, all low-acid foods should be sterilized at temperatures of 240 to 250°F, attainable with pressure canners operated at 10 to 15 PSIG. (PSIG means pounds per square inch of pressure as measured by a gauge.) At these temperatures, the time needed to destroy bacteria in low-acid canned foods ranges from twenty to 100 minutes. The exact time depends on the kind of food being canned, the way it is packed into jars, and the size of jars. The time needed to safely process low-acid foods in boiling water ranges from seven to eleven hours; the time needed to process acid foods in boiling water varies from five to eighty-five minutes.

Know Your Altitude

It is important to know your approximate elevation or altitude above sea level in order to determine a safe processing time for canned foods. Since the boiling temperature of liquid

Food Acidity and Processing Methods

Whether food should be processed in a pressure canner or boiling-water canner to control botulism bacteria depends on the acidity in the food. Acidity may be natural, as in most fruits, or added, as in pickled food. Low-acid canned foods contain too little acidity to prevent the growth of these bacteria. Other foods may contain enough acidity to block their growth or to destroy them rapidly when heated. The term "pH" is a measure of acidity: the lower its value, the more acidic the food. The acidity level in foods can be increased by adding lemon juice, citric acid, or vinegar.

Low-acid foods have pH values higher than 4.6. They include red meats, seafood, poultry, milk, and all fresh vegetables except for most tomatoes. Most products that are mixtures of low-acid and acid foods also have pH values above 4.6 unless their ingredients include enough lemon juice, citric acid, or vinegar to make them acid foods. Acid foods have a pH of 4.6 or lower. They include

is lower at higher elevations, it is critical that additional time be given for the safe processing of foods at altitudes above sea level.

What Not to Do

Open-kettle canning and the processing of freshly filled jars in conventional ovens, microwave ovens, and dishwashers are not recommended because these practices do not prevent all risks of spoilage. Steam canners are not recommended because processing times for use with current models have not been adequately researched. Because steam canners may not heat foods in the same manner as boiling-water canners, their use with boiling-water processing times may result in spoilage. So-called canning powders are useless as preservatives and do not replace the need for proper heat processing.

It is not recommended that pressures in excess of 15 PSIG be applied when using new pressure-canning equipment.

Ensuring High-Quality Canned Foods

Examine food carefully for freshness and wholesomeness. Discard diseased and moldy food. Trim small diseased lesions or spots from food.

Can fruits and vegetables picked from your garden or purchased from nearby producers when the products are at their peak of quality—within six to twelve hours after harvest for most vegetables. However, apricots, nectarines, peaches, pears, and plums should be ripened one or more days between harvest and canning. If you must delay the canning of other fresh produce, keep it in a shady, cool place.

Fresh, home-slaughtered red meats and poultry should be chilled and canned without delay. Do not can meat from sickly or diseased animals. Put fish and seafood on ice after harvest, eviscerate immediately, and can them within two days.

Maintaining Color and Flavor in Canned Food

To maintain good natural color and flavor in stored canned food, you must:

- Remove oxygen from food tissues and jars
- Quickly destroy the food enzymes, and
- Obtain high jar vacuums and airtight jar seals.

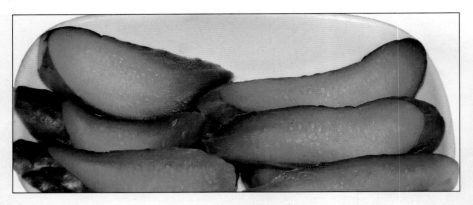

Pure powdered form—Seasonally available among canning supplies in supermarkets. One level teaspoon of pure powder weighs about 3 grams. Use 1 teaspoon per gallon of water as a treatment solution.

Vitamin C tablets—Economical and available year-round in many stores. Buy 500-milligram tablets; crush and dissolve six tablets per gallon of water as a treatment solution.

Commercially prepared mixes of ascorbic and citric acid—Seasonally available among canning supplies in supermarkets. Sometimes citric acid powder is sold in supermarkets, but it is less effective in controlling discoloration. If you choose to use these products, follow the manufacturer's directions.

- Fill hot foods into jars and adjust headspace as specified in recipes
- Tighten screw bands securely, but if you are especially strong, not as tightly as possible
- Process and cool jars
- Store the jars in a relatively cool, dark place, preferably between 50 and 70°F
- Can no more food than you will use within a year.

Advantages of Hot Packing

Many fresh foods contain from 10 percent to more than 30 percent air. The length of time that food will last at premium quality depends on how much air is removed from the food before jars are sealed. The more air that is removed, the higher the quality of the canned product.

Raw packing is the practice of filling jars tightly with freshly prepared but unheated food. Such foods, especially fruit, will float in the jars. The entrapped air in and around the food may cause discoloration within two to three months of storage. Raw-packing is

Follow these guidelines to ensure that your canned foods retain optimal colors and flavors during processing and storage:

- Use only high-quality foods that are at the proper maturity and are free of diseases and bruises
- Use the hot-pack method, especially with acid foods to be processed in boiling water
- Don't unnecessarily expose prepared foods to air; can them as soon as possible

- While preparing a canner load of jars, keep peeled, halved, quartered, sliced or diced apples, apricots, nectarines, peaches, and pears in a solution of 3 grams (3,000 milligrams) ascorbic acid to 1 gallon of cold water. This procedure is also useful in maintaining the natural color of mushrooms and potatoes and for preventing stem-end discoloration in cherries and grapes.

You can get ascorbic acid in several forms:

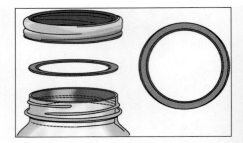

more suitable for vegetables processed in a pressure canner.

Hot packing is the practice of heating freshly prepared food to boiling, simmering it three to five minutes, and promptly filling jars loosely with the boiled food. Hot packing is the best way to remove air and is the preferred pack style for foods processed in a boiling-water canner. At first, the color of hot-packed foods may appear no better than that of raw-packed foods, but within a short storage period both color and flavor of hot-packed foods will be superior.

Whether food has been hot packed or raw packed, the juice, syrup, or water to be added to the foods should be heated to boiling before it is added to the jars. This practice helps to remove air from food tissues, shrinks food, helps keep the food from floating in the jars, increases vacuum in sealed jars, and improves shelf life. Preshrinking food allows you to add more food to each jar.

Controlling Headspace

The unfilled space above the food in a jar and below its lid is termed headspace. It is best to leave a ¼-inch headspace for jams and jellies, ½-inch for fruits and tomatoes to be processed in boiling water, and from 1 to 1¼ inches in low-acid foods to be processed in a pressure canner.

This space is needed for expansion of food as jars are processed and for forming

vacuums in cooled jars. The extent of expansion is determined by the air content in the food and by the processing temperature. Air expands greatly when heated to high temperatures—the higher the temperature, the greater the expansion. Foods expand less than air when heated.

Jars and Lids

Food may be canned in glass jars or metal containers. Metal containers can be used only once. They require special sealing equipment and are much more costly than jars.

Mason-type jars designed for home canning are ideal for preserving food by pressure or boiling-water canning. Regular and wide-mouthed threaded mason jars with self-sealing lids are the best choices. They are available in half-pint, pint, 1½-pint, and quart sizes. The standard jar mouth opening is about 2⅜ inches. Wide-mouthed jars have openings of about 3 inches, making them more easily filled and emptied. Regular-mouth decorative

jelly jars are available in eight-ounce and 12-ounce sizes.

With careful use and handling, mason jars may be reused many times, requiring only new lids each time. When lids are used properly, jar seals and vacuums are excellent.

Jar Cleaning

Before reuse, wash empty jars in hot water with detergent and rinse well by hand, or wash in a dishwasher. Rinse thoroughly, as detergent residue may cause unnatural flavors and colors. Scale or hard-water films on jars are easily removed by soaking jars several hours in a solution containing 1 cup of vinegar (5 percent acid) per gallon of water.

Sterilization of Empty Jars

Use sterile jars for all jams, jellies, and pickled products processed less than ten minutes. To sterilize empty jars, put them right side up on the rack in a boiling-water canner. Fill the canner and jars with hot (not boiling) water to 1 inch above the tops of

or deformed lids or lids with gaps or other defects in the sealing gasket.

After filling jars with food, release air bubbles by inserting a flat plastic (not metal) spatula between the food and the jar. Slowly turn the jar and move the spatula up and down to allow air bubbles to escape. Adjust the headspace and then clean the jar rim (sealing surface) with a dampened paper towel. Place the lid, gasket down, onto the cleaned jar-sealing surface. Uncleaned jar-sealing surfaces may cause seal failures.

Then fit the metal screw band over the flat lid. Follow the manufacturer's guidelines enclosed with or on the box for tightening the jar lids properly.

- If screw bands are too tight, air cannot vent during processing, and food will discolor during storage. Overtightening also may cause lids to buckle and jars to break, especially with raw-packed, pressure-processed food.
- If screw bands are too loose, liquid may escape from jars during processing, seals may fail, and the food will need to be reprocessed.

Do not retighten lids after processing jars. As jars cool, the contents in the jar contract, pulling the self-sealing lid firmly against the jar to form a high vacuum. Screw bands are not needed on stored jars. They can be removed easily after jars are cooled. When removed, washed, dried, and stored in a dry

the jars. Boil ten minutes. Remove and drain hot sterilized jars one at a time. Save the hot water for processing filled jars. Fill jars with food, add lids, and tighten screw bands.

Empty jars used for vegetables, meats, and fruits to be processed in a pressure canner need not be sterilized beforehand. It is also unnecessary to sterilize jars for fruits, tomatoes, and pickled or fermented foods that will be processed ten minutes or longer in a boiling-water canner.

Lid Selection, Preparation, and Use

The common self-sealing lid consists of a flat metal lid held in place by a metal screw band during processing. The flat lid is crimped around its bottom edge to form a trough, which is filled with a colored gasket material. When jars are processed, the lid gasket softens and flows slightly to cover the jar-sealing surface, yet allows air to escape from the jar. The gasket then forms an airtight seal as the jar cools. Gaskets in unused lids work well for at least five years from date of manufacture. The gasket material in older unused lids may fail to seal on jars.

It is best to buy only the quantity of lids you will use in a year. To ensure a good seal, carefully follow the manufacturer's directions in preparing lids for use. Examine all metal lids carefully. Do not use old, dented,

(PSIG) you wish to use and match it with your pack style (raw or hot) and jar size to find the correct processing time.

Recommended Canners

There are two main types of canners for heat-processing home-canned food: boiling-water canners and pressure canners. Most are designed to hold seven one-quart jars or eight to nine one-pint jars. Small pressure canners hold four one-quart jars; some large pressure canners hold eighteen 1-pint jars in two layers but hold only seven quart jars. Pressure saucepans with smaller volume capacities are not recommended for use in canning. Treat small pressure canners the same as standard larger canners; they should be vented using the typical venting procedures.

Low-acid foods must be processed in a pressure canner to be free of botulism risks. Although pressure canners also may be used for processing acid foods, boiling-water canners are recommended because they are faster. A pressure canner would require from fifty-five to 100 minutes to can a load of jars; the total time for canning most acid foods in boiling water varies from twenty-five to sixty minutes.

A boiling-water canner loaded with filled jars requires about twenty to thirty minutes of heating before its water begins to boil. A loaded pressure canner requires about twelve to fifteen minutes of heating before it begins

area, screw bands may be used many times. If left on stored jars, they become difficult to remove, often rust, and may not work properly again.

Selecting the Correct Processing Time

When food is canned in boiling water, more processing time is needed for most raw-packed foods and for quart jars than is needed for hot-packed foods and pint jars.

To destroy microorganisms in acid foods processed in a boiling-water canner, you must:

- Process jars for the correct number of minutes in boiling water;
- Cool the jars at room temperature.

To destroy microorganisms in low-acid foods processed with a pressure canner, you must:

- Process the jars for the correct number of minutes at 240°F (10 PSIG) or 250°F (15 PSIG);
- Allow canner to cool at room temperature until it is completely depressurized.

The food may spoil if you fail to use the proper processing times, fail to vent steam from canners properly, process at lower pressure than specified, process for fewer minutes than specified, or cool the canner with water.

Processing times for haft-pint and pint jars are the same, as are times for 1 ½-pint and quart jars. For some products, you have a choice of processing at 5, 10, or 15 PSIG. In these cases, choose the canner pressure

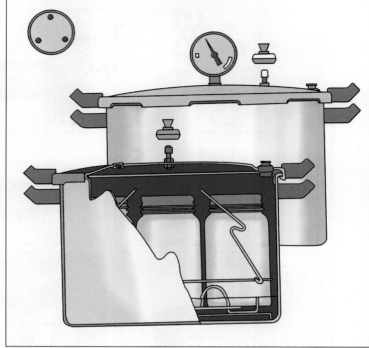

to vent, another ten minutes to vent the canner, another five minutes to pressurize the canner, another eight to ten minutes to process the acid food, and, finally, another twenty to sixty minutes to cool the canner before removing jars.

Boiling-Water Canners

These canners are made of aluminum or porcelain-covered steel. They have removable perforated racks and fitted lids. The canner must be deep enough so that at least 1 inch of briskly boiling water will cover the tops of jars during processing. Some boiling-water canners do not have flat bottoms. A flat bottom must be used on an electric range. Either a flat or ridged bottom can be used on a gas burner. To ensure uniform processing of all jars with an electric range, the canner should be no more than 4 inches wider in diameter than the element on which it is heated.

Using a Boiling-Water Canner

Follow these steps for successful boiling-water canning:

1. Fill the canner halfway with water.
2. Preheat water to 140°F for raw-packed foods and to 180°F for hot-packed foods.
3. Load filled jars, fitted with lids, into the canner rack and use the handles to lower the rack into the water; or fill the canner, one jar at a time, with a jar lifter.
4. Add more boiling water, if needed, so the water level is at least 1 inch above jar tops.
5. Turn heat to its highest position until water boils vigorously.
6. Set a timer for the minutes required for processing the food.
7. Cover with the canner lid and lower the heat setting to maintain a gentle boil throughout the processing time.
8. Add more boiling water, if needed, to keep the water level above the jars.
9. When jars have been boiled for the recommended time, turn off the heat and remove the canner lid.
10. Using a jar lifter, remove the jars and place them on a towel, leaving at least 1 inch of space between the jars during cooling.

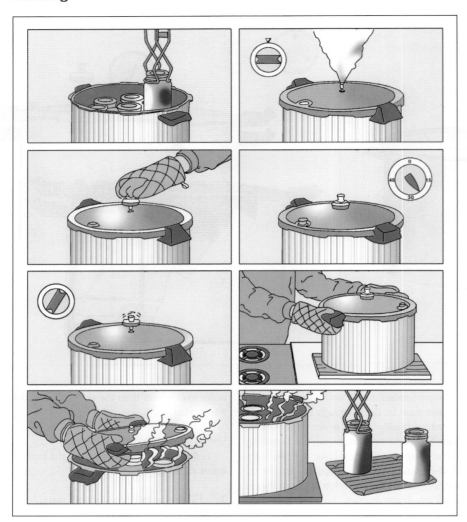

The highest volume of air trapped in a canner occurs in processing raw-packed foods in dial-gauge canners. These canners do not vent air during processing. To be safe, all types of pressure canners must be vented ten minutes before they are pressurized.

To vent a canner, leave the vent port uncovered on newer models or manually open petcocks on some older models. Heating the filled canner with its lid locked into place boils water and generates steam that escapes through the petcock or vent port. When steam first escapes, set a timer for ten minutes. After venting ten minutes, close the petcock or place the counterweight or weighted gauge over the vent port to pressurize the canner.

Weighted-gauge models exhaust tiny amounts of air and steam each time their gauge rocks or jiggles during processing. The sound of the weight rocking or jiggling indicates that the canner is maintaining the recommended pressure and needs no further attention until the load has been processed for the set time. Weighted-gauge canners cannot correct precisely for higher altitudes, and at altitudes above 1,000 feet must be operated at a pressure of 15.

Check dial gauges for accuracy before use each year and replace if they read high by more than 1 pound at 5, 10, or 15 pounds of pressure. Low readings cause over-processing and may indicate that the accuracy of the gauge is unpredictable. If

Pressure Canners

Pressure canners for use in the home have been extensively redesigned in recent years. Models made before the 1970s were heavy-walled kettles with clamp-on lids. They were fitted with a dial gauge, a vent port in the form of a petcock or counterweight, and a safety fuse. Modern pressure canners are lightweight, thin-walled kettles; most have turn-on lids. They have a jar rack, gasket, dial or weighted gauge, an automatic vent or cover lock, a vent port (steam vent) that is closed with a counterweight or weighted gauge, and a safety fuse.

Pressure does not destroy microorganisms, but high temperatures applied for a certain period of time do. The success of destroying all microorganisms capable of growing in canned food is based on the temperature obtained in pure steam, free of air, at sea level. At sea level, a canner operated at a gauge pressure of 10 pounds provides an internal temperature of 240°F.

Air trapped in a canner lowers the inside temperature and results in under-processing.

a gauge is consistently low, you may adjust the processing pressure. For example, if the directions call for 12 pounds of pressure and your dial gauge has tested 1 pound low, you can safely process at 11 pounds of pressure. If the gauge is more than 2 pounds low, it is unpredictable, and it is best to replace it. Gauges may be checked at most USDA county extension offices, which are located in every state across the country. To find one near you, visit www.csrees.usda.gov.

Handle gaskets of canner lids carefully and clean them according to the manufacturer's directions. Nicked or dried gaskets will allow steam leaks during pressurization of canners. Gaskets of older canners may need to be lightly coated with vegetable oil once per year, but newer models are pre-lubricated. Check your canner's instructions.

Lid safety fuses are thin metal inserts or rubber plugs designed to relieve excessive pressure from the canner. Do not pick at or scratch fuses while cleaning lids. Use only canners that have Underwriter's Laboratory (UL) approval to ensure their safety.

Replacement gauges and other parts for canners are often available at stores offering canner equipment or from canner manufacturers. To order parts, list canner model number and describe the parts needed.

Using a Pressure Canner

Follow these steps for successful pressure canning:

1. Put 2 to 3 inches of hot water in the canner. Place filled jars on the rack, using a jar lifter. Fasten canner lid securely.
2. Open petcock or leave weight off vent port. Heat at the highest setting until steam flows from the petcock or vent port.
3. Maintain high heat setting, exhaust steam ten minutes, and then place weight on vent port or close petcock. The canner will pressurize during the next three to five minutes.
4. Start timing the process when the pressure reading on the dial gauge indicates that the recommended pressure has been reached or when the weighted gauge begins to jiggle or rock.
5. Regulate heat under the canner to maintain a steady pressure at or slightly above the correct gauge pressure. Quick and large pressure variations during processing may cause unnecessary liquid losses from jars. Weighted gauges on Mirro canners should jiggle about two or three times per minute. On Presto canners, they should rock slowly throughout the process.

When processing time is completed, turn off the heat, remove the canner from heat if possible, and let the canner depressurize. Do not force-cool the canner. If you cool it with cold running water in a sink or open the vent port before the canner depressurizes by itself, liquid will spurt from jars, causing low liquid levels and jar seal failures. Force-cooling also may warp the canner lid of older model canners, causing steam leaks.

Depressurization of older models should be timed. Standard size heavy-walled canners require about thirty minutes when loaded with pints and forty-five minutes with quarts. Newer thin-walled canners cool more rapidly and are equipped with vent locks. These canners are depressurized when their vent lock piston drops to a normal position.

1. After the vent port or petcock has been open for two minutes, unfasten the lid and carefully remove it. Lift the lid away from you so that the steam does not burn your face.
2. Remove jars with a lifter, and place on towel or cooling rack, if desired.

Cooling Jars

Cool the jars at room temperature for twelve to twenty-four hours. Jars may be cooled on racks or towels to minimize heat damage to counters. The food level and liquid volume of raw-packed jars will be noticeably lower after cooling because air is exhausted during processing and food shrinks. If a jar loses excessive liquid during processing, do not open it to add more liquid. As long as the seal is good, the product is still usable.

⌃ Testing jar seals.

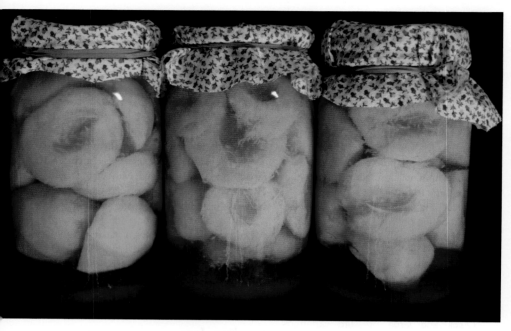

Storing Canned Foods

If lids are tightly vacuum-sealed on cooled jars, remove screw bands, wash the lid and jar to remove food residue, then rinse and dry jars. Label and date the jars and store them in a clean, cool, dark, dry place. Do not store jars at temperatures above 95°F or near hot pipes, a range, a furnace, in an un-insulated attic, or in direct sunlight. Under these conditions, food will lose quality in a few weeks or months and may spoil. Dampness may corrode metal lids, break seals, and allow recontamination and spoilage.

Accidental freezing of canned foods will not cause spoilage unless jars become unsealed and re-contaminated. However, freezing and thawing may soften food. If jars must be stored where they may freeze, wrap them in newspapers, place them in heavy cartons, and cover them with more newspapers and blankets.

Testing Jar Seals

After cooling jars for twelve to twenty-four hours, remove the screw bands and test seals with one of the following methods:

Method 1: Press the middle of the lid with a finger or thumb. If the lid springs up when you release your finger, the lid is unsealed and reprocessing will be necessary.

Method 2: Tap the lid with the bottom of a teaspoon. If it makes a dull sound, the lid is not sealed. If food is in contact with the underside of the lid, it will also cause a dull sound. If the jar lid is sealed correctly, it will make a ringing, high-pitched sound.

Method 3: Hold the jar at eye level and look across the lid. The lid should be concave (curved down slightly in the center). If center of the lid is either flat or bulging, it may not be sealed.

Reprocessing Unsealed Jars

If a jar fails to seal, remove the lid and check the jar-sealing surface for tiny nicks. If necessary, change the jar, add a new, properly prepared lid, and reprocess within twenty-four hours using the same processing time.

Another option is to adjust headspace in unsealed jars to 1½ inches and freeze jars and contents instead of reprocessing. However, make sure jars have straight sides. Freezing may crack jars with "shoulders."

Foods in single unsealed jars could be stored in the refrigerator and consumed within several days.

should completely cover the containers with a minimum of 1 inch of water above the containers. Avoid splashing the water. Place a lid on the pot and heat the water to boiling. Boil thirty minutes to ensure detoxifying the food and all container components. Cool and discard lids and food in the trash or bury in the soil.

Thoroughly clean all counters, containers, and equipment including can opener, clothing, and hands that may have come in contact with the food or the containers. Discard any sponges or washcloths that were used in the cleanup. Place them in a plastic bag and discard in the trash.

Canned Foods for Special Diets

The cost of commercially canned special diet food often prompts interest in preparing these products at home. Some low-sugar and low-salt foods may be easily and safely canned at home. However, it may take some experimentation to create a product with the desired color, flavor, and texture. Start with a small batch and then make appropriate adjustments before producing large quantities.

How much should you can?

The amount of food to preserve for your family, either by canning or freezing, should be based on individual choices. The following table can serve as a worksheet to plan how much food you should can for use within a year.

Identifying and Handling Spoiled Canned Food

Growth of spoilage bacteria and yeast produces gas, which pressurizes the food, swells lids, and breaks jar seals. As each stored jar is selected for use, examine its lid for tightness and vacuum. Lids with concave centers have good seals.

Next, while holding the jar upright at eye level, rotate the jar and examine its outside surface for streaks of dried food originating at the top of the jar. Look at the contents for rising air bubbles and unnatural color.

While opening the jar, smell for unnatural odors and look for spurting liquid and cotton-like mold growth (white, blue, black, or green) on the top food surface and underside of lid. Do not taste food from a stored jar you discover to have an unsealed lid or that otherwise shows signs of spoilage.

All suspect containers of spoiled low-acid foods should be treated as having produced botulinum toxin and should be handled carefully as follows:

- If the suspect glass jars are unsealed, open, or leaking, they should be detoxified before disposal.
- If the suspect glass jars are sealed, remove lids and detoxify the entire jar, contents, and lids.

Detoxification Process

Carefully place the suspect containers and lids on their sides in an eight-quart-volume or larger stockpot, pan, or boiling-water canner. Wash your hands thoroughly. Carefully add water to the pot. The water

PROCESS TIMES FOR FRUITS AND FRUIT PRODUCTS IN A DIAL-GAUGE PRESSURE CANNER*

Type of Fruit	Style of Pack	Jar Size	Process Time	Canner Pressure (PSI) at Altitudes of:			
				0–2,000 ft	2,001–4,000 ft	4,001–6,000 ft	6,001–8,000 ft
Applesauce	Hot	Pints	8 minutes	6 lbs	7 lbs	8 lbs	9 lbs
	Hot	Quarts	ten minutes	6 lbs	7 lbs	8 lbs	9 lbs
Apples, sliced	Hot	Pints or Quarts	8 minutes	6 lbs	7 lbs	8 lbs	9 lbs
Berries, whole	Hot	Pints or Quarts	8 minutes	6 lbs	7 lbs	8 lbs	9 lbs
	Raw	Pints	8 minutes	6 lbs	7 lbs	8 lbs	9 lbs
	Raw	Quarts	ten minutes	6 lbs	7 lbs	8 lbs	9 lbs
Cherries, sour or sweet	Hot	Pints	8 minutes	6 lbs	7 lbs	8 lbs	9 lbs
	Hot	Quarts	ten minutes	6 lbs	7 lbs	8 lbs	9 lbs
	Raw	Pints or Quarts	ten minutes	6 lbs	7 lbs	8 lbs	9 lbs
Fruit purées	Hot	Pints or Quarts	8 minutes	6 lbs	7 lbs	8 lbs	9 lbs
Grapefruit or orange sections	Hot	Pints or Quarts	8 minutes	6 lbs	7 lbs	8 lbs	9 lbs
	Raw	Pints	8 minutes	6 lbs	7 lbs	8 lbs	9 lbs
	Raw	Quarts	ten minutes	6 lbs	7 lbs	8 lbs	9 lbs
Peaches, apricots, or nectarines	Hot or Raw	Pints or Quarts	ten minutes	6 lbs	7 lbs	8 lbs	9 lbs
Pears	Hot	Pints or Quarts	ten minutes	6 lbs	7 lbs	8 lbs	9 lbs
Plums	Hot or Raw	Pints or Quarts	ten minutes	6 lbs	7 lbs	8 lbs	9 lbs
Rhubarb	Hot	Pints or Quarts	8 minutes	6 lbs	7 lbs	8 lbs	9 lbs

*After the process is complete, turn off the heat and remove the canner lid. Wait five to ten minutes before removing jars.

Fruit

There's nothing quite like opening a jar of home- preserved strawberries in the middle of a winter snowstorm. It takes you right back to the warm early-summer sunshine, the smell of the strawberry patch's damp earth, and the feel of the firm berries as you snipped them from the vines. Best of all, you get to indulge in the sweet, summery flavor even as the snow swirls outside the windows.

Preserving fruit is simple, safe, and it allows you to enjoy the fruits of your summer's labor all year round. On the next pages you will find reference charts for processing various fruits and fruit products in a dial-gauge pressure canner or a weighted-gauge pressure canner. The same information is also included with each recipe's directions. In some cases a boiling-water canner will serve better; for these instances, directions for its use are offered instead.

Adding syrup to canned fruit helps to retain its flavor, color, and shape, although it does not prevent spoilage. To maintain the most natural flavor, use the Very Light Syrup listed in the table found on page 181. Many fruits that are typically packed in heavy syrup are just as good—and a lot better for you—when packed in lighter syrups. However, if you're preserving fruit that's on the sour side, like cherries or tart apples, you might want to splurge on one of the sweeter versions.

Syrups

Adding syrup to canned fruit helps to retain its flavor, color, and shape, although jars still need to be processed to prevent spoilage. Follow the chart below for syrups of varying sweetness. Light corn syrups or mild-flavored honey may be used to replace up to half the table sugar called for in syrups.

For hot packs, bring water and sugar to a boil, add fruit, reheat to a boil, and fill into jars immediately.

PROCESS TIMES FOR FRUITS AND FRUIT PRODUCTS IN A WEIGHTED-GAUGE PRESSURE CANNER*

Type of Fruit	Style of Pack	Jar Size	Process Time	Canner Pressure (PSI) at Altitudes of: 0–1,000 ft	Above 1,000 ft
Applesauce	Hot	Pints	8 minutes	5 lbs	10 lbs
	Hot	Quarts	ten minutes	5 lbs	10 lbs
Apples, sliced	Hot	Pints or Quarts	8 minutes	5 lbs	10 lbs
Berries, whole	Hot	Pints or Quarts	8 minutes	5 lbs	10 lbs
	Raw	Pints	8 minutes	5 lbs	10 lbs
	Raw	Quarts	ten minutes	5 lbs	10 lbs
Cherries, sour or sweet	Hot	Pints	8 minutes	5 lbs	10 lbs
	Hot	Quarts	ten minutes	5 lbs	10 lbs
	Raw	Pints or Quarts	ten minutes	5 lbs	10 lbs
Fruit purées	Hot	Pints or Quarts	8 minutes	5 lbs	10 lbs
Grapefruit or orange sections	Hot	Pints or Quarts	8 minutes	5 lbs	10 lbs
	Raw	Pints	8 minutes	5 lbs	10 lbs
	Raw	Quarts	ten minutes	5 lbs	10 lbs
Peaches, apricots, or nectarines	Hot or Raw	Pints or Quarts	ten minutes	5 lbs	10 lbs
Pears	Hot	Pints or Quarts	ten minutes	5 lbs	10 lbs
Plums	Hot or Raw	Pints or Quarts	ten minutes	5 lbs	10 lbs
Rhubarb	Hot	Pints or Quarts	8 minutes	5 lbs	10 lbs

*After the process is complete, turn off the heat and remove the canner lid. Wait five to ten minutes before removing jars.

SUGAR AND WATER IN SYRUP

Syrup Type	Approx. % Sugar	Measures of Water and Sugar For 9-Pt Load* Cups Water	Cups Sugar	For 7-Qt Load Cups Water	Cups Sugar	Fruits Commonly Packed in Syrup
Very Light	10	6½	¾	10½	1¼	Approximates natural sugar levels in most fruits and adds the fewest calories.
Light	20	5¾	1½	9	2¼	Very sweet fruit. Try a small amount the first time to see if your family likes it.
Medium	30	5¼	2¼	8¼	3¾	Sweet apples, sweet cherries, berries, grapes.
Heavy	40	5	3¼	7¾	5¼	Tart apples, apricots, sour cherries, gooseberries, nectarines, peaches, pears, plums.
Very Heavy	50	4¼	4¼	6½	6¾	Very sour fruit. Try a small amount the first time to see if your family likes it.

*This amount is also adequate for a four-quart load.

Canning Without Sugar

In canning regular fruits without sugar, it is very important to select fully ripe but firm fruits of the best quality. It is generally best to can fruit in its own juice, but blends of unsweetened apple, pineapple, and white grape juice are also good for pouring over solid fruit pieces. Adjust headspaces and lids and use the processing recommendations for regular fruits. Add sugar substitutes, if desired, when serving.

PROCESS TIMES FOR APPLE JUICE IN A BOILING-WATER CANNER*

Style of Pack	Jar Size	Process Time at Altitudes of:		
		0–1,000 ft	1,001–6,000 ft	Above 6,000 ft
Hot	Pints or Quarts	5 minutes	ten minutes	15 minutes
	Half-Gallons	ten minutes	15 minutes	20 minutes

*After the process is complete, turn off the heat and remove the canner lid. Wait five minutes before removing jars.

Directions

1. Bring water and sugar to a boil in a medium saucepan.
2. Pour over raw fruits in jars.

Apple Juice

The best apple juice is made from a blend of varieties. If you don't have your own apple press, try to buy fresh juice from a local cider maker within twenty-four hours after it has been pressed.

Directions

1. Refrigerate juice for twenty-four to forty-eight hours.
2. Without mixing, carefully pour off clear liquid and discard sediment. Strain the clear liquid through a paper coffee filter or double layers of damp cheesecloth.
3. Heat quickly in a saucepan, stirring occasionally, until juice begins to boil.
4. Fill immediately into sterile pint or quart jars or into clean half-gallon jars, leaving ¼-inch headspace.
5. Adjust lids and process. See below for recommended times for a boiling-water canner.

PROCESS TIMES FOR APPLE BUTTER IN A BOILING-WATER CANNER*

Style of Pack	Jar Size	Process Time at Altitudes of:		
		0–1,000 ft	1,001–6,000 ft	Above 6,000 ft
Hot	Half-pints or Pints	5 minutes	ten minutes	15 minutes
	Quarts	ten minutes	15 minutes	20 minutes

*After the process is complete, turn off the heat and remove the canner lid. Wait five minutes before removing jars.

Apple Butter

The best apple varieties to use for apple butter include Jonathan, Winesap, Stayman, Golden Delicious, and Macintosh apples, but any of your favorite varieties will work. Don't bother to peel the apples, as you will strain the fruit before cooking it anyway. This recipe will yield eight to nine pints.

Ingredients

* 8 lbs. apples
* 2 cups cider
* 2 cups vinegar
* 2¼ cups white sugar
* 2¼ cups packed brown sugar
* 2 tbsp. ground cinnamon
* 1 tbsp. ground cloves

Directions

1. Wash, stem, quarter, and core apples.
2. Cook slowly in cider and vinegar until soft. Press fruit through a colander, food mill, or strainer.
3. Cook fruit pulp with sugar and spices, stirring frequently. To test for doneness, remove a spoonful and hold it away from steam for 2 minutes. If the butter remains mounded on the spoon, it is done. If you're still not sure, spoon a small quantity onto a plate. When a rim of liquid does not separate around the edge of the butter, it is ready for canning.
4. Fill hot into sterile half-pint or pint jars, leaving ¼-inch headspace. Quart jars need not be pre-sterilized.

Applesauce

Besides being delicious on its own or paired with dishes like pork chops or latkes, applesauce can be used as a butter substitute in many baked goods. Select apples that are sweet, juicy, and crisp. For a tart flavor, add one to two pounds of tart apples to each three pounds of sweeter fruit.

Directions

1. Wash, peel, and core apples. Slice apples into water containing a little lemon juice to prevent browning.
2. Place drained slices in an 8- to 10-quart pot. Add ½ cup water. Stirring occasionally to prevent burning, heat quickly

PROCESS TIMES FOR APPLESAUCE IN A BOILING-WATER CANNER*

Style of Pack	Jar Size	Process Time at Altitudes of:			
		0–1,000 ft	1,001–3,000 ft	3,001–6,000 ft	Above 6,000 ft
Hot	Pints	15 minutes	20 minutes	20 minutes	25 minutes
	Quarts	20 minutes	25 minutes	30 minutes	35 minutes
*After the process is complete, turn off the heat and remove the canner lid. Wait five minutes before removing jars.					

PROCESS TIMES FOR APPLESAUCE IN A DIAL-GAUGE PRESSURE CANNER*

Style of Pack	Jar Size	Process Time	Canner Pressure (PSI) at Altitudes of:			
			0–2,000 ft	2,001–4,000 ft	4,001–6,000 ft	6,001–8,000 ft
Hot	Pints	8 minutes	6 lbs	7 lbs	8 lbs	9 lbs
	Quarts	ten minutes	6 lbs	7 lbs	8 lbs	9 lbs

*After the canner is completely depressurized, remove the weight from the vent port or open the petcock. Wait ten minutes; then unfasten the lid and remove it carefully. Lift the lid with the underside away from you so that the steam coming out of the canner does not burn your face.

PROCESS TIMES FOR APPLESAUCE IN A WEIGHTED-GAUGE PRESSURE CANNER*

Style of Pack	Jar Size	Process Time	Canner Pressure (PSI) at Altitudes of:	
			0–1,000 ft	Above 1,000 ft
Hot	Pints	8 minutes	5 lbs	10 lbs
	Quarts	ten minutes	5 lbs	10 lbs

*After the canner is completely depressurized, remove the weight from the vent port or open the petcock. Wait ten minutes; then unfasten the lid and remove it carefully. Lift the lid with the underside away from you so that the steam coming out of the canner does not burn your face.

until tender (5 to 20 minutes, depending on maturity and variety).

3. Press through a sieve or food mill, or skip the pressing step if you prefer chunky-style sauce. Sauce may be packed without sugar, but if desired, sweeten to taste (start with 1/8 cup sugar per quart of sauce).

4. Reheat sauce to boiling. Fill jars with hot sauce, leaving 1/2-inch headspace. Adjust lids and process.

Spiced Apple Rings

- 12 lbs. firm tart apples (maximum diameter 2-1/2 inches)
- 12 cups sugar
- 6 cups water
- 1 1/4 cups white vinegar (5%)
- 3 tbsp whole cloves
- 3/4 cup red hot cinnamon candies or 8 cinnamon sticks
- 1 tsp red food coloring (optional)

Yield: About 8 to 9 pints

Directions

1. Wash apples. To prevent discoloration, peel and slice one apple at a time. Immediately cut crosswise into

QUANTITY

1. An average of 21 pounds of apples is needed per canner load of seven quarts.

2. An average of 13½ pounds of apples is needed per canner load of nine pints.

3. A bushel weighs 48 pounds and yields 14 to 19 quarts of sauce— an average of three pounds per quart.

1/2-inch slices, remove core area with a melon baller and immerse in ascorbic acid solution.

2. To make flavored syrup, combine sugar water, vinegar, cloves, cinnamon candies, or cinnamon sticks and food coloring in a 6-qt saucepan. Stir, heat to boil, and simmer 3 minutes.

3. Drain apples, add to hot syrup, and cook 5 minutes. Fill jars (preferably wide-mouth) with apple rings and hot flavored syrup, leaving 1/2-inch headspace. Adjust lids and process according to the chart below.

PROCESS TIME FOR SPICED APPLE RINGS IN A BOILING-WATER CANNER.

Style of Pack	Jar Size	Process Time at Altitudes of		
		0–1,000 ft	1,001–6,000 ft	Above 6,000 ft
Hot	Half-pints or Pints	10 minutes	15 minutes	20 minutes

Table 1. Recommended process time for **Spiced Apple Rings** in a boiling-water canner.

Apricots, Halved or Sliced

Apricots are excellent in baked goods, stuffing, chutney, or on their own. Choose firm, well-colored mature fruit for best results.

Directions

1. Dip fruit in boiling water for 30 to 60 seconds until skins loosen. Dip quickly in cold water and slip off skins.
2. Cut in half, remove pits, and slice if desired. To prevent darkening, keep peeled fruit in water with a little lemon juice.
3. Prepare and boil a very light, light, or medium syrup (see page 181) or pack apricots in water, apple juice, or white grape juice.

QUANTITY

- An average of 16 pounds is needed per canner load of seven quarts.
- An average of 10 pounds is needed per canner load of nine pints.
- A bushel weighs 50 pounds and yields 20 to 25 quarts—an average of 2¼ pounds per quart.

Berries, Whole

Preserved berries are perfect for use in pies, muffins, pancakes, or in poultry or pork dressings. Nearly every berry preserves well, including blackberries, blueberries, currants, dewberries, elderberries, gooseberries, huckleberries, loganberries, mulberries, and raspberries. Choose ripe, sweet berries with uniform color.

Directions

1. Wash 1 or 2 quarts of berries at a time. Drain, cap, and stem if necessary. For gooseberries, snip off heads and tails with scissors.
2. Prepare and boil preferred syrup, if desired (see page 181). Add ½ cup syrup, juice, or water to each clean jar.

PROCESS TIMES FOR HALVED OR SLICED APRICOTS IN A DIAL-GAUGE PRESSURE CANNER*

Style of Pack	Jar Size	Process Time	Canner Pressure (PSI) at Altitudes of:			
			0–2,000 ft	2,001–4,000 ft	4,001–6,000 ft	6,001–8,000 ft
Hot or Raw	Pints or Quarts	ten minutes	6 lbs	7 lbs	8 lbs	9 lbs
*After the process is complete, turn off the heat and remove the canner lid. Wait five minutes before removing jars.						

PROCESS TIMES FOR HALVED OR SLICED APRICOTS IN A WEIGHTED-GAUGE PRESSURE CANNER*

Style of Pack	Jar Size	Process Time	Canner Pressure (PSI) at Altitudes of:	
			0–1,000 ft	Above 1,000 ft
Hot or Raw	Pints or Quarts	ten minutes	5 lbs	10 lbs
*After the process is complete, turn off the heat and remove the canner lid. Wait five minutes before removing jars.				

RECOMMENDED PROCESS TIMES FOR WHOLE BERRIES IN A BOILING-WATER CANNER*

Style of Pack	Jar Size	Process Time at Altitudes of:			
		0–1,000 ft	1,001–3,000 ft	3,001–6,000 ft	Above 6,000 ft
Hot	Pints or Quarts	15 minutes	20 minutes	20 minutes	25 minutes
Raw	Pints	15 minutes	20 minutes	20 minutes	25 minutes
Raw	Quarts	20 minutes	25 minutes	30 minutes	35 minutes

*After the process is complete, turn off the heat and remove the canner lid. Wait five minutes before removing jars.

QUANTITY

- An average of 12 pounds is needed per canner load of seven quarts.
- An average of 8 pounds is needed per canner load of nine pints.
- A 24-quart crate weighs 36 pounds and yields 18 to 24 quarts—an average of 1¾ pounds per quart.

PROCESS TIMES FOR WHOLE BERRIES IN A DIAL-GAUGE PRESSURE CANNER*

Style of Pack	Jar Size	Process Time	Canner Pressure (PSI) at Altitudes of:			
			0–2,000 ft	2,001–4,000 ft	4,001–6,000 ft	6,001–8,000 ft
Hot	Pints or Quarts	8 minutes	6 lbs	7 lbs	8 lbs	9 lbs
Raw	Pints	8 minutes	6 lbs	7 lbs	8 lbs	9 lbs
Raw	Quarts	ten minutes	6 lbs	7 lbs	8 lbs	9 lbs

*After the process is complete, turn off the heat and remove the canner lid. Wait five minutes before removing jars.

PROCESS TIMES FOR WHOLE BERRIES IN A WEIGHTED-GAUGE PRESSURE CANNER*

Style of Pack	Jar Size	Process Time	Canner Pressure (PSI) at Altitudes of:	
			0–1,000 ft	Above 1,000 ft
Hot	Pints or Quarts	8 minutes	5 lbs	10 lbs
Raw	Pints	8 minutes	5 lbs	10 lbs
Raw	Quarts	ten minutes	5 lbs	10 lbs

*After the process is complete, turn off the heat and remove the canner lid. Wait five minutes before removing jars.

PROCESS TIMES FOR BERRY SYRUP IN A BOILING-WATER CANNER*

Style of Pack	Jar Size	Process Time at Altitudes of:		
		0–1,000 ft	1,001–6,000 ft	Above 6,000 ft
Hot	Half-pints or Pints	ten minutes	15 minutes	20 minutes
*After the process is complete, turn off the heat and remove the canner lid. Wait five minutes before removing jars.				

To make syrup with whole berries, rather than crushed, save 1 or 2 cups of the fresh or frozen fruit, combine these with the sugar, and simmer until soft. Remove from heat, skim off foam, and fill into clean jars, following processing directions for regular berry syrup.

Hot pack—(Best for blueberries, currants, elderberries, gooseberries, and huckleberries) Heat berries in boiling water for thirty seconds and drain. Fill jars and cover with hot juice, leaving ½-inch headspace.

Raw pack—Fill jars with any of the raw berries, shaking down gently while filling. Cover with hot syrup, juice, or water, leaving ½-inch headspace.

Berry Syrup

Juices from fresh or frozen blueberries, cherries, grapes, raspberries (black or red), and strawberries are easily made into toppings for use on ice cream and pastries. For an elegant finish to cheesecakes or pound cakes, drizzle a thin stream in a zigzag across the top just before serving. Berry syrups are also great additions to smoothies or milkshakes. This recipe makes about nine half-pints.

Directions

1. Select 6½ cups of fresh or frozen berries of your choice. Wash, cap, and stem berries and crush in a saucepan.
2. Heat to boiling and simmer until soft (5 to ten minutes). Strain hot through a colander placed in a large pan and drain until cool enough to handle.
3. Strain the collected juice through a double layer of cheesecloth or jelly bag. Discard the dry pulp. The yield of the pressed juice should be about 4½ to 5 cups.
4. Combine the juice with 6¾ cups of sugar in a large saucepan, bring to a boil, and simmer 1 minute.
5. Fill into clean half-pint or pint jars, leaving ½-inch headspace. Adjust lids and process.

Fruit Purées

Almost any fruit can be puréed for use as baby food, in sauces, or just as a nutritious snack. Puréed prunes and apples can be used as a butter replacement in many baked goods. Use this recipe for any fruit except figs and tomatoes.

Directions

1. Stem, wash, drain, peel, and remove pits if necessary. Measure fruit into large saucepan, crushing slightly if desired.
2. Add 1 cup hot water for each quart of fruit. Cook slowly until fruit is soft, stirring frequently. Press through sieve or food mill. If desired, add sugar to taste.
3. Reheat pulp to boil, or until sugar dissolves (if added). Fill hot into clean jars, leaving ¼-inch headspace. Adjust lids and process.

PROCESS TIMES FOR FRUIT PURÉES IN A BOILING-WATER CANNER*

Style of Pack	Jar Size	Process Time at Altitudes of:		
		0–1,000 ft	1,001–6,000 ft	Above 6,000 ft
Hot	Pints or Quarts	15 minutes	20 minutes	25 minutes

*After the process is complete, turn off the heat and remove the canner lid. Wait five minutes before removing jars.

PROCESS TIMES FOR FRUIT PURÉES IN A DIAL-GAUGE PRESSURE CANNER*

Style of Pack	Jar Size	Process Time	Canner Pressure (PSI) at Altitudes of:			
			0–2,000 ft	2,001–4,000 ft	4,001–6,000 ft	6,001–8,000 ft
Hot	Pints or Quarts	8 minutes	6 lbs	7 lbs	8 lbs	9 lbs

*After the canner is completely depressurized, remove the weight from the vent port or open the petcock. Wait ten minutes; then unfasten the lid and remove it carefully. Lift the lid with the underside away from you so that the steam coming out of the canner does not burn your face.

PROCESS TIMES FOR FRUIT PURÉES IN A WEIGHTED-GAUGE PRESSURE CANNER*

Style of Pack	Jar Size	Process Time (Min)	Canner Pressure (PSI) at Altitudes of:	
			0–1,000 ft	Above 1,000 ft
Hot	Pints or Quarts	8 minutes	5 lbs	10 lbs

*After the canner is completely depressurized, remove the weight from the vent port or open the petcock. Wait ten minutes; then unfasten the lid and remove it carefully. Lift the lid with the underside away from you so that the steam coming out of the canner does not burn your face.

QUANTITY

- An average of 24½ pounds is needed per canner load of seven quarts.
- An average of 16 pounds per canner load of nine pints.
- A lug weighs 26 pounds and yields seven to nine quarts of juice—an average of 3½ pounds per quart.

Grape Juice

Purple grapes are full of antioxidants and help to reduce the risk of heart disease, cancer, and Alzheimer's disease. For juice, select sweet, well-colored, firm, mature fruit.

Directions

1. Wash and stem grapes. Place grapes in a saucepan and add boiling water to cover. Heat and simmer slowly until skin is soft.

2. Strain through a damp jelly bag or double layers of cheesecloth, and discard solids. Refrigerate juice for 24 to 48 hours.

3. Without mixing, carefully pour off clear liquid and save; discard sediment. If desired, strain through a paper coffee filter for a clearer juice.

4. Add juice to a saucepan and sweeten to taste. Heat and stir until sugar is dissolved. Continue heating with occasional stirring until juice begins to boil. Fill into jars immediately, leaving ¼-inch headspace. Adjust lids and process.

Peaches, Halved or Sliced

Peaches are delicious in cobblers, crisps, and muffins, or grilled for a unique cake topping. Choose ripe, mature fruit with minimal bruising.

Directions

1. Dip fruit in boiling water for 30 to 60 seconds until skins loosen. Dip quickly in cold water and slip off skins. Cut in half, remove pits, and slice if desired. To prevent darkening, keep peeled fruit in ascorbic acid solution.

2. Prepare and boil a very light, light, or medium syrup or pack peaches in water, apple juice, or white grape juice. Raw packs make poor quality peaches.

Hot pack—In a large saucepan, place drained fruit in syrup, water, or juice and bring to boil. Fill jars with hot fruit and cooking liquid, leaving ½-inch headspace. Place halves in layers, cut side down.

Raw pack—Fill jars with raw fruit, cut side down, and add hot water, juice, or syrup, leaving ½-inch headspace.

3. Adjust lids and process.

PROCESS TIMES FOR GRAPE JUICE IN A BOILING-WATER CANNER*

Style of Pack	Jar Size	Process Time at Altitudes of:		
		0–1,000 ft	1,001–6,000 ft	Above 6,000 ft
Hot	Pints or Quarts	5 minutes	ten minutes	15 minutes
	Half-gallons	ten minutes	15 minutes	20 minutes

*After the process is complete, turn off the heat and remove the canner lid. Wait five minutes before removing jars.

PROCESS TIMES FOR HALVED OR SLICED PEACHES IN A BOILING-WATER CANNER*

Style of Pack	Jar Size	Process Time at Altitudes of:			
		0–1,000 ft	1,001–3,000 ft	3,001–6,000 ft	Above 6,000 ft
Hot	Pints	20 minutes	25 minutes	30 minutes	35 minutes
	Quarts	25 minutes	30 minutes	35 minutes	40 minutes
Raw	Pints	25 minutes	30 minutes	35 minutes	40 minutes
	Quarts	30 minutes	35 minutes	40 minutes	45 minutes

*After the process is complete, turn off the heat and remove the canner lid. Wait five minutes before removing jars.

PROCESS TIMES FOR HALVED OR SLICED PEACHES IN A DIAL-GAUGE PRESSURE CANNER*

Style of Pack	Jar Size	Process Time	Canner Pressure (PSI) at Altitudes of:			
			0–2,000 ft	2,001–4,000 ft	4,001–6,000 ft	6,001–8,000 ft
Hot or Raw	Pints or Quarts	ten minutes	6 lbs	7 lbs	8 lbs	9 lbs

*After the canner is completely depressurized, remove the weight from the vent port or open the petcock. Wait ten minutes; then unfasten the lid and remove it carefully. Lift the lid with the underside away from you so that the steam coming out of the canner does not burn your face.

PROCESS TIMES FOR HALVED OR SLICED PEACHES IN A WEIGHTED-GAUGE PRESSURE CANNER*

Style of Pack	Jar Size	Process Time	Canner Pressure (PSI) at Altitudes of:	
			0–1,000 ft	Above 1,000 ft
Hot or Raw	Pints or Quarts	ten minutes	5 lbs	10 lbs

*After the canner is completely depressurized, remove the weight from the vent port or open the petcock. Wait ten minutes; then unfasten the lid and remove it carefully. Lift the lid with the underside away from you so that the steam coming out of the canner does not burn your face.

Canning

Pears, Halved

Choose ripe, mature fruit for best results. For a special treat, filled halved pears with a mixture of chopped dried apricots, pecans, brown sugar, and butter; bake or microwave until warm and serve with vanilla ice cream.

Directions

1. Wash and peel pears. Cut lengthwise in halves and remove core. A melon baller or metal measuring spoon works well for coring pears. To prevent discoloration, keep pears in water with a little lemon juice.
2. Prepare a very light, light, or medium syrup (see page 181) or use apple juice, white grape juice, or water. Raw packs make poor quality pears. Boil drained pears 5 minutes in syrup, juice, or water. Fill jars with hot fruit and cooking liquid, leaving ½-inch headspace. Adjust lids and process.

QUANTITY

- An average of 17½ pounds is needed per canner load of seven quarts.
- An average of 11 pounds is needed per canner load of nine pints.
- A bushel weighs 50 pounds and yields 16 to 25 quarts—an average of 2½ pounds per quart.

PROCESS TIMES FOR HALVED PEARS IN A BOILING-WATER CANNER*

Style of Pack	Jar Size	Process Time at Altitudes of:			
		0–1,000 ft	1,001–3,000 ft	3,001–6,000 ft	Above 6,000 ft
Hot	Pints	20 minutes	25 minutes	30 minutes	35 minutes
	Quarts	25 minutes	30 minutes	35 minutes	40 minutes
*After the process is complete, turn off the heat and remove the canner lid. Wait five minutes before removing jars.					

PROCESS TIMES FOR HALVED PEARS IN A DIAL-GAUGE PRESSURE CANNER*

Style of Pack	Jar Size	Process Time	Canner Pressure (PSI) at Altitudes of:			
			0–2,000 ft	2,001–4,000 ft	4,001–6,000 ft	6,001–8,000 ft
Hot	Pints or Quarts	ten minutes	6 lbs	7 lbs	8 lbs	9 lbs
*After the canner is completely depressurized, remove the weight from the vent port or open the petcock. Wait ten minutes; then unfasten the lid and remove it carefully. Lift the lid with the underside away from you so that the steam coming out of the canner does not burn your face.						

PROCESS TIMES FOR HALVED PEARS IN A WEIGHTED-GAUGE PRESSURE CANNER*

Style of Pack	Jar Size	Process Time	Canner Pressure (PSI) at Altitudes of:	
			0–1,000 ft	Above 1,000 ft
Hot	Pints or Quarts	ten minutes	5 lbs	10 lbs
*After the canner is completely depressurized, remove the weight from the vent port or open the petcock. Wait ten minutes; then unfasten the lid and remove it carefully. Lift the lid with the underside away from you so that the steam coming out of the canner does not burn your face.				

Rhubarb, Stewed

Rhubarb in the garden is a sure sign that spring has sprung and summer is well on its way. But why not enjoy rhubarb all year round? The brilliant red stalks make it as appropriate for a holiday table as for an early summer feast. Rhubarb is also delicious in crisps, cobblers, or served hot over ice cream. Select young, tender, well-colored stalks from the spring or, if available, late fall crop.

Directions

1. Trim off leaves. Wash stalks and cut into ½-inch to 1-inch pieces.
2. Place rhubarb in a large saucepan, and add ½ cup sugar for each quart of fruit. Let stand until juice appears. Heat gently to boiling. Fill jars without delay, leaving ½-inch headspace. Adjust lids and process.

QUANTITY

- An average of 10½ pounds is needed per canner load of seven quarts.
- An average of 7 pounds is needed per canner load of nine pints.
- A lug weighs 28 pounds and yields 14 to 28 quarts—an average of 1½ pounds per quart.

PROCESS TIMES FOR STEWED RHUBARB IN A BOILING-WATER CANNER*

Style of Pack	Jar Size	Process Time at Altitudes of:		
		0–1,000 ft	1,001–6,000 ft	Above 6,000 ft
Hot	Pints or Quarts	15 minutes	20 minutes	25 minutes

*After the process is complete, turn off the heat and remove the canner lid. Wait five minutes before removing jars.

PROCESS TIMES FOR STEWED RHUBARB IN A DIAL-GAUGE PRESSURE CANNER*

Style of Pack	Jar Size	Process Time	Canner Pressure (PSI) at Altitudes of			
			0–2,000 ft	2,001–4,000 ft	4,001–6,000 ft	6,001–8,000 ft
Hot	Pints or Quarts	8 minutes	6 lbs	7 lbs	8 lbs	9 lbs

*After the canner is completely depressurized, remove the weight from the vent port or open the petcock. Wait ten minutes; then unfasten the lid and remove it carefully. Lift the lid with the underside away from you so that the steam coming out of the canner does not burn your face.

PROCESS TIMES FOR STEWED RHUBARB IN A WEIGHTED-GAUGE PRESSURE CANNER*

Style of Pack	Jar Size	Process Time	Canner Pressure (PSI) at Altitudes of:	
			0–1,000 ft	Above 1,000 ft
Hot	Pints or Quarts	8 minutes	5 lbs	10 lbs

*After the canner is completely depressurized, remove the weight from the vent port or open the petcock. Wait ten minutes; then unfasten the lid and remove it carefully. Lift the lid with the underside away from you so that the steam coming out of the canner does not burn your face.

Canned Pie Fillings

Using a pre-made pie filling will cut your pie preparation time by more than half, but most commercially produced fillings are oozing with high fructose corn syrup and all manner of artificial coloring and flavoring. (Food coloring is not at all necessary, but if you're really concerned about how the inside of your pie will look, appropriate amounts are added to each recipe as an optional ingredient.) Making and preserving your own pie fillings means that you can use your own fresh ingredients and adjust the sweetness to your taste. Because some folks like their pies rich and sweet and others prefer a natural tart flavor, you might want to first make a single quart, make a pie with it, and see how you like it. Then you can adjust the sugar and spices in the recipe to suit your personal preferences before making a large batch. Experiment with combining fruits or adding different spices, but the amount of lemon juice should not be altered, as it aids in controlling the safety and storage stability of the fillings.

These recipes use Clear Jel® (sometimes sold as Clear Jel A®), a chemically modified cornstarch that produces excellent sauce consistency even after fillings are canned and baked. By using Clear Jel® you can lower the sugar content of your fillings without sacrificing safety, flavor, or texture. (Note: Instant Clear Jel® is not meant to be cooked and should not be used for these recipes. Sure-Gel® is a natural fruit pectin and is not a suitable substitute for Clear Jel®. Cornstarch, tapioca starch, or arrowroot starch can be used in place of Clear Jel®, but the finished

Ingredients

	1 Quart	7 Quarts
Blanched, sliced fresh apples	3½ cups	6 quarts
Granulated sugar	¾ cup + 2 tbsp	5 ½ cups
Clear Jel®	¼ cup	1 ½ cup
Cinnamon	½ tsp	1 tbsp
Cold water	½ cup	2 ½ cups
Apple juice	¾ cup	5 cups
Bottled lemon juice	2 tbsp	¾ cup
Nutmeg (optional)	⅛ tsp	1 tsp

product is likely to be runny.) One pound of Clear Jel® costs less than five dollars and is enough to make fillings for about fourteen pies. It will keep for at least a year if stored in a cool, dry place. Clear Jel® is increasingly available among canning and freezing supplies in some stores. Alternately, you can order it by the pound at any of the following online stores:

- www.barryfarm.com
- www.kitchenkrafts.com
- www.theingredientstore.com

Apple Pie Filling

Use firm, crisp apples, such as Stayman, Golden Delicious, or Rome varieties for the best results. If apples lack tartness, use an additional ¼ cup of lemon juice for each six quarts of slices. Ingredients are included for a one-quart (enough for one 8-inch pie) or a seven-quart recipe.

Directions

1. Wash, peel, and core apples. Prepare slices ½ inch wide and place in water containing a little lemon juice to prevent browning.

2. For fresh fruit, place 6 cups at a time in 1 gallon of boiling water. Boil each batch 1 minute after the water returns to a boil. Drain, but keep heated fruit in a covered bowl or pot.

When using frozen cherries and blueberries, select unsweetened fruit. If sugar has been added, rinse it off while fruit is frozen. Thaw fruit, then collect, measure, and use juice from fruit to partially replace the water specified in the recipe.

Directions

1. Wash and drain blueberries. Place 6 cups at a time in 1 gallon boiling water. Allow water to return to a boil and cook each batch for 1 minute. Drain but keep heated fruit in a covered bowl or pot.
2. Combine sugar and Clear Jel® in a large kettle. Stir. Add water and food coloring if desired. Cook on medium-high heat until mixture thickens and begins to bubble.
3. Add lemon juice and boil 1 minute, stirring constantly. Fold in drained berries immediately and fill jars with mixture without delay, leaving 1-inch headspace. Adjust lids and process immediately.

Ingredients

	1 Quart	7 Quarts
Fresh or thawed blueberries	3½ cups	6 quarts
Granulated sugar	¾ cup + 2 tbsp	6 cups
Clear Jel®	¼ cup + 1 tbsp	2¼ cups
Cold water	1 cup	7 cups
Bottled lemon juice	3½ cups	½ cup
Blue food coloring (optional)	3 drops	20 drops
Red food coloring (optional)	1 drop	7 drops

PROCESS TIMES FOR APPLE PIE FILLING IN A BOILING-WATER CANNER*

		Process Time at Altitudes of:			
Style of Pack	Jar Size	0–1,000 ft	1,001–3,000 ft	3,001–6,000 ft	Above 6,000 ft
Hot	Pints or Quarts	25 minutes	30 minutes	35 minutes	40 minutes

*After the process is complete, turn off the heat and remove the canner lid. Wait five minutes before removing jars.

3. Combine sugar, Clear Jel®, and cinnamon in a large kettle with water and apple juice. Add nutmeg, if desired. Stir and cook on medium-high heat until mixture thickens and begins to bubble.
4. Add lemon juice and boil 1 minute, stirring constantly. Fold in drained apple slices immediately and fill jars with mixture without delay, leaving 1-inch headspace. Adjust lids and process immediately.

Blueberry Pie Filling

Select fresh, ripe, and firm blueberries. Unsweetened frozen blueberries may be used. If sugar has been added, rinse it off while fruit is still frozen. Thaw fruit, then collect, measure, and use juice from fruit to partially replace the water specified in the recipe. Ingredients are included for a one-quart (enough for one 8-inch pie) or seven-quart recipe.

PROCESS TIMES FOR BLUEBERRY PIE FILLING IN A BOILING-WATER CANNER*

Style of Pack	Jar Size	Process Time at Altitudes of:			
		0–1,000 ft	1,001–3,000 ft	3,001–6,000 ft	Above 6,000 ft
Hot	Pints or Quarts	30 minutes	35 minutes	40 minutes	45 minutes

*After the process is complete, turn off the heat and remove the canner lid. Wait five minutes before removing jars.

Cherry Pie Filling

Select fresh, very ripe, and firm cherries. Unsweetened frozen cherries may be used. If sugar has been added, rinse it off while the fruit is still frozen. Thaw fruit, then collect, measure, and use juice from fruit to partially replace the water specified in the recipe. Ingredients are included for a one-quart (enough for one 8-inch pie) or seven-quart recipe.

Directions

1. Rinse and pit fresh cherries, and hold in cold water. To prevent stem end browning, use water with a little lemon juice. Place 6 cups at a time in 1 gallon boiling water. Boil each batch 1 minute after the water returns to a boil. Drain but keep heated fruit in a covered bowl or pot.

Ingredients

	1 Quart	7 Quarts
Fresh or thawed sour cherries	3⅓ cups	6 quarts
Granulated sugar	1 cup	7 cups
Clear Jel®	¼ cup + 1 tbsp	1¾ cups
Cold water	1⅓ cups	9⅓ cups
Bottled lemon juice	1 tbsp + 1 tsp	½ cup
Cinnamon (optional)	⅛ tsp	1 tsp
Almond extract (optional)	¼ tsp	2 tsps
Red food coloring (optional)	6 drops	¼ tsp

PROCESS TIMES FOR CHERRY PIE FILLING IN A BOILING-WATER CANNER*

Style of Pack	Jar Size	Process Time at Altitudes of:			
		0–1,000 ft	1,001–3,000 ft	3,001–6,000 ft	Above 6,000 ft
Hot	Pints or Quarts	30 minutes	35 minutes	40 minutes	45 minutes

*After the process is complete, turn off the heat and remove the canner lid. Wait five minutes before removing jars.

2. Combine sugar and Clear Jel® in a large saucepan and add water. If desired, add cinnamon, almond extract, and food coloring. Stir mixture and cook over medium-high heat until mixture thickens and begins to bubble.

3. Add lemon juice and boil 1 minute, stirring constantly. Fold in drained cherries immediately and fill jars with mixture without delay, leaving 1-inch headspace. Adjust lids and process immediately.

Festive Mincemeat Pie Filling

Mincemeat pie originated as "Christmas Pie" in the eleventh century, when the English crusaders returned from the Holy Land bearing oriental spices. They added three of these spices—cinnamon, cloves, and nutmeg—to their meat pies to represent the three gifts that the magi brought to the Christ child. Mincemeat pies are traditionally small and are perfect paired with a mug of hot buttered rum. Walnuts or pecans can be used in place of meat if preferred. This recipe yields about seven quarts.

Ingredients

- 2 cups finely chopped suet
- 4 lbs. ground beef or 4 lbs. ground venison and 1 lb. sausage
- 5 qts. chopped apples
- 2 lbs. dark seedless raisins
- 1 lb. white raisins
- 2 qts. apple cider
- 2 tbsps ground cinnamon
- 2 tsps ground nutmeg
- ½ tsp cloves
- 5 cups sugar
- 2 tbsps salt

Directions

1. Cook suet and meat in water to avoid browning. Peel, core, and quarter apples. Put suet, meat, and apples

through food grinder using a medium blade.

2. Combine all ingredients in a large saucepan, and simmer 1 hour or until slightly thickened. Stir often.

3. Fill jars with mixture without delay, leaving 1-inch headspace. Adjust lids and process.

PROCESS TIMES FOR FESTIVE MINCEMEAT PIE FILLING IN A DIAL-GAUGE PRESSURE CANNER*

Style of Pack	Jar Size	Process Time	Canner Pressure (PSI) at Altitudes of:			
			0–2,000 ft	2,001–4,000 ft	4,001–6,000 ft	6,000–8,000 ft
Hot	Quarts	90 minutes	11 lbs	12 lbs	13 lbs	14 lbs

*After the canner is completely depressurized, remove the weight from the vent port or open the petcock. Wait ten minutes; then unfasten the lid and remove it carefully. Lift the lid with the underside away from you so that the steam coming out of the canner does not burn your face.

PROCESS TIMES FOR FESTIVE MINCEMEAT PIE FILLING IN A WEIGHTED-GAUGE PRESSURE CANNER*

Style of Pack	Jar Size	Process Time	Canner Pressure (PSI) at Altitudes of:	
			0–1,000 ft	Above 1,000 ft
Hot	Quarts	90 minutes	10 lbs	15 lbs

*After the canner is completely depressurized, remove the weight from the vent port or open the petcock. Wait ten minutes; then unfasten the lid and remove it carefully. Lift the lid with the underside away from you so that the steam coming out of the canner does not burn your face.

Jams, Jellies, and Other Fruit Spreads

Homemade jams and jellies have lots more flavor than store-bought, over-processed varieties. The combinations of fruits and spices are limitless, so have fun experimenting with these recipes. If you can bear to part with your creations when you're all done, they make wonderful gifts for any occasion.

Pectin is what makes jams and jellies thicken and gel. Many fruits, such as crab apples, citrus fruits, sour plums, currants, quinces, green apples, or Concord grapes, have plenty of their own natural pectin, so there's no need to add more pectin to your recipes. You can use less sugar when you don't add pectin, but you will have to boil the fruit for longer. Still, the process is relatively simple and you don't have to worry about having store-bought pectin on hand.

To use fresh fruits with a low pectin content or canned or frozen fruit juice, powdered or liquid pectin must be added for your jams and jellies to thicken and set properly. Jelly or jam made with added pectin requires less cooking and generally gives a larger yield. These products have more natural fruit flavors, too. In addition, using added pectin eliminates the need to test hot jellies and jams for proper gelling.

Beginning this section are descriptions of the differences between methods and tips for success with whichever you use.

Making Jams and Jellies without Added Pectin

Jelly without Added Pectin

Making jelly without added pectin is not an exact science. You can add a little more or less sugar according to your taste, substitute honey for up to ½ of the sugar, or experiment with combining small amounts of low-pectin fruits with other high-pectin fruits. The Ingredients table below shows you the basics for common high-pectin fruits. Use it as a guideline as you experiment with other fruits.

As fruit ripens, its pectin content decreases, so use fruit that has recently been picked, and mix ¾ ripe fruit with ¼ under-ripe. Cooking cores and peels along with the fruit will also increase the pectin level. Avoid using canned or frozen fruit as they contain very little pectin. Be sure to wash all fruit thoroughly before cooking. One pound of fruit should yield at least 1 cup of clear juice.

Directions

1. Crush soft fruits or berries; cut firmer fruits into small pieces (there is no need to peel or core the fruits, as cooking all the parts adds pectin).

2. Add water to fruits that require it, as listed in the Ingredients table above. Put fruit and water in large saucepan and bring to a boil. Then simmer according to the times below until fruit is soft, while stirring to prevent scorching.

3. When fruit is tender, strain through a colander, then strain through a double layer of cheesecloth or a jelly bag. Allow juice to drip through, using a stand or colander to hold the bag. Avoid pressing or squeezing the bag or cloth as it will cause cloudy jelly.

4. Using no more than 6 to 8 cups of extracted fruit juice at a time, measure fruit juice, sugar, and lemon juice according to the Ingredients table, and heat to boiling.

5. Stir until the sugar is dissolved. Boil over high heat to the jellying point. To test jelly for doneness, follow these steps:

6. Remove from heat and quickly skim off foam. Fill sterile jars with jelly. Use a measuring cup or ladle the jelly through a wide-mouthed funnel, leaving ¼-inch headspace. Adjust lids and process.

Ingredients

Fruit	Water to be Added per Pound of Fruit	Minutes to Simmer Fruit before Extracting Juice	Ingredients Added to Each Cup of Strained Juice		Yield from 4 Cups of Juice (Half-pints)
			Sugar (Cups)	Lemon Juice (Tsp)	
Apples	1 cup	20 to 25	¾	1½ (opt)	4 to 5
Blackberries	None or ¼ cup	5 to 10	¾ to 1	None	7 to 8
Crab apples	1 cup	20 to 25	1	None	4 to 5
Grapes	None or ¼ cup	5 to 10	¾ to 1	None	8 to 9
Plums	½ cup	15 to 20	¾	None	8 to 9

Temperature test—Use a jelly or candy thermometer and boil until mixture reaches the following temperatures:

Sea Level	1,000 ft	2,000 ft	3,000ft	4,000 ft	5,000ft	6,000ft	7,000 ft	8,000 ft
220°F	218°F	216°F	214°F	212°F	211°F	209°F	207°F	205°F

Sheet or spoon test—Dip a cool metal spoon into the boiling jelly mixture. Raise the spoon about 12 inches above the pan (out of steam). Turn the spoon so the liquid runs off the side. The jelly is done when the syrup forms two drops that flow together and sheet or hang off the edge of the spoon.

PROCESS TIMES FOR JELLY WITHOUT ADDED PECTIN IN A BOILING WATER CANNER*

Style of Pack	Jar Size	Process Time at Altitudes of:		
		0–1,000 ft	1,001–6,000 ft	Above 6,000 ft
Hot	Half-pints or pints	5 minutes	ten minutes	15 minutes

*After the process is complete, turn off the heat and remove the canner lid. Wait five minutes before removing jars.

Preventing spoilage

Even though sugar helps preserve jellies and jams, molds can grow on the surface of these products. Research now indicates that the mold which people usually scrape off the surface of jellies may not be as harmless as it seems. Mycotoxins have been found in some jars of jelly having surface mold growth. Mycotoxins are known to cause cancer in animals; their effects on humans are still being researched. Because of possible mold contamination, paraffin or wax seals are no longer recommended for any sweet spread, including jellies. To prevent growth of molds and loss of good flavor or color, fill products hot into sterile Mason jars, leaving ¼-inch headspace, seal with self-sealing lids, and process five minutes in a boiling-water canner. Correct process time at higher elevations by adding one additional minute per 1,000 feet above sea level. If unsterile jars are used, the filled jars should be processed ten minutes. Use of sterile jars is preferred, especially when fruits are low in pectin, since the added five-minute process time may cause weak gels.

Lemon Curd

Lemon curd is a rich, creamy spread that can be used on (or in) a variety of teatime treats—crumpets, scones, cake fillings, tartlets, or meringues are all enhanced by its tangy-sweet flavor. Follow the recipe carefully, as variances in ingredients, order, and temperatures may lead to a poor texture or flavor. For Lime Curd, use the same recipe but substitute 1 cup bottled lime juice and ¼ cup fresh lime zest for the lemon juice and zest. This recipe yields about three to four half-pints.

Ingredients

- 2½ cups superfine sugar*
- ½ cup lemon zest (freshly zested), optional
- 1 cup bottled lemon juice**
- ¾ cup unsalted butter, chilled, cut into approximately ¾-inch pieces
- 7 large egg yolks
- 4 large whole eggs

Directions

1. Wash 4 half-pint canning jars with warm, soapy water. Rinse well; keep hot until ready to fill. Prepare canning lids according to manufacturer's directions.

2. Fill boiling water canner with enough water to cover the filled jars by 1 to 2 inches. Use a thermometer to preheat the water to 180°F by the time filled jars are ready to be added. **Caution:** Do not heat the water in the canner to more than 180°F before jars are added. If the water in the canner is too hot when jars are added, the process time will not be long enough. The time it takes for the canner to reach boiling after the jars are added is expected to be 25 to 30 minutes for this product. Process time starts after the water in the canner comes to a full boil over the tops of the jars.

3. Combine the sugar and lemon zest in a small bowl, stir to mix, and set aside about 30 minutes. Pre-measure the lemon juice and prepare the chilled butter pieces.

4. Heat water in the bottom pan of a double boiler until it boils gently. The water should not boil vigorously or touch the bottom of the top double boiler pan or bowl in which the curd is to be cooked. Steam produced will be sufficient for the cooking process to occur.

5. In the top of the double boiler, on the counter top or table, whisk the egg yolks and whole eggs together until thoroughly mixed. Slowly whisk in the sugar and zest, blending until well mixed and smooth. Blend in the lemon juice and then add the butter pieces to the mixture.

6. Place the top of the double boiler over boiling water in the bottom pan. Stir gently but continuously with a silicone spatula or cooking spoon, to prevent the mixture from sticking to the bottom of the pan. Continue cooking until the mixture reaches a temperature of 170°F. Use a food thermometer to monitor the temperature.

7. Remove the double boiler pan from the stove and place on a protected surface, such as a dishcloth or towel on the counter top. Continue to stir gently until the curd thickens (about 5 minutes). Strain curd through a mesh strainer into a glass or stainless steel bowl; discard collected zest.

8. Fill hot strained curd into the clean, hot half-pint jars, leaving ½-inch headspace. Remove air bubbles and adjust headspace if needed. Wipe rims of jars with a dampened, clean paper towel; apply two-piece metal canning lids. Process. Let cool, undisturbed, for twelve to twenty-four hours and check for seals.

---

* If superfine sugar is not available, run granulated sugar through a grinder or food processor for 1 minute, let settle, and use in place of superfine sugar. Do not use powdered sugar.

** Bottled lemon juice is used to standardize acidity. Fresh lemon juice can vary in acidity and is not recommended.

*** If a double boiler is not available, a substitute can be made with a large bowl or saucepan that can fit partway down into a saucepan of a smaller diameter. If the bottom pan has a larger diameter, the top bowl or pan should have a handle or handles that can rest on the rim of the lower pan.

PROCESS TIMES FOR LEMON CURD IN A BOILING-WATER CANNER*

Style of Pack	Jar Size	Process Time at Altitudes of:		
		0–1,000 ft	1,001–6,000 ft	Above 6,000 ft
Hot	Half-pints	15 minutes	20 minutes	25 minutes

*After the process is complete, turn off the heat and remove the canner lid. Wait five minutes before removing jars.

Jam without Added Pectin

Making jam is even easier than making jelly, as you don't have to strain the fruit. However, you'll want to be sure to remove all stems, skins, and pits. Be sure to wash and rinse all fruits thoroughly before cooking, but don't let them soak. For best flavor, use fully ripe fruit. Use the ingredients table below as a guideline as you experiment with less common fruits.

1. Remove stems, skins, seeds, and pits; cut into pieces and crush. For berries, remove stems and blossoms and crush. Seedy berries may be put through a sieve or food mill. Measure crushed fruit into large saucepan using the ingredient quantities specified above.

2. Add sugar and bring to a boil while stirring rapidly and constantly. Continue to boil until mixture thickens. Use one of the following tests to determine when jams and jellies are ready to fill. Remember that the jam will thicken as it cools.

3. Remove from heat and skim off foam quickly. Fill sterile jars with jam. Use a measuring cup or ladle the jam through a wide-mouthed funnel, leaving ¼-inch headspace. Adjust lids and process.

Ingredients

Fruit	Quantity (Crushed)	Sugar	Lemon Juice	Yield (Half-pints)
Apricots	4 to 4½ cups	4 cups	2 tbsps	5 to 6
Berries*	4 cups	4 cups	None	3 to 4
Peaches	5½ to 6 cups	4 to 5 cups	2 tbsps	6 to 7

* Includes blackberries, boysenberries, dewberries, gooseberries, loganberries, raspberries, and strawberries.

Temperature test—Use a jelly or candy thermometer and boil until mixture reaches the temperature for your altitude.

Sea Level	1,000 ft	2,000 ft	3,000 ft	4,000 ft	5,000 ft	6,000 ft	7,000 ft	8,000 ft
220°F	218°F	216°F	214°F	212°F	211°F	209°F	207°F	205°F

Refrigerator test—Remove the jam mixture from the heat. Pour a small amount of boiling jam on a cold plate and put it in the freezer compartment of a refrigerator for a few minutes. If the mixture gels, it is ready to fill.

PROCESS TIMES FOR JAMS WITHOUT ADDED PECTIN IN A BOILING-WATER CANNER*

Style of Pack	Jar Size	Process Time at Altitudes of:		
		0–1,000 ft	1,001–6,000 ft	Above 6,000 ft
Hot	Half-pints	5 minutes	ten minutes	15 minutes

*After the process is complete, turn off the heat and remove the canner lid. Wait five minutes before removing jars.

Jams and Jellies with Added Pectin

To use fresh fruits with a low pectin content or canned or frozen fruit juice, powdered or liquid pectin must be added for your jams and jellies to thicken and set properly. Jelly or jam made with added pectin requires less cooking and generally gives a larger yield. These products have more natural fruit flavors, too. In addition, using added pectin eliminates the need to test hot jellies and jams for proper gelling.

Commercially produced pectin is a natural ingredient, usually made from apples and available at most grocery stores. There are several types of pectin now commonly available; liquid, powder, low-sugar, and no-sugar pectins each have their own advantages and downsides. Pomona's Universal Pectin is a citrus pectin that allows you to make jams and jellies with little or no sugar. Because the order of combining ingredients depends on the type of pectin used, it is best to follow the common jam and jelly recipes that are included right on most pectin packages. How ever, if you want to try something a little different, follow one of the following recipes for mixed fruit and spiced fruit jams and jellies.

PROCESS TIMES FOR JAMS AND JELLIES WITH ADDED PECTIN IN A BOILING-WATER CANNER*

Style of Pack	Jar Size	Process Time at Altitudes of:		
		0–1,000 ft	1,001–6,000 ft	Above 6,000 ft
Hot	Half-pints	5 minutes	ten minutes	15 minutes

*After the process is complete, turn off the heat and remove the canner lid. Wait five minutes before removing jars.

Pear-Apple Jam

This is a delicious jam perfect for making at the end of autumn, just before the frost gets the last apples. For a warming, spicy twist add a teaspoon of fresh grated ginger along with the cinnamon. This recipe yields seven to eight half-pints.

Ingredients

- 2 cups peeled, cored, and finely chopped pears (about 2 lbs.)
- 1 cup peeled, cored, and finely chopped apples
- ¼ tsp ground cinnamon
- 6½ cups sugar
- ⅓ cup bottled lemon juice
- 6 oz liquid pectin

Directions

1. Peel, core, and slice apples and pears into a large saucepan and stir in cinnamon. Thoroughly mix sugar and lemon juice with fruits and bring to a boil over high heat, stirring constantly and crushing fruit with a potato masher as it softens.

2. Once boiling, immediately stir in pectin. Bring to a full rolling boil and boil hard 1 minute, stirring constantly.

3. Remove from heat, quickly skim off foam, and fill sterile jars, leaving ¼ inch headspace. Adjust lids and process.

PROCESS TIMES FOR PEAR-APPLE JAM IN A BOILING WATER CANNER*

Style of Pack	Jar Size	Process Time at Altitudes of:		
		0–1,000 ft	1,001–6,000 ft	Above 6,000 ft
Hot	Half-pints	5 minutes	ten minutes	15 minutes

*After the process is complete, turn off the heat and remove the canner lid. Wait five minutes before removing jars.

TIP

- Adding ½ teaspoon of butter or margarine with the juice and pectin will reduce foaming. However, these may cause off-flavor in a long-term storage of jellies and jams.
- Purchase fresh fruit pectin each year. Old pectin may result in poor gels.
- Be sure to use mason canning jars, self-sealing two-piece lids, and a five-minute process (corrected for altitude, as necessary) in boiling water.

Canning

Strawberry-Rhubarb Jelly

Strawberry-rhubarb jelly will turn any ordinary piece of bread into a delightful treat. You can also spread it on shortcake or pound cake for a simple and unique dessert. This recipe yields about seven half-pints.

Ingredients

- 1½ lbs. red stalks of rhubarb
- 1½ qts ripe strawberries
- ½ tsp butter or margarine to reduce foaming (optional)
- 6 cups sugar
- 6 oz liquid pectin

Directions

1. Wash and cut rhubarb into 1-inch pieces and blend or grind. Wash, stem, and crush strawberries, one layer at a time, in a saucepan. Place both fruits in a jelly bag or double layer of cheesecloth and gently squeeze juice into a large measuring cup or bowl.
2. Measure 3½ cups of juice into a large saucepan. Add butter and sugar, thoroughly mixing into juice. Bring to a boil over high heat, stirring constantly.
3. As soon as mixture begins to boil, stir in pectin. Bring to a full rolling boil and boil hard 1 minute, stirring constantly. Remove from heat, quickly skim off foam, and fill sterile jars, leaving ¼-inch headspace. Adjust lids and process.

Blueberry-Spice Jam

This is a summery treat that is delicious spread over waffles with a little butter. Using wild blueberries results in a stronger flavor, but cultivated blueberries also work well. This recipe yields about five half-pints.

PROCESS TIMES FOR STRAWBERRY-RHUBARB JELLY IN A BOILING-WATER CANNER*

Style of Pack	Jar Size	Process Time at Altitudes of:		
		0–1,000 ft	1,001–6,000 ft	Above 6,000 ft
Hot	Half-pints or pints	5 minutes	ten minutes	15 minutes
*After the process is complete, turn off the heat and remove the canner lid. Wait five minutes before removing jars.				

PROCESS TIMES FOR BLUEBERRY-SPICE JAM IN A BOILING-WATER CANNER*

Style of Pack	Jar Size	Process Time at Altitudes of:		
		0–1,000 ft	1,001–6,000 ft	Above 6,000 ft
Hot	Half-pints or pints	5 minutes	ten minutes	15 minutes
*After the process is complete, turn off the heat and remove the canner lid. Wait five minutes before removing jars.				

Ingredients

- 2½ pints ripe blueberries
- 1 tbsp lemon juice
- ½ tsp ground nutmeg or cinnamon
- ¾ cup water
- 5½ cups sugar
- 1 box (1¾ oz) powdered pectin

Directions

1. Wash and thoroughly crush blueberries, adding one layer at a time, in a saucepan. Add lemon juice, spice, and water. Stir pectin and bring to a full, rolling boil over high heat, stirring frequently.
2. Add the sugar and return to a full rolling boil. Boil hard for 1 minute, stirring constantly. Remove from heat, quickly skim off foam, and fill sterile jars, leaving ¼-inch headspace. Adjust lids and process.

Grape-Plum Jelly

If you think peanut butter and jelly sandwiches are only for kids, try grape-plum jelly spread with a natural nut butter over a thick slice of whole wheat bread. You'll change your mind. This recipe yields about 10 half-pints.

Ingredients

- 3½ lbs. ripe plums
- 3 lbs. ripe Concord grapes
- 8½ cups sugar
- 1 cup water
- ½ tsp butter or margarine to reduce foaming (optional)
- 1 box (1¾ oz) powdered pectin

Directions

1. Wash and pit plums; do not peel. Thoroughly crush the plums and grapes, adding one layer at a time, in a saucepan

PROCESS TIMES FOR GRAPE-PLUM JELLY IN A BOILING-WATER CANNER*

Style of Pack	Jar Size	Process Time at Altitudes of:		
		0–1,000 ft	1,001–6,000 ft	Above 6,000 ft
Hot	Half-pints or pints	5 minutes	10 minutes	15 minutes
*After the process is complete, turn off the heat and remove the canner lid. Wait five minutes before removing jars.				

with water. Bring to a boil, cover, and simmer ten minutes.

2. Strain juice through a jelly bag or double layer of cheesecloth. Measure sugar and set aside. Combine 6½ cups of juice with butter and pectin in large saucepan. Bring to a hard boil over high heat, stirring constantly.

3. Add the sugar and return to a full rolling boil. Boil hard for 1 minute, stirring constantly. Remove from heat, quickly skim off foam, and fill sterile jars, leaving ¼-inch headspace. Adjust lids and process.

Making Reduced-Sugar Fruit Spreads

A variety of fruit spreads may be made that are tasteful, yet lower in sugars and calories than regular jams and jellies. The most straightforward method is probably to buy low-sugar pectin and follow the directions on the package, but the recipes below show alternate methods of using gelatin or fruit pulp as thickening agents. Gelatin recipes should not be processed and should be refrigerated and used within four weeks.

Peach-Pineapple Spread

This recipe may be made with any combination of peaches, nectarines, apricots, and plums. You can use no sugar, up to two cups of sugar, or a combination of sugar and another sweetener (such as honey, Splenda , or agave nectar). Note that if you use aspartame, the spread may lose its sweetness within three to four weeks. Add cinnamon or star anise if desired. This recipe yields five to six half-pints.

Ingredients

- 4 cups drained peach pulp (follow directions below)
- 2 cups drained unsweetened crushed pineapple
- ¼ cup bottled lemon juice
- 2 cups sugar (optional)

Directions

1. Thoroughly wash 4 to 6 pounds of firm, ripe peaches. Drain well. Peel and remove pits. Grind fruit flesh with a medium or coarse blade, or crush with a fork (do not use a blender).

2. Place ground or crushed peach pulp in a 2-quart saucepan. Heat slowly to release juice, stirring constantly, until fruit is tender. Place cooked fruit in a jelly bag or strainer lined with four layers of cheesecloth. Allow juice to drip about 15 minutes. Save the juice for jelly or other uses.

3. Measure 4 cups of drained peach pulp for making spread. Combine the 4

PROCESS TIMES FOR PEACH-PINEAPPLE SPREAD IN A BOILING-WATER CANNER*

Style of Pack	Jar Size	Process Time at Altitudes of:			
		0–1,000 ft	1,001–3,000 ft	3,001–6,000 ft	Above 6,000 ft
Hot	Half-pints	15 minutes	20 minutes	20 minutes	25 minutes
	Pints	20 minutes	25 minutes	30 minutes	35 minutes
*After the process is complete, turn off the heat and remove the canner lid. Wait five minutes before removing jars.					

cups of pulp, pineapple, and lemon juice in a 4-quart saucepan. Add up to 2 cups of sugar or other sweetener, if desired, and mix well.

4. Heat and boil gently for 10 to 15 minutes, stirring enough to prevent sticking. Fill jars quickly, leaving ¼-inch headspace. Adjust lids and process.

Refrigerated Apple Spread

This recipe uses gelatin as a thickener, so it does not require processing but it should be refrigerated and used within four weeks. For spiced apple jelly, add two sticks of cinnamon and four whole cloves to mixture before boiling. Remove both spices before adding the sweetener and food coloring (if desired). This recipe yields four half-pints.

Ingredients

- 2 tbsps unflavored gelatin powder
- 1 qt bottle unsweetened apple juice
- 2 tbsps bottled lemon juice
- 2 tbsps liquid low-calorie sweetener (e.g., sucralose, honey, or 1–2 tsps liquid stevia)

Directions

1. In a saucepan, soften the gelatin in the apple and lemon juices. To dissolve gelatin, bring to a full rolling boil and boil 2 minutes. Remove from heat.

2. Stir in sweetener and food coloring (if desired). Fill jars, leaving ¼-inch head-space. Adjust lids. Refrigerate (do not process or freeze).

Refrigerated Grape Spread

This is a simple, tasty recipe that doesn't require processing. Be sure to refrigerate and use within four weeks. This recipe makes three half-pints.

Ingredients

- 2 tbsps unflavored gelatin powder
- 1 bottle (24 oz) unsweetened grape juice
- 2 tbsps bottled lemon juice
- 2 tbsps liquid low-calorie sweetener (e.g., sucralose, honey, or 1–2 tsps liquid stevia)

Directions

1. In a saucepan, heat the gelatin in the grape and lemon juices until mixture is soft. Bring to a full rolling boil to dissolve gelatin. Boil 1 minute and remove from heat. Stir in sweetener.

2. Fill jars quickly, leaving ¼-inch head-space. Adjust lids. Refrigerate (do not process or freeze).

Remaking Soft Jellies

Sometimes jelly just doesn't turn out right the first time. Jelly that is too soft can be used as a sweet sauce to drizzle over ice cream, cheesecake, or angel food cake, but

it can also be re-cooked into the proper consistency.

To Remake with Powdered Pectin

1. Measure jelly to be re-cooked. Work with no more than 4 to 6 cups at a time. For each quart (4 cups) of jelly, mix ¼ cup sugar, ½ cup water, 2 tablespoons bottled lemon juice, and 4 teaspoons powdered pectin. Bring to a boil while stirring.

2. Add jelly and bring to a rolling boil over high heat, stirring constantly. Boil hard ½ minute. Remove from heat, quickly skim foam off jelly, and fill sterile jars, leaving ¼-inch headspace. Adjust new lids and process as recommended (see page 200).

To Remake with Liquid Pectin

1. Measure jelly to be re-cooked. Work with no more than 4 to 6 cups at a time. For each quart (4 cups) of jelly, measure into a bowl ¾ cup sugar, 2 tablespoons bottled lemon juice, and 2 tablespoons liquid pectin.

2. Bring jelly only to boil over high heat, while stirring. Remove from heat and quickly add the sugar, lemon juice, and pectin. Bring to a full rolling boil, stirring constantly. Boil hard for 1 minute. Quickly skim off foam and fill sterile jars, leaving ¼-inch headspace. Adjust new lids and process as recommended (see page 201)

To Remake without Added Pectin

1. For each quart of jelly, add 2 table-spoons bottled lemon juice. Heat to boiling and continue to boil for 3 to 4 minutes.

Temperature test—Use a jelly or candy thermometer and boil until mixture reaches the following temperatures at the altitudes below:

Sea Level	1,000 ft	2,000 ft	3,000 ft	4,000 ft	5,000 ft	6,000 ft	7,000 ft	8,000 ft
220°F	218°F	216°F	214°F	212°F	211°F	209°F	207°F	205°F

Sheet or spoon test—Dip a cool metal spoon into the boiling jelly mixture. Raise the spoon about 12 inches above the pan (out of steam). Turn the spoon so the liquid runs off the side. The jelly is done when the syrup forms two drops that flow together and sheet or hang off the edge of the spoon.

PROCESS TIMES FOR REMADE SOFT JELLIES IN A BOILING-WATER CANNER*

Style of Pack	Jar Size	Process Time at Altitudes of:		
		0–1,000 ft	1,001–6,000 ft	Above 6,000 ft
Hot	Half-pints or pints	5 minutes	10 minutes	15 minutes

*After the process is complete, turn off the heat and remove the canner lid. Wait five minutes before removing jars.

QUANTITY

- An average of five pounds is needed per canner load of seven quarts.
- An average of 3¼ pounds is needed per canner load of nine pints—an average of ¾ pounds per quart.

2. To test jelly for doneness, use one of the following methods:
3. Remove from heat, quickly skim off foam, and fill sterile jars, leaving ¼-inch headspace. Adjust new lids and process.

Vegetables, Pickles, and Tomatoes

Beans or Peas, Shelled or Dried (All Varieties)

Shelled or dried beans and peas are inexpensive and easy to buy or store in bulk, but they are not very convenient when it comes to preparing them to eat. Hydrating and canning beans or peas enable you to simply open a can and use them rather than waiting for them to soak. Sort and discard discolored seeds before rehydrating.

Directions

1. Place dried beans or peas in a large pot and cover with water. Soak 12 to 18 hours in a cool place. Drain water. To quickly hydrate beans, you may cover sorted and washed beans with boiling water in a saucepan. Boil 2 minutes, remove from heat, soak 1 hour, and drain.
2. Cover beans soaked by either method with fresh water and boil 30 minutes. Add ½ teaspoon of salt per pint or 1 teaspoon per quart to each jar, if

PROCESS TIMES FOR BEANS OR PEAS IN A DIAL-GAUGE PRESSURE CANNER*

Style of Pack	Jar Size	Process Time	Canner Pressure (PSI) at Altitudes of:			
			0–2,000 ft	2,001–4,000 ft	4,001–6,000 ft	6,001–8,000 ft
Hot	Pints	75 minutes	11 lbs	12 lbs	13 lbs	14 lbs
	Quarts	90 minutes	11 lbs	12 lbs	13 lbs	14 lbs

*After the canner is completely depressurized, remove the weight from the vent port or open the petcock. Wait ten minutes; then unfasten the lid and remove it carefully. Lift the lid with the underside away from you so that the steam coming out of the canner does not burn your face.

PROCESS TIMES FOR BEANS OR PEAS IN A WEIGHTED-GAUGE PRESSURE CANNER*

Style of pack	Jar Size	Process Time	Canner Pressure (PSI) at Altitudes of:	
			0–1,000 ft	Above 1,000 ft
Hot	Pints	75 minutes	10 lbs	15 lbs
	Quarts	90 minutes	10 lbs	15 lbs

*After the canner is completely depressurized, remove the weight from the vent port or open the petcock. Wait ten minutes; then unfasten the lid and remove it carefully. Lift the lid with the underside away from you so that the steam coming out of the canner does not burn your face.

desired. Fill jars with beans or peas and cooking water, leaving 1-inch head-space. Adjust lids and process.

Baked Beans

Baked beans are an old New England favorite, but every cook has his or her favorite variation. Two recipes are included here, but feel free to alter them to your own taste.

Directions

1. Sort and wash dry beans. Add 3 cups of water for each cup of dried beans. Boil 2 minutes, remove from heat, soak 1 hour, and drain.

QUANTITY

- An average of five pounds of beans is needed per canner load of seven quarts.
- An average of 3¼ pounds is needed per canner load of nine pints—an average of ¾ pounds per quart.

---

2. Heat to boiling in fresh water, and save liquid for making sauce. Make your choice of the following sauces:

Tomato Sauce—Mix 1 quart tomato juice, 3 tablespoons sugar, 2 teaspoons salt, 1 tablespoon chopped onion, and ¼ teaspoon each of ground cloves, allspice, mace, and cayenne pepper. Heat to boiling. Add 3 quarts cooking liquid from beans and bring back to boiling.

Molasses Sauce—Mix 4 cups water or cooking liquid from beans, 3 tablespoons dark molasses, 1 tablespoon vinegar, 2 teaspoons salt, and ¾ teaspoon powdered dry mustard. Heat to boiling.

3. Place seven ¾-inch pieces of pork, ham, or bacon in an earthenware crock, a large casserole, or a pan. Add beans and enough molasses sauce to cover beans.

PROCESS TIMES FOR BAKED BEANS IN A DIAL-GAUGE PRESSURE CANNER*

Style of Pack	Jar Size	Process Time	Canner Pressure (PSI) at Altitudes of:			
			0–2,000 ft	2,001–4,000 ft	4,001–6,000 ft	6,001–8,000 ft
Hot	Pints	65 minutes	11 lbs	12 lbs	13 lbs	14 lbs
	Quarts	75 minutes	11 lbs	12 lbs	13 lbs	14 lbs

*After the canner is completely depressurized, remove the weight from the vent port or open the petcock. Wait ten minutes; then unfasten the lid and remove it carefully. Lift the lid with the underside away from you so that the steam coming out of the canner does not burn your face.

PROCESS TIMES FOR BAKED BEANS IN A WEIGHTED-GAUGE PRESSURE CANNER*

Style of pack	Jar Size	Process Time	Canner Pressure (PSI) at Altitudes of:	
			0–1,000 ft	Above 1,000 ft
Hot	Pints	65 minutes	10 lbs	15 lbs
	Quarts	75 minutes	10 lbs	15 lbs

*After the canner is completely depressurized, remove the weight from the vent port or open the petcock. Wait ten minutes; then unfasten the lid and remove it carefully. Lift the lid with the underside away from you so that the steam coming out of the canner does not burn your face.

4. Cover and bake 4 to 5 hours at 350°F. Add water as needed—about every hour. Fill jars, leaving 1-inch headspace. Adjust lids and process.

Green Beans

This process will work equally well for snap, Italian, or wax beans. Select filled but tender, crisp pods, removing any diseased or rusty pods.

Directions

1. Wash beans and trim ends. Leave whole, or cut or break into 1-inch pieces.
2. Adjust lids and process.

Hot pack—Cover with boiling water; boil 5 minutes. Fill jars loosely, leaving 1-inch headspace.

Raw pack—Fill jars tightly with raw beans, leaving 1-inch headspace. Add 1 teaspoon of salt per quart to each jar, if desired. Add boiling water, leaving 1-inch headspace.

PROCESS TIMES FOR GREEN BEANS IN A DIAL-GAUGE PRESSURE CANNER*

Style of Pack	Jar Size	Process Time	Canner Pressure (PSI) at Altitudes of:			
			0–2,000 ft	2,001–4,000 ft	4,001–6,000 ft	6,001–8,000 ft
Hot or Raw	Pints	20 minutes	11 lbs	12 lbs	13 lbs	14 lbs
	Quarts	25 minutes	11 lbs	12 lbs	13 lbs	14 lbs

*After the canner is completely depressurized, remove the weight from the vent port or open the petcock. Wait ten minutes; then unfasten the lid and remove it carefully. Lift the lid with the underside away from you so that the steam coming out of the canner does not burn your face.

PROCESS TIMES FOR GREEN BEANS IN A WEIGHTED-GAUGE PRESSURE CANNER*

Style of Pack	Jar Size	Process Time	Canner Pressure (PSI) at Altitudes of:	
			0–1,000 ft	Above 1,000 ft
Hot or Raw	Pints	20 minutes	10 lbs	15 lbs
	Quarts	25 minutes	10 lbs	15 lbs

*After the canner is completely depressurized, remove the weight from the vent port or open the petcock. Wait ten minutes; then unfasten the lid and remove it carefully. Lift the lid with the underside away from you so that the steam coming out of the canner does not burn your face.

QUANTITY

- An average of 14 pounds is needed per canner load of seven quarts.
- An average of nine pounds is needed per canner load of nine pints.
- A bushel weighs 30 pounds and yields 12 to 20 quarts—an average of 2 pounds per quart.

Canning

Beets

You can preserve beets whole, cubed, or sliced, according to your preference. Beets that are 1 to 2 inches in diameter are the best, as larger ones tend to be too fibrous.

Directions

1. Trim off beet tops, leaving an inch of stem and roots to reduce bleeding of color. Scrub well. Cover with boiling water. Boil until skins slip off easily, about 15 to 25 minutes depending on size.
2. Cool, remove skins, and trim off stems and roots. Leave baby beets whole. Cut medium or large beets into ½-inch cubes or slices. Halve or quarter very large slices. Add 1 teaspoon of salt per quart to each jar, if desired.
3. Fill jars with hot beets and fresh hot water, leaving 1-inch headspace. Adjust lids and process.

Carrots

Carrots can be preserved sliced or diced according to your preference. Choose small carrots, preferably 1 to 1¼ inches in diameter, as larger ones are often too fibrous.

Directions

1. Wash, peel, and rewash carrots. Slice or dice.

Hot pack—Cover with boiling water; bring to boil and simmer for 5 minutes. Fill jars with carrots, leaving 1-inch headspace.

QUANTITY

- An average of 21 pounds (without tops) is needed per canner load of seven quarts.
- An average of 13½ pounds is needed per canner load of nine pints.
- A bushel (without tops) weighs 52 pounds and yields 15 to 20 quarts—an average of three pounds per quart.

PROCESS TIMES FOR BEETS IN A DIAL-GAUGE PRESSURE CANNER*

Style of Pack	Jar Size	Process Time	Canner Pressure (PSI) at Altitudes of:			
			0–2,000 ft	2,001–4,000 ft	4,001–6,000 ft	6,001–8,000 ft
Hot	Pints	30 minutes	11 lbs	12 lbs	13 lbs	14 lbs
	Quarts	35 minutes	11 lbs	12 lbs	13 lbs	14 lbs

*After the canner is completely depressurized, remove the weight from the vent port or open the petcock. Wait ten minutes; then unfasten the lid and remove it carefully. Lift the lid with the underside away from you so that the steam coming out of the canner does not burn your face.

PROCESS TIMES FOR BEETS IN A WEIGHTED-GAUGE PRESSURE CANNER*

Style of Pack	Jar Size	Process Time	Canner Pressure (PSI) at Altitudes of:	
			0–1,000 ft	Above 1,000 ft
Hot or Raw	Pints	30 minutes	10 lbs	15 lbs
	Quarts	35 minutes	10 lbs	15 lbs

*After the canner is completely depressurized, remove the weight from the vent port or open the petcock. Wait ten minutes; then unfasten the lid and remove it carefully. Lift the lid with the underside away from you so that the steam coming out of the canner does not burn your face.

QUANTITY

- An average of 17½ pounds (without tops) is needed per canner load of seven quarts.
- An average of 11 pounds is needed per canner load of nine pints.
- A bushel (without tops) weighs 50 pounds and yields 17 to 25 quarts—an average of 2½ pounds per quart.

Raw pack—Fill jars tightly with raw carrots, leaving 1-inch headspace.

2. Add 1 teaspoon of salt per quart to the jar, if desired. Add hot cooking liquid or water, leaving 1-inch headspace. Adjust lids and process.

Corn, Cream Style

The creamy texture comes from scraping the corncobs thoroughly and including the juices and corn pieces with the kernels. If you want to add milk or cream, butter, or other ingredients, do so just before serving (do not add dairy products before canning). Select ears containing slightly immature kernels for this recipe.

PROCESS TIMES FOR CARROTS IN A DIAL-GAUGE PRESSURE CANNER*

Style of Pack	Jar Size	Process Time	Canner Pressure (PSI) at Altitudes of:			
			0–2,000 ft	2,001–4,000 ft	4,001–6,000 ft	6,001–8,000 ft
Hot or Raw	Pints	25 minutes	11 lbs	12 lbs	13 lbs	14 lbs
	Quarts	30 minutes	11 lbs	12 lbs	13 lbs	14 lbs

*After the canner is completely depressurized, remove the weight from the vent port or open the petcock. Wait ten minutes; then unfasten the lid and remove it carefully. Lift the lid with the underside away from you so that the steam coming out of the canner does not burn your face.

PROCESS TIMES FOR CARROTS IN A WEIGHTED-GAUGE PRESSURE CANNER*

Style of Pack	Jar Size	Process Time	Canner Pressure (PSI) at Altitudes of:	
			0–1,000 ft	Above 1,000 ft
Hot or Raw	Pints	25 minutes	10 lbs	15 lbs
	Quarts	30 minutes	10 lbs	15 lbs

*After the canner is completely depressurized, remove the weight from the vent port or open the petcock. Wait ten minutes; then unfasten the lid and remove it carefully. Lift the lid with the underside away from you so that the steam coming out of the canner does not burn your face.

Directions

1. Husk corn, remove silk, and wash ears. Cut corn from cob at about the center of kernel. Scrape remaining corn from cobs with a table knife.

Hot pack—To each quart of corn and scrapings in a saucepan, add 2 cups of boiling water. Heat to boiling. Add ½ teaspoon salt to each jar, if desired. Fill pint jars with hot corn mixture, leaving 1-inch headspace.

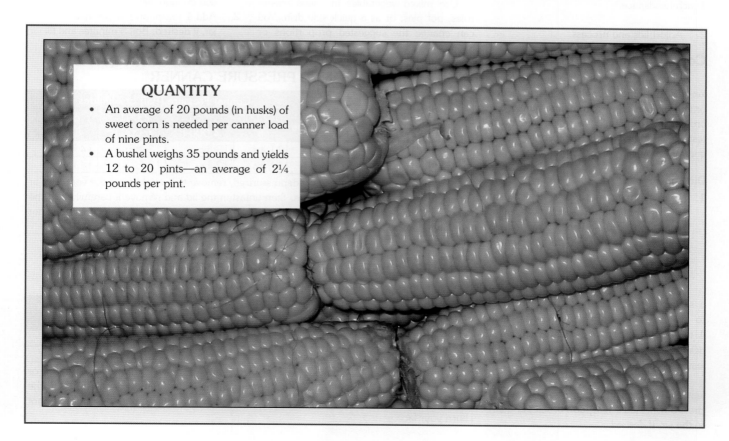

QUANTITY

- An average of 20 pounds (in husks) of sweet corn is needed per canner load of nine pints.
- A bushel weighs 35 pounds and yields 12 to 20 pints—an average of 2¼ pounds per pint.

PROCESS TIMES FOR CREAM STYLE CORN IN A DIAL-GAUGE PRESSURE CANNER

Style of pack	Jar Size	Process Time	Canner Pressure (PSI) at Altitudes of:			
			0–2,000 ft	2,001–4,000 ft	4,001–6,000 ft	6,001–8,000 ft
Hot	Pints	85 minutes	11 lbs	12 lbs	13 lbs	14 lbs
Raw	Pints	95 minutes	11 lbs	12 lbs	13 lbs	14 lbs

*After the canner is completely depressurized, remove the weight from the vent port or open the petcock. Wait ten minutes; then unfasten the lid and remove it carefully. Lift the lid with the underside away from you so that the steam coming out of the canner does not burn your face.

PROCESS TIMES FOR CREAM STYLE CORN IN A WEIGHTED-GAUGE PRESSURE CANNER*

Style of Pack	Jar Size	Process Time	Canner Pressure (PSI) at Altitudes of:	
			0–1,000 ft	Above 1,000 ft
Hot	Pints	85 minutes	10 lbs	15 lbs
Raw	Pints	95 minutes	10 lbs	15 lbs

*After the canner is completely depressurized, remove the weight from the vent port or open the petcock. Wait ten minutes; then unfasten the lid and remove it carefully. Lift the lid with the underside away from you so that the steam coming out of the canner does not burn your face.

Raw pack—Fill pint jars with raw corn, leaving 1-inch headspace. Do not shake or press down. Add ½ teaspoon salt to each jar, if desired. Add fresh boiling water, leaving 1-inch headspace.

2. Adjust lids and process.

Corn, Whole Kernel

Select ears containing slightly immature kernels. Canning of some sweeter varieties or kernels that are too immature may cause browning. Try canning a small amount to test color and flavor before canning large quantities.

Directions

1. Husk corn, remove silk, and wash. Blanch 3 minutes in boiling water. Cut corn from cob at about three-fourths the depth of kernel. Do not scrape cob, as it will create a creamy texture.

Hot pack —To each quart of kernels in a saucepan, add 1 cup of hot water, heat to boiling, and simmer 5 minutes. Add 1 teaspoon of salt per quart to each jar, if desired. Fill jars with corn and cooking liquid, leaving 1-inch headspace.

Raw pack—Fill jars with raw kernels, leaving 1-inch headspace. Do not shake or press down. Add 1 teaspoon of salt per quart to the jar, if desired.

2. Add fresh boiling water, leaving 1-inch headspace. Adjust lids and process.

Mixed Vegetables

Use mixed vegetables in soups, casseroles, pot pies, or as a quick side dish. You can change the suggested proportions or substitute other favorite vegetables, but avoid leafy greens, dried beans, cream-style corn, winter squash, and sweet potatoes as they will ruin the consistency of the other vegetables. This recipe yields about seven quarts.

QUANTITY

- An average of 31½ pounds (in husks) of sweet corn is needed per canner load of seven quarts.
- An average of 20 pounds is needed per canner load of nine pints.
- A bushel weighs 35 pounds and yields 6 to 11 quarts—an average of 4½ pounds per quart.

Ingredients

- 6 cups sliced carrots
- 6 cups cut, whole-kernel sweet corn
- 6 cups cut green beans
- 6 cups shelled lima beans
- 4 cups diced or crushed tomatoes
- 4 cups diced zucchini

Directions

1. Carefully wash, peel, de-shell, and cut vegetables as necessary. Combine all vegetables in a large pot or kettle, and add enough water to cover pieces.

2. Add 1 teaspoon salt per quart to each jar, if desired. Boil 5 minutes and fill jars

PROCESS TIMES FOR WHOLE KERNEL CORN IN A DIAL-GAUGE PRESSURE CANNER*

Style of Pack	Jar Size	Process Time	Canner Pressure (PSI) at Altitudes of:			
			0–2,000 ft	2,001–4,000 ft	4,001–6,000 ft	6,001–8,000 ft
Hot or Raw	Pints	55 minutes	11 lbs	12 lbs	13 lbs	14 lbs
	Quarts	85 minutes	11 lbs	12 lbs	13 lbs	14 lbs

*After the canner is completely depressurized, remove the weight from the vent port or open the petcock. Wait ten minutes; then unfasten the lid and remove it carefully. Lift the lid with the underside away from you so that the steam coming out of the canner does not burn your face.

PROCESS TIMES FOR WHOLE KERNEL CORN IN A WEIGHTED-GAUGE PRESSURE CANNER*

Style of Pack	Jar Size	Process Time	Canner Pressure (PSI) at Altitudes of:	
			0–1,000 ft	Above 1,000 ft
Hot or Raw	Pints	55 minutes	10 lbs	15 lbs
	Quarts	85 minutes	10 lbs	15 lbs

*After the canner is completely depressurized, remove the weight from the vent port or open the petcock. Wait ten minutes; then unfasten the lid and remove it carefully. Lift the lid with the underside away from you so that the steam coming out of the canner does not burn your face.

with hot pieces and liquid, leaving 1-inch headspace. Adjust lids and process.

Peas, Green or English, Shelled

Green and English peas preserve well when canned, but sugar snap and Chinese edible pods are better frozen. Select filled pods containing young, tender, sweet seeds, and discard any diseased pods.

Directions

1. Shell and wash peas. Add 1 teaspoon of salt per quart to each jar, if desired.

   Hot pack—Cover with boiling water. Bring to a boil in a saucepan, and boil 2 minutes. Fill jars loosely with hot peas, and add cooking liquid, leaving 1-inch headspace.

   Raw pack—Fill jars with raw peas, and add boiling water, leaving 1-inch headspace. Do not shake or press down peas.

2. Adjust lids and process.

PROCESS TIMES FOR MIXED VEGETABLES IN A DIAL-GAUGE PRESSURE CANNER*

Style of Pack	Jar Size	Process Time	Canner Pressure (PSI) at Altitudes of:			
			0–2,000 ft	2,001–4,000 ft	4,001–6,000 ft	6,001–8,000 ft
Hot	Pints	75 minutes	11 lbs	12 lbs	13 lbs	14 lbs
	Quarts	90 minutes	11 lbs	12 lbs	13 lbs	14 lbs

*After the canner is completely depressurized, remove the weight from the vent port or open the petcock. Wait ten minutes; then unfasten the lid and remove it carefully. Lift the lid with the underside away from you so that the steam coming out of the canner does not burn your face.

PROCESS TIMES FOR MIXED VEGETABLES IN A WEIGHTED-GAUGE PRESSURE CANNER*

Style of Pack	Jar Size	Process Time	Canner Pressure (PSI) at Altitudes of:	
			0–1,000 ft	Above 1,000 ft
Hot	Pints	75 minutes	10 lbs	15 lbs
	Quarts	90 minutes	10 lbs	15 lbs

*After the canner is completely depressurized, remove the weight from the vent port or open the petcock. Wait ten minutes; then unfasten the lid and remove it carefully. Lift the lid with the underside away from you so that the steam coming out of the canner does not burn your face.

QUANTITY

- An average of 31½ pounds (in pods) is needed per canner load of seven quarts.
- An average of 20 pounds is needed per canner load of nine pints.
- A bushel weighs 30 pounds and yields 5 to 10 quarts—an average of 4½ pounds per quart.

Potatoes, Sweet

Sweet potatoes can be preserved whole, in chunks, or in slices, according to your preference. Choose small to medium-sized potatoes that are mature and not too fibrous. Can within one to two months after harvest.

Directions

1. Wash potatoes and boil or steam until partially soft (15 to 20 minutes). Remove skins. Cut medium potatoes, if needed, so that pieces are uniform in size. Do not mash or purée pieces.
2. Fill jars, leaving 1-inch headspace. Add 1 teaspoon salt per quart to each jar, if desired. Cover with your choice of fresh boiling water or syrup, leaving 1-inch headspace. Adjust lids and process.

PROCESS TIMES FOR PEAS IN A DIAL-GAUGE PRESSURE CANNER*

Style of Pack	Jar Size	Process Time	Canner Pressure (PSI) at Altitudes of:			
			0–2,000 ft	2,001–4,000 ft	4,001–6,000 ft	6,001–8,000 ft
Hot or Raw	Pints or Quarts	40 minutes	11 lbs	12 lbs	13 lbs	14 lbs

*After the canner is completely depressurized, remove the weight from the vent port or open the petcock. Wait ten minutes; then unfasten the lid and remove it carefully. Lift the lid with the underside away from you so that the steam coming out of the canner does not burn your face.

PROCESS TIMES FOR PEAS IN A WEIGHTED-GAUGE PRESSURE CANNER*

Style of Pack	Jar Size	Process Time	Canner Pressure (PSI) at Altitudes of:	
			0–1,000 ft	Above 1,000 ft
Hot or Raw	Pints or Quarts	40 minutes	10 lbs	15 lbs

*After the canner is completely depressurized, remove the weight from the vent port or open the petcock. Wait ten minutes; then unfasten the lid and remove it carefully. Lift the lid with the underside away from you so that the steam coming out of the canner does not burn your face.

QUANTITY

- An average of 17½ pounds is needed per canner load of seven quarts.
- An average of 11 pounds is needed per canner load of nine pints.
- A bushel weighs 50 pounds and yields 17 to 25 quarts—an average of 2½ pounds per quart.

PROCESS TIMES FOR SWEET POTATOES IN A DIAL-GAUGE PRESSURE CANNER*

Style of Pack	Jar Size	Process Time	Canner Pressure (PSI) at Altitudes of:			
			0–2,000 ft	2,001–4,000 ft	4,001–6,000 ft	6,001–8,000 ft
Hot	Pints	65 minutes	11 lbs	12 lbs	13 lbs	14 lbs
	Quarts	90 minutes	11 lbs	12 lbs	13 lbs	14 lbs

*After the canner is completely depressurized, remove the weight from the vent port or open the petcock. Wait ten minutes; then unfasten the lid and remove it carefully. Lift the lid with the underside away from you so that the steam coming out of the canner does not burn your face.

PROCESS TIMES FOR SWEET POTATOES IN A WEIGHTED-GAUGE PRESSURE CANNER*

Style of Pack	Jar Size	Process Time	Canner Pressure (PSI) at Altitudes of:	
			0–1,000 ft	Above 1,000 ft
Hot	Pints	65 minutes	10 lbs	15 lbs
	Quarts	90 minutes	10 lbs	15 lbs

*After the canner is completely depressurized, remove the weight from the vent port or open the petcock. Wait ten minutes; then unfasten the lid and remove it carefully. Lift the lid with the underside away from you so that the steam coming out of the canner does not burn your face.

PROCESS TIMES FOR PUMPKIN AND WINTER SQUASH IN A DIAL-GAUGE PRESSURE CANNER*

Style of Pack	Jar Size	Process Time	Canner Pressure (PSI) at Altitudes of:			
			0–2,000 ft	2,001–4,000 ft	4,001–6,000 ft	6,001–8,000 ft
Hot	Pints	55 minutes	11 lbs	12 lbs	13 lbs	14 lbs
	Quarts	90 minutes	11 lbs	12 lbs	13 lbs	14 lbs

*After the canner is completely depressurized, remove the weight from the vent port or open the petcock. Wait ten minutes; then unfasten the lid and remove it carefully. Lift the lid with the underside away from you so that the steam coming out of the canner does not burn your face.

PROCESS TIMES FOR PUMPKIN AND WINTER SQUASH IN A WEIGHTED-GAUGE PRESSURE CANNER*

Style of Pack	Jar Size	Process Time	Canner Pressure (PSI) at Altitudes of:	
			0–1,000 ft	Above 1,000 ft
Hot	Pints	55 minutes	10 lbs	15 lbs
	Quarts	90 minutes	10 lbs	15 lbs

*After the canner is completely depressurized, remove the weight from the vent port or open the petcock. Wait ten minutes; then unfasten the lid and remove it carefully. Lift the lid with the underside away from you so that the steam coming out of the canner does not burn your face.

Pumpkin and Winter Squash

Pumpkin and squash are great to have on hand for use in pies, soups, quick breads, or as side dishes. They should have a hard rind and stringless, mature pulp. Small pumpkins (sugar or pie varieties) are best. Before using for pies, drain jars and strain or sieve pumpkin or squash cubes.

Directions

1. Wash, remove seeds, cut into 1-inch-wide slices, and peel. Cut flesh into 1-inch cubes. Boil 2 minutes in water. Do not mash or purée.
2. Fill jars with cubes and cooking liquid, leaving 1-inch headspace. Adjust lids and process.

Succotash

To spice up this simple, satisfying dish, add a little paprika and celery salt before serving. It is also delicious made into a pot pie, with or without added chicken, turkey, or beef. This recipe yields seven quarts.

Ingredients

- 1 lb unhusked sweet corn or 3 qts cut whole kernels
- 14 lbs mature green podded lima beans or 4 qts shelled lima beans
- 2 qts crushed or whole tomatoes (optional)

Directions

1. Husk corn, remove silk, and wash. Blanch 3 minutes in boiling water. Cut corn from cob at about three-fourths the depth of kernel. Do not scrape cob, as it will create a creamy texture. Shell lima beans and wash thoroughly.

Hot pack—Combine all prepared vegetables in a large kettle with enough water to cover the pieces. Add 1 teaspoon salt to each quart jar, if desired. Boil gently 5

Canning

QUANTITY

- An average of 16 pounds is needed per canner load of seven quarts.
- An average of 10 pounds is needed per canner load of nine pints—an average of 2¼ pounds per quart.

minutes and fill jars with pieces and cooking liquid, leaving 1-inch headspace.

Raw pack—Fill jars with equal parts of all prepared vegetables, leaving 1-inch headspace. Do not shake or press down pieces. Add 1 teaspoon salt to each quart jar, if desired. Add fresh boiling water, leaving 1-inch headspace.

2. Adjust lids and process.

Soups

Vegetable, dried bean or pea, meat, poultry, or seafood soups can all be canned. Add pasta, rice, or other grains to soup just prior to serving, as grains tend to get soggy when canned. If dried beans or peas are used, they *must* be fully rehydrated first.

Dairy products should also be avoided in the canning process.

Directions

1. Select, wash, and prepare vegetables
2. Cook vegetables. For each cup of dried beans or peas, add 3 cups of water, boil 2 minutes, remove from heat, soak 1 hour, and heat to boil. Drain and combine with meat broth, tomatoes, or water to cover. Boil 5 minutes.
3. Salt to taste, if desired. Fill jars halfway with solid mixture. Add remaining liquid, leaving 1-inch headspace. Adjust lids and process.

Meat Stock (Broth)

"Good broth will resurrect the dead," says a South American proverb. Bones contain calcium, magnesium, phosphorus, and other trace minerals, while cartilage and tendons hold glucosamine, which is important for joints and muscle health. When simmered for extended periods, these nutrients are released into the water and broken down into a form that our bodies can absorb. Not to mention that good broth is the secret to deli-

cious risotto, reduction sauces, gravies, and dozens of other gourmet dishes.

Beef

1. Saw or crack fresh trimmed beef bones to enhance extraction of flavor. Rinse bones and place in a large stockpot or kettle, cover bones with water, add pot cover, and simmer 3 to 4 hours.
2. Remove bones, cool broth, and pick off meat. Skim off fat, add meat removed from bones to broth, and reheat to boiling. Fill jars, leaving 1-inch headspace. Adjust lids and process.

Chicken or Turkey

1. Place large carcass bones in a large stockpot, add enough water to cover bones, cover pot, and simmer 30 to 45 minutes or until meat can be easily stripped from bones.
2. Remove bones and pieces, cool broth, strip meat, discard excess fat, and return meat to broth. Reheat to boiling and fill jars, leaving 1-inch headspace. Adjust lids and process.

PROCESS TIMES FOR SUCCOTASH IN A DIAL-GAUGE PRESSURE CANNER*

Style of Pack	Jar Size	Process Time	Canner Pressure (PSI) at Altitudes of:			
			0–2,000 ft	2,001–4,000 ft	4,001–6,000 ft	6,001–8,000 ft
Hot or Raw	Pints	60 minutes	11 lbs	12 lbs	13 lbs	14 lbs
	Quarts	85 minutes	11 lbs	12 lbs	13 lbs	14 lbs

*After the canner is completely depressurized, remove the weight from the vent port or open the petcock. Wait ten minutes; then unfasten the lid and remove it carefully. Lift the lid with the underside away from you so that the steam coming out of the canner does not burn your face.

PROCESS TIMES FOR SUCCOTASH IN A WEIGHTED-GAUGE PRESSURE CANNER*

Style of Pack	Jar Size	Process Time	Canner Pressure (PSI) at Altitudes of:	
			0–1,000 ft	Above 1,000 ft
Hot or Raw	Pints	60 minutes	10 lbs	15 lbs
	Quarts	85 minutes	10 lbs	15 lbs

*After the canner is completely depressurized, remove the weight from the vent port or open the petcock. Wait ten minutes; then unfasten the lid and remove it carefully. Lift the lid with the underside away from you so that the steam coming out of the canner does not burn your face.

Fermented Foods and Pickled Vegetables

Pickled vegetables play a vital role in Italian antipasto dishes, Chinese stir-fries, British piccalilli, and much of Russian and Finnish cuisine. And, of course, the Germans love their sauerkraut, kimchee is found on nearly every Korean dinner table, and many

an American won't eat a sandwich without a good strong dill pickle on the side.

Fermenting vegetables is not complicated, but you'll want to have the proper containers, covers, and weights ready before you begin. For containers, keep the following in mind:

PROCESS TIMES FOR SOUPS IN A DIAL-GAUGE PRESSURE CANNER*

Style of Pack	Jar Size	Process Time	Canner Pressure (PSI) at Altitudes of:			
			0–2,000 ft	2,001–4,000 ft	4,001–6,000 ft	6,001–8,000 ft
Hot	Pints	6 0 * minutes	11 lbs	12 lbs	13 lbs	14 lbs
	Quarts	7 5 * minutes	11 lbs	12 lbs	13 lbs	14 lbs

*Caution: Process 100 minutes if soup contains seafood.

*After the canner is completely depressurized, remove the weight from the vent port or open the petcock. Wait ten minutes; then unfasten the lid and remove it carefully. Lift the lid with the underside away from you so that the steam coming out of the canner does not burn your face.

PROCESS TIMES FOR SOUPS IN A WEIGHTED-GAUGE PRESSURE CANNER*

Style of Pack	Jar Size	Process Time	Canner Pressure (PSI) at Altitudes of:	
			0–1,000 ft	Above 1,000 ft
Hot	Pints	60* minutes	10 lbs	15 lbs
	Quarts	75* minutes	10 lbs	15 lbs

*Caution: Process 100 minutes if soup contains seafood.

*After the canner is completely depressurized, remove the weight from the vent port or open the petcock. Wait ten minutes; then unfasten the lid and remove it carefully. Lift the lid with the underside away from you so that the steam coming out of the canner does not burn your face.

- A one-gallon container is needed for each five pounds of fresh vegetables. Therefore, a five-gallon stone crock is of ideal size for fermenting about 25 pounds of fresh cabbage or cucumbers.
- Food-grade plastic and glass containers are excellent substitutes for stone crocks. Other one- to three-gallon non-food-grade plastic containers may be used if lined inside with a clean food-grade plastic bag. **Caution: Be certain**

PROCESS TIMES FOR MEAT STOCK IN A DIAL-GAUGE PRESSURE CANNER*

Style of Pack	Jar Size	Process Time	Canner Pressure (PSI) at Altitudes of:			
			0–2,000 ft	2,001–4,000 ft	4,001–6,000 ft	6,001–8,000 ft
Hot	Pints	20 minutes	11 lbs	12 lbs	13 lbs	14 lbs
	Quarts	25 minutes	11 lbs	12 lbs	13 lbs	14 lbs

*After the canner is completely depressurized, remove the weight from the vent port or open the petcock. Wait ten minutes; then unfasten the lid and remove it carefully. Lift the lid with the underside away from you so that the steam coming out of the canner does not burn your face.

PROCESS TIMES FOR MEAT STOCK IN A WEIGHTED-GAUGE PRESSURE CANNER*

Style of Pack	Jar Size	Process Time	Canner Pressure (PSI) at Altitudes of:	
			0–1,000 ft	Above 1,000 ft
Hot	Pints	20 minutes	10 lbs	15 lbs
	Quarts	25 minutes	10 lbs	15 lbs

*After the canner is completely depressurized, remove the weight from the vent port or open the petcock. Wait ten minutes; then unfasten the lid and remove it carefully. Lift the lid with the underside away from you so that the steam coming out of the canner does not burn your face.

that foods contact only food-grade plastics. Do not use garbage bags or trash liners.
- Fermenting sauerkraut in quart and half-gallon mason jars is an acceptable practice, but may result in more spoilage losses.

Some vegetables, like cabbage and cucumbers, need to be kept 1 to 2 inches under brine while fermenting. If you find them floating to top of the container, here are some suggestions:

- After adding prepared vegetables and brine, insert a suitably sized dinner plate or glass pie plate inside the fermentation container. The plate must be slightly smaller than the container opening, yet large enough

some are brined several hours or overnight, then drained and covered with vinegar and seasonings. Fruit pickles usually are prepared by heating fruit in a seasoned syrup acidified with either lemon juice or vinegar. Relishes are made from chopped fruits and vegetables that are cooked with seasonings and vinegar.

Be sure to remove and discard a $1/16$-inch slice from the blossom end of fresh cucumbers. Blossoms may contain an enzyme which causes excessive softening of pickles.

Caution: The level of acidity in a pickled product is as important to its safety as it is to taste and texture.

- **Do not alter vinegar, food, or water proportions in a recipe or use a vinegar with unknown acidity.**
- **Use only recipes with tested proportions of ingredients.**
- **There must be a minimum, uniform level of acid throughout the mixed product to prevent the growth of botulinum bacteria.**

Ingredients

Select fresh, firm fruits or vegetables free of spoilage. Measure or weigh amounts carefully, because the proportion of fresh food to other ingredients will affect flavor and, in many instances, safety.

Use canning or pickling salt. Noncaking material added to other salts may make the brine cloudy. Since flake salt varies in density, it is not recommended for making pickled and fermented foods. White granulated and brown sugars are most often used. Corn syrup and honey, unless called for in reliable recipes, may produce undesirable flavors. White distilled and cider vinegars of 5 percent acidity (50 grain) are recommended. White vinegar is usually preferred when light color is desirable, as is the case with fruits and cauliflower.

Pickles with reduced salt content

In the making of fresh-pack pickles, cucumbers are acidified quickly with vinegar. Use only tested recipes formulated to produce the proper acidity. While these pickles may be prepared safely with reduced or no salt, their quality may be noticeably lower. Both texture and flavor may be slightly, but noticeably, different than expected. You may wish to make small quantities first to determine if you like them.

However, the salt used in making fermented sauerkraut and brined pickles not only provides characteristic flavor but also

to cover most of the shredded cabbage or cucumbers.

- To keep the plate under the brine, weight it down with two to three sealed quart jars filled with water. Covering the container opening with a clean, heavy bath towel helps to prevent contamination from insects and molds while the vegetables are fermenting.
- Fine quality fermented vegetables are also obtained when the plate is weighted down with a very large, clean, plastic bag filled with three quarts of water containing $4\frac{1}{2}$ table-spoons of salt. Be sure to seal the plastic bag. Freezer bags sold for packaging turkeys are suitable for use with five-gallon containers.

Be sure to wash the fermentation container, plate, and jars in hot sudsy water, and rinse well with very hot water before use.

Regular dill pickles and sauerkraut are fermented and cured for about three weeks. Refrigerator dills are fermented for about one week. During curing, colors and flavors change and acidity increases. Fresh-pack or quick-process pickles are not fermented;

is vital to safety and texture. In fermented foods, salt favors the growth of desirable bacteria while inhibiting the growth of others. **Caution: Do not attempt to make sauer-kraut or fermented pickles by cutting back on the salt required.**

Preventing spoilage

Pickle products are subject to spoilage from microorganisms, particularly yeasts and molds, as well as enzymes that may affect flavor, color, and texture. Processing the pickles in a boiling-water canner will prevent both of these problems. Standard canning jars and self-sealing lids are recommended. Processing times and procedures will vary according to food acidity and the size of food pieces.

Dill Pickles

Feel free to alter the spices in this recipe, but stick to the same proportion of cucumbers, vinegar, and water. Check the label of your vinegar to be sure it contains 5 percent acetic acid. Fully fermented pickles may be stored in the original container for about four

to six months, provided they are refrigerated and surface scum and molds are removed regularly, but canning is a better way to store fully fermented pickles.

Ingredients

- Use the following quantities for each gallon capacity of your container:
- 4 lbs. of 4-inch pickling cucumbers
- 2 tbsps dill seed or 4 to 5 heads fresh or dry dill weed
- ½ cup salt
- ¼ cup vinegar (5 percent acetic acid)
- 8 cups water and one or more of the following ingredients:
- 2 cloves garlic (optional)
- 2 dried red peppers (optional)
- 2 tsp whole mixed pickling spices (optional)

Directions

1. Wash cucumbers. Cut ¹⁄₁₆-inch slice off blossom end and discard. Leave ¼ inch of stem attached. Place half of dill and spices on bottom of a clean, suitable container.
2. Add cucumbers, remaining dill, and spices. Dissolve salt in vinegar and water and pour over cucumbers. Add suitable cover and weight. Store where temperature is between 70 and 75°F for about 3 to 4 weeks while fermenting. Temperatures of 55 to 65°F are acceptable, but the fermentation will take 5 to 6 weeks. Avoid temperatures above 80°F, or pickles will become too soft during fermentation. Fermenting pickles cure slowly. Check the container several times a week and promptly remove surface scum or mold. **Caution: If the pickles become soft, slimy, or develop a disagreeable odor, discard them.**
3. Once fully fermented, pour the brine into a pan, heat slowly to a boil, and

PROCESS TIMES FOR DILL PICKLES IN A BOILING-WATER CANNER*

Style of Pack	Jar Size	Process Time at Altitudes of:		
		0–1,000 ft	1,001–6,000 ft	Above 6,000 ft
Raw	Pints	10 minutes	15 minutes	20 minutes
	Quarts	15 minutes	20 minutes	25 minutes

*After the process is complete, turn off the heat and remove the canner lid. Wait five minutes before removing jars.

simmer 5 minutes. Filter brine through paper coffee filters to reduce cloudiness, if desired. Fill jars with pickles and hot brine, leaving ½-inch headspace. Adjust lids and process in a boiling water canner, or use the low-temperature pasteurization treatment described below.

Low-Temperature Pasteurization Treatment

The following treatment results in a better product texture but must be carefully managed to avoid possible spoilage.

1   Place jars in a canner filled halfway with warm (120 to 140°F) water. Then, add hot water to a level 1 inch above jars.
2.   Heat the water enough to maintain 180 to 185°F water temperature for 30 minutes. Check with a candy or jelly thermometer to be certain that the water temperature is at least 180°F during the entire 30 minutes. Temperatures higher than 185°F may cause unnecessary softening of pickles.

Sauerkraut

For the best sauerkraut, use firm heads of fresh cabbage. Shred cabbage and start kraut between twenty-four and forty-eight hours after harvest. This recipe yields about nine quarts.

Ingredients

- 25 lbs. cabbage
- ¾ cup canning or pickling salt

Directions

1. Work with about 5 pounds of cabbage at a time. Discard outer leaves. Rinse heads under cold running water and drain. Cut heads in quarters and remove cores. Shred or slice to the thickness of a quarter.
2. Put cabbage in a suitable fermentation container, and add 3 tablespoons of salt. Mix thoroughly, using clean hands. Pack firmly until salt draws juices from cabbage.
3. Repeat shredding, salting, and packing until all cabbage is in the container. Be sure it is deep enough so that its rim is at least 4 or 5 inches above the cabbage. If juice does not cover cabbage, add boiled and cooled brine (1½ tablespoons of salt per quart of water).
4. Add plate and weights; cover container with a clean bath towel. Store at 70 to 75°F while fermenting. At temperatures between 70 and 75°F, kraut will be fully fermented in about 3 to 4 weeks; at 60° to 65°F, fermentation may take 5 to 6 weeks. At temperatures lower than 60°F, kraut may not ferment. Above 75°F, kraut may become soft.

Note: If you weigh the cabbage down with a brine-filled bag, do not disturb the crock until normal fermentation is completed (when bubbling ceases). If you use jars as weight, you will have to check the kraut 2 to 3 times each week and remove scum if it forms. Fully fermented kraut may be kept tightly covered in the refrigerator for several months or it may be canned as follows:

Hot pack—Bring kraut and liquid slowly to a boil in a large kettle, stirring frequently. Remove from heat and fill jars rather firmly with kraut and juices, leaving ½-inch headspace.

Raw pack—Fill jars firmly with kraut and cover with juices, leaving ½-inch headspace.

5.   Adjust lids and process.

Pickled Three-Bean Salad

This is a great side dish to bring to a summer picnic or potluck. Feel free to add or adjust spices to your taste. This recipe yields about five to six half-pints.

Ingredients

- 1½ cups cut and blanched green or yellow beans (prepared as below)
- 1½ cups canned, drained red kidney beans
- 1 cup canned, drained garbanzo beans
- ½ cup peeled and thinly sliced onion (about 1 medium onion)
- ½ cup trimmed and thinly sliced celery (1½ medium stalks)
- ½ cup sliced green peppers (½ medium pepper)
- ½ cup white vinegar (5 percent acetic acid)
- ¼ cup bottled lemon juice
- ¾ cup sugar
- 1¼ cups water
- ¼ cup oil
- ½ tsp canning or pickling salt

Directions

1. Wash and snap off ends of fresh beans. Cut or snap into 1- to 2-inch pieces. Blanch 3 minutes and cool immediately. Rinse kidney beans with tap water and drain again. Prepare and measure all other vegetables.
2. Combine vinegar, lemon juice, sugar, and water and bring to a boil. Remove from heat. Add oil and salt and mix well. Add beans, onions, celery, and green pepper to solution and bring to a simmer.
3. Marinate 12 to 14 hours in refrigerator, then heat entire mixture to a boil. Fill clean jars with solids. Add hot liquid, leaving ½-inch headspace. Adjust lids and process.

Pickled Horseradish Sauce

Select horseradish roots that are firm and have no mold, soft spots, or green spots. Avoid roots that have begun to sprout. The pungency of fresh horseradish fades within one to two months, even when refrigerated, so make only small quantities at a time. This recipe yields about two half-pints.

Ingredients

- 2 cups (¾ lb.) freshly grated horse-radish
- 1 cup white vinegar (5 percent acetic acid)

PROCESS TIMES FOR SAUERKRAUT IN A BOILING-WATER CANNER*

Style of Pack	Jar Size	Process Time at Altitudes of:			
		0–1,000 ft	1,001-3,000 ft	3,001-6,000 ft	Above 6,000 ft
Hot	Pints	10 minutes	15 minutes	15 minutes	20 minutes
	Quarts	15 minutes	20 minutes	20 minutes	25 minutes
Raw	Pints	20 minutes	25 minutes	30 minutes	35 minutes
	Quarts	25 minutes	30 minutes	35 minutes	40 minutes

*After the process is complete, turn off the heat and remove the canner lid. Wait five minutes before removing jars.

PROCESS TIMES FOR PICKLED THREE-BEAN SALAD IN A BOILING WATER CANNER*

Style of Pack	Jar Size	Process Time at Altitudes of:		
		0–1,000 ft	1,001–6,000 ft	Above 6,000 ft
Hot	Half-pints or Pints	15 minutes	20 minutes	25 minutes

- ½ tsp canning or pickling salt
- ¼ tsp powdered ascorbic acid

Directions

1. Wash horseradish roots thoroughly and peel off brown outer skin. Grate the peeled roots in a food processor or cut them into small cubes and put through a food grinder.
2. Combine ingredients and fill into sterile jars, leaving ¼-inch headspace. Seal jars tightly and store in a refrigerator.

Marinated Peppers

Any combination of bell, Hungarian, banana, or jalapeño peppers can be used in this recipe. Use more jalapeño peppers if you want your mix to be hot, but remember to wear rubber or plastic gloves while handling them or wash hands thoroughly with soap and water before touching your face. This recipe yields about nine half-pints.

Ingredients

- 4 lbs. firm peppers
- 1 cup bottled lemon juice
- 2 cups white vinegar (5 percent acetic acid)

- 1 tbsp oregano leaves
- 1 cup olive or salad oil
- ½ cup chopped onions
- 2 tbsp. prepared horseradish (optional)
- 2 cloves garlic, quartered (optional)
- 2¼ tsp salt (optional)

Directions

1. Select your favorite pepper. Peppers may be left whole or quartered. Wash, slash two to four slits in each pepper, and blanch in boiling water or blister

PROCESS TIMES FOR MARINATED PEPPERS IN A BOILING-WATER CANNER*

Style of Pack	Jar Size	Process Time at Altitudes of:			
		0–1,000 ft	1,001–3,000 ft	3,001–6,000 ft	Above 6,000 ft
Raw	Half-pints and pints	15 minutes	20 minutes	20 minutes	25 minutes

*After the process is complete, turn off the heat and remove the canner lid. Wait five minutes before removing jars.

PROCESS TIMES FOR PICCALILLI IN A BOILING-WATER CANNER

Style of Pack	Jar Size	Process Time at Altitudes of:		
		0–1,000 ft	1,001–6,000 ft	Above 6,000 ft
Hot	Half-pints or Pints	5 minutes	10 minutes	15 minutes

*After the process is complete, turn off the heat and remove the canner lid. Wait five minutes before removing jars.

in order to peel tough-skinned hot peppers. Blister peppers using one of the following methods:

Oven or broiler method—Place peppers in a hot oven (400°F) or broiler for 6 to 8 minutes or until skins blister.

Range-top method—Cover hot burner, either gas or electric, with heavy wire mesh. Place peppers on burner for several minutes until skins blister.

2. Allow peppers to cool. Place in pan and cover with a damp cloth. This will make peeling the peppers easier. After several minutes of cooling, peel each pepper. Flatten whole peppers.
3. Mix all remaining ingredients except garlic and salt in a saucepan and heat to boiling. Place ¼ garlic clove (optional) and ¼ teaspoon salt in each half-pint or ½ teaspoon per pint. Fill jars with peppers, and add hot, well-mixed oil/pickling solution over peppers, leaving ½-inch headspace. Adjust lids and process.

Piccalilli

Piccalilli is a nice accompaniment to roasted or braised meats and is common in British and Indian meals. It can also be mixed with mayonnaise or crème fraîche as the basis of a French remoulade. This recipe yields nine half-pints.

Ingredients

- 6 cups chopped green tomatoes
- 1½ cups chopped sweet red peppers
- 1½ cups chopped green peppers
- 2¼ cups chopped onions
- 7½ cups chopped cabbage

- ½ cup canning or pickling salt
- 3 tbsps whole mixed pickling spice
- 4½ cups vinegar (5 percent acetic acid)
- 3 cups brown sugar

Directions

1. Wash, chop, and combine vegetables with salt. Cover with hot water and let stand 12 hours. Drain and press in a clean white cloth to remove all possible liquid.
2. Tie spices loosely in a spice bag and add to combined vinegar and brown sugar and heat to a boil in a saucepan. Add vegetables and boil gently 30 minutes or until the volume of the mixture is reduced by one-half. Remove spice bag.
3. Fill hot sterile jars with hot mixture, leaving ½-inch headspace. Adjust lids and process.

Bread-and-Butter Pickles

These slightly sweet, spiced pickles will add flavor and crunch to any sandwich. If desired, slender (1 to 1½ inches in diameter) zucchini or yellow summer squash can be substituted for cucumbers. After processing and cooling, jars should be stored four to five weeks to develop ideal flavor. This recipe yields about eight pints.

Ingredients

- 6 lbs. of 4- to 5-inch pickling cucumbers
- 8 cups thinly sliced onions (about 3 pounds)
- ½ cup canning or pickling salt
- 4 cups vinegar (5 percent acetic acid)
- 4½ cups sugar
- 2 tbsp mustard seed
- 1½ tbsp celery seed
- 1 tbsp ground turmeric
- 1 cup pickling lime (optional—for use in variation below for making firmer pickles)

Directions

1. Wash cucumbers. Cut 1/16 inch off blossom end and discard. Cut into 3/16-inch slices. Combine cucumbers and onions in a large bowl. Add salt. Cover with 2 inches crushed or cubed ice. Refrigerate 3 to 4 hours, adding more ice as needed.
2. Combine remaining ingredients in a large pot. Boil ten minutes. Drain cucumbers and onions, add to pot, and slowly reheat to boiling. Fill jars with slices and cooking syrup, leaving ½-inch headspace.
3. Adjust lids and process in boiling-water canner, or use the low-temperature pasteurization treatment described below.

Low-Temperature Pasteurization Treatment

The following treatment results in a better product texture but must be carefully managed to avoid possible spoilage.

PROCESS TIMES FOR BREAD-AND-BUTTER PICKLES IN A BOILING-WATER CANNER*

Style of Pack	Jar Size	Process Time at Altitudes of:		
		0–1,000 ft	1,001–6,000 ft	Above 6,000 ft
Hot	Pints or Quarts	10 minutes	15 minutes	20 minutes

*After the process is complete, turn off the heat and remove the canner lid. Wait five minutes before removing jars.

Canning

1. Place jars in a canner filled halfway with warm (120 to 140°F) water. Then, add hot water to a level 1 inch above jars.
2. Heat the water enough to maintain 180 to 185°F water temperature for 30 minutes. Check with a candy or jelly thermometer to be certain that the water temperature is at least 180°F during the entire 30 minutes. Temperatures higher than 185°F may cause unnecessary softening of pickles.

Variation for firmer pickles: Wash cucumbers. Cut ¹⁄₁₆ inch off blossom end and discard. Cut into ³⁄₁₆-inch slices. Mix 1 cup pickling lime and ½ cup salt to 1 gallon water in a 2- to 3-gallon crock or enamelware container. Avoid inhaling lime dust while mixing the lime-water solution. Soak cucumber slices in lime water for twelve to twenty-four hours, stirring occasionally. Remove from lime solution, rinse, and resoak one hour in fresh cold water. Repeat the rinsing and soaking steps two more times. Handle carefully, as slices will be brittle. Drain well.

PROCESS TIMES FOR QUICK FRESH-PACK DILL PICKLES IN A BOILING-WATER CANNER*

Style of Pack	Jar Size	Process Time at Altitudes of:		
		0–1,000 ft	1,001–6,000 ft	Above 6,000 ft
Raw	Pints	10 minutes	15 minutes	20 minutes
	Quarts	15 minutes	20 minutes	25 minutes

*After the process is complete, turn off the heat and remove the canner lid. Wait five minutes before removing jars.

PROCESS TIMES FOR PICKLE RELISH IN A BOILING-WATER CANNER*

Style of Pack	Jar Size	Process Time at Altitudes of:		
		0–1,000 ft	1,001–6,000 ft	Above 6,000 ft
Hot	Half-pints or Pints	10 minutes	15 minutes	20 minutes

*After the process is complete, turn off the heat and remove the canner lid. Wait five minutes before removing jars.

Quick Fresh-Pack Dill Pickles

For best results, pickle cucumbers within twenty-four hours of harvesting, or immediately after purchasing. This recipe yields seven to nine pints.

Ingredients

- 8 lbs. of 3- to 5-inch pickling cucumbers
- 2 gallons water
- 1¼ to 1½ cups canning or pickling salt
- 1½ qts vinegar (5 percent acetic acid)
- ¼ cup sugar
- 2 to 2¼ quarts water
- 2 tbsp whole mixed pickling spice
- 3 to 5 tbsp whole mustard seed (2 tsps to 1 tsp per pint jar)
- 14 to 21 heads of fresh dill (1½ to 3 heads per pint jar) *or*
- 4½ to 7 tbsps dill seed (1½ tsp to 1 tbsp per pint jar)

Directions

1. Wash cucumbers. Cut ¹⁄₁₆-inch slice off blossom end and discard, but leave ¼-inch of stem attached. Dissolve ¾ cup salt in 2 gallons water. Pour over cucumbers and let stand twelve hours. Drain.
2. Combine vinegar, ½ cup salt, sugar and 2 quarts water. Add mixed pickling spices tied in a clean white cloth. Heat to boiling. Fill jars with cucumbers. Add 1 tsp mustard seed and 1½ heads fresh dill per pint.
3. Cover with boiling pickling solution, leaving ½-inch headspace. Adjust lids and process.

Pickle Relish

A food processor will make quick work of chopping the vegetables in this recipe. Yields about nine pints.

Ingredients

- 3 qts chopped cucumbers
- 3 cups each of chopped sweet green and red peppers
- 1 cup chopped onions
- ¾ cup canning or pickling salt
- 4 cups ice
- 8 cups water
- 4 tsp each of mustard seed, turmeric, whole allspice, and whole cloves
- 2 cups sugar
- 6 cups white vinegar (5 percent acetic acid)

Directions

1. Add cucumbers, peppers, onions, salt, and ice to water and let stand 4 hours. Drain and re-cover vegetables with fresh ice water for another hour. Drain again.
2. Combine spices in a spice or cheese-cloth bag. Add spices to sugar and vinegar. Heat to boiling and pour mixture over vegetables. Cover and refrigerate 24 hours.
3. Heat mixture to boiling and fill hot into clean jars, leaving ½-inch headspace. Adjust lids and process.

Quick Sweet Pickles

Quick and simple to prepare, these are the sweet pickles to make when you're short on time. After processing and cooling, jars should be stored four to five weeks to develop ideal flavor. If desired, add two slices of raw whole onion to each jar before filling with cucumbers. This recipe yields about seven to nine pints.

Ingredients

- 8 lbs of 3- to 4-inch pickling cucumbers
- 1/3 cup canning or pickling salt
- 4½ cups sugar
- 3½ cups vinegar (5 percent acetic acid)
- 2 tsp celery seed
- 1 tbsp whole allspice
- 2 tbsp mustard seed
- 1 cup pickling lime (optional)

Directions

1. Wash cucumbers. Cut 1/16 inch off blossom end and discard, but leave ¼

PROCESS TIMES FOR QUICK SWEET PICKLES IN A BOILING-WATER CANNER*

		0–1,000 ft	1,001–6,000 ft	Above 6,000 ft
Hot	Pints or Quarts	5 minutes	10 minutes	15 minutes
Raw	Pints	10 minutes	15 minutes	20 minutes
	Quarts	15 minutes	20 minutes	25 minutes

*After the process is complete, turn off the heat and remove the canner lid. Wait five minutes before removing jars.

PROCESS TIMES FOR REDUCED-SODIUM SLICED SWEET PICKLES IN A BOILING-WATER CANNER*

Style of Pack	Jar Size	Process Time at Altitudes of:		
		0–1,000 ft	1,001–6,000 ft	Above 6,000 ft
Hot	Pints	ten minutes	15 minutes	20 minutes

*After the process is complete, turn off the heat and remove the canner lid. Wait five minutes before removing jars.

inch of stem attached. Slice or cut in strips, if desired.

2. Place in bowl and sprinkle with salt. Cover with 2 inches of crushed or cubed ice. Refrigerate 3 to 4 hours. Add more ice as needed. Drain well.

3. Combine sugar, vinegar, celery seed, allspice, and mustard seed in 6-quart kettle. Heat to boiling.

Hot pack—Add cucumbers and heat slowly until vinegar solution returns to boil. Stir occasionally to make sure mixture heats evenly. Fill sterile jars, leaving ½-inch headspace.

Raw pack—Fill jars, leaving ½-inch headspace.

4. Add hot pickling syrup, leaving ½-inch headspace. Adjust lids and process.

Variation for firmer pickles: Wash cucumbers. Cut 1/16 inch off blossom end and discard, but leave ¼ inch of stem attached. Slice or strip cucumbers. Mix 1 cup pickling lime and 1/3 cup salt with 1 gallon water in a 2- to 3-gallon crock or enamelware container. **Caution: Avoid inhaling lime dust while mixing the lime-water solution.** Soak cucumber slices or strips in lime-water solution for twelve to twenty-four hours, stirring occasionally. Remove from lime solution, rinse, and soak 1 hour in fresh cold water. Repeat the rinsing and soaking two more times. Handle carefully, because slices or strips will be brittle. Drain well.

Reduced-Sodium Sliced Sweet Pickles

Whole allspice can be tricky to find. If it's not available at your local grocery store, it can be ordered at www.spicebarn.com or at www.gourmetsleuth.com. This recipe yields about four to five pints.

Ingredients

- 4 lbs (3- to 4-inch) pickling cucumbers

Canning syrup:
- 1²/3 cups distilled white vinegar (5 percent acetic acid)
- 3 cups sugar
- 1 tbsp whole allspice
- 2¼ tsp celery seed

Brining solution:
- 1 qt distilled white vinegar (5 percent acetic acid)
- 1 tbsp canning or pickling salt
- 1 tbsp mustard seed
- ½ cup sugar

Directions

1. Wash cucumbers and cut 1/1 inch off blossom end, and discard. Cut cucumbers into ¼-inch slices. Combine all ingredients for canning syrup in a saucepan and bring to boiling. Keep syrup hot until used.

2. In a large kettle, mix the ingredients for the brining solution. Add the cut cucumbers, cover, and simmer until the cucumbers change color from bright to dull green (about 5 to 7 minutes). Drain the cucumber slices.

3. Fill jars, and cover with hot canning syrup leaving ½-inch headspace. Adjust lids and process.

Tomatoes

Canned tomatoes should be a staple in every cook's pantry. They are easy to prepare and, when made with garden-fresh produce, make ordinary soups, pizza, or pastas into five-star meals. Be sure to select only disease-free, preferably vine-ripened, firm fruit. Do not can tomatoes from dead or frost-killed vines.

Green tomatoes are more acidic than ripened fruit and can be canned safely with the following recommendations.

- To ensure safe acidity in whole, crushed, or juiced tomatoes, add two tablespoons of bottled lemon juice or ½ teaspoon of citric acid per quart of tomatoes. For pints, use one tablespoon bottled lemon juice or ¼ teaspoon citric acid.
- Acid can be added directly to the jars before filling with product. Add sugar to offset acid taste, if desired. Four tablespoons of 5 percent acidity vinegar per quart may be used instead of lemon juice or citric acid. However, vinegar may cause undesirable flavor changes.
- Using a pressure canner will result in higher quality and more nutritious canned tomato products. If your pressure canner cannot be operated above 15 PSI, select a process time at a lower pressure.

Tomato Juice

Tomato juice is a good source of vitamin A and C and is tasty on its own or in a cocktail. It's also the secret ingredient in some very delicious cakes. If desired, add carrots, celery, and onions, or toss in a few jalapeños for a little kick.

Directions

1. Wash tomatoes, remove stems, and trim off bruised or discolored portions. To prevent juice from separating, quickly cut about 1 pound of fruit into quarters and put directly into saucepan. Heat immediately to boiling while crushing.
2. Continue to slowly add and crush freshly cut tomato quarters to the boiling mixture. Make sure the mixture

boils constantly and vigorously while you add the remaining tomatoes. Simmer 5 minutes after you add all pieces.
3. Press heated juice through a sieve or food mill to remove skins and seeds. Add bottled lemon juice or citric acid to jars (see page 170). Heat juice again to boiling.
4. Add 1 teaspoon of salt per quart to the jars, if desired. Fill jars with hot tomato juice, leaving ½-inch headspace. Adjust lids and process.

QUANTITY

- An average of 23 pounds is needed per canner load of seven quarts, or an average of 14 pounds per canner load of nine pints.
- A bushel weighs 53 pounds and yields 15 to 18 quarts of juice—an average of 3¼ pounds per quart.

PROCESS TIMES FOR TOMATO JUICE IN A DIAL-GAUGE PRESSURE CANNER*

Style of Pack	Jar Size	Process Time	Canner Gauge Pressure (PSI) at Altitudes of:			
			0–2,000 ft	2,001–4,000 ft	4,001–6,000 ft	6,001–8,000 ft
Hot	Pints or Quarts	20 minutes	6 lbs	7 lbs	8 lbs	9 lbs
		15 minutes	11 lbs	12 lbs	13 lbs	14 lbs

*After the canner is completely depressurized, remove the weight from the vent port or open the petcock. Wait ten minutes; then unfasten the lid and remove it carefully. Lift the lid with the underside away from you so that the steam coming out of the canner does not burn your face.

PROCESS TIMES FOR TOMATO JUICE IN A BOILING-WATER CANNER*

Style of Pack	Jar Size	Process Time at Altitudes of:			
		0–1,000 ft	1,001–3,000 ft	3,001–6,000 ft	Above 6,000 ft
Hot	Pints	35 minutes	40 minutes	45 minutes	50 minutes
	Quarts	40 minutes	45 minutes	50 minutes	55 minutes

*After the process is complete, turn off the heat and remove the canner lid. Wait five minutes before removing jars.

PROCESS TIMES FOR TOMATO JUICE IN A WEIGHTED-GAUGE PRESSURE CANNER

Style of Pack	Jar Size	Process Time	Canner Gauge Pressure (PSI) at Altitudes of:	
			0–1,000 ft	Above 1,000 ft
Hot	Pints or Quarts	20 minutes	5 lbs	10 lbs
		15 minutes	10 lbs	15 lbs

Crushed Tomatoes with No Added Liquid

Crushed tomatoes are great for use in soups, stews, thick sauces, and casseroles. Simmer crushed tomatoes with kidney beans, chili powder, sautéed onions, and garlic to make an easy pot of chili.

Directions

1. Wash tomatoes and dip in boiling water for 30 to 60 seconds or until skins split. Then dip in cold water, slip off skins, and remove cores. Trim off any bruised or discolored portions and quarter.
2. Heat $1/6$ of the quarters quickly in a large pot, crushing them with a wooden mallet or spoon as they are added to the pot. This will exude juice. Continue heating the tomatoes, stirring to prevent burning.
3. Once the tomatoes are boiling, gradually add remaining quartered tomatoes, stirring constantly. These remaining tomatoes do not need to be crushed; they will soften with heating and stirring. Continue until all tomatoes are added. Then boil gently 5 minutes.
4. Add bottled lemon juice or citric acid to jars (see page 170). Add 1 teaspoon of salt per quart to the jars, if desired. Fill jars immediately with hot tomatoes, leaving ½-inch headspace. Adjust lids and process.

QUANTITY

- An average of 22 pounds is needed per canner load of seven quarts.
- An average of 14 fresh pounds is needed per canner load of nine pints.
- A bushel weighs 53 pounds and yields 17 to 20 quarts of crushed tomatoes—an average of 2¾ pounds per quart.

PROCESS TIMES FOR CRUSHED TOMATOES IN A DIAL-GAUGE PRESSURE CANNER*

Style of Pack	Jar Size	Process Time	Canner Gauge Pressure (PSI) at Altitudes of:			
			0–2,000 ft	2,001–4,000 ft	4,001–6,000 ft	6,001–8,000 ft
Hot	Pints or Quarts	20 minutes	6 lbs	7 lbs	8 lbs	9 lbs
		15 minutes	11 lbs	12 lbs	13 lbs	14 lbs

*After the canner is completely depressurized, remove the weight from the vent port or open the petcock. Wait ten minutes; then unfasten the lid and remove it carefully. Lift the lid with the underside away from you so that the steam coming out of the canner does not burn your face.

PROCESS TIMES FOR CRUSHED TOMATOES IN A WEIGHTED-GAUGE PRESSURE CANNER*

Style of Pack	Jar Size	Process Time	Canner Gauge Pressure (PSI) at Altitudes of:	
			0–1,000 ft	Above 1,000 ft
Hot	Pints or Quarts	20 minutes	5 lbs	10 lbs
		15 minutes	10 lbs	15 lbs

*After the canner is completely depressurized, remove the weight from the vent port or open the petcock. Wait ten minutes; then unfasten the lid and remove it carefully. Lift the lid with the underside away from you so that the steam coming out of the canner does not burn your face.

Tomato Sauce

This plain tomato sauce can be spiced up before using in soups or in pink or red sauces. The thicker you want your sauce, the more tomatoes you'll need.

Directions

1. Prepare and press as for making tomato juice (see page 225). Simmer in a large saucepan until sauce reaches desired consistency. Boil until volume is reduced by about one-third for thin sauce, or by one-half for thick sauce.

QUANTITY

For thin sauce:
- An average of 35 pounds is needed per canner load of seven quarts.
- An average of 21 pounds is needed per canner load of nine pints.
- A bushel weighs 53 pounds and yields 10 to 12 quarts of sauce—an average of five pounds per quart.

For thick sauce:
- An average of 46 pounds is needed per canner load of seven quarts.
- An average of 28 pounds is needed per canner load of nine pints.
- A bushel weighs 53 pounds and yields seven to nine quarts of sauce—an average of 6½ pounds per quart.

PROCESS TIMES FOR CRUSHED TOMATOES IN A BOILING-WATER CANNER*

Style of Pack	Jar Size	Process Time at Altitudes of:			
		0–1,000 ft	1,001–3,000 ft	3,001–6,000 ft	Above 6,000 ft
Hot	Pints	35 minutes	40 minutes	45 minutes	50 minutes
	Quarts	45 minutes	50 minutes	55 minutes	60 minutes
*After the process is complete, turn off the heat and remove the canner lid. Wait five minutes before removing jars.					

PROCESS TIMES FOR TOMATO SAUCE IN A BOILING-WATER CANNER*

Style of Pack	Jar Size	Process Time at Altitudes of:			
		0–1,000 ft	1,001–3,000 ft	3,001–6,000 ft	Above 6,000 ft
Hot	Pints	35 minutes	40 minutes	45 minutes	50 minutes
	Quarts	40 minutes	45 minutes	50 minutes	55 minutes
*After the process is complete, turn off the heat and remove the canner lid. Wait five minutes before removing jars.					

PROCESS TIMES FOR TOMATO SAUCE IN A DIAL-GAUGE PRESSURE CANNER*

Style of Pack	Jar Size	Process Time	Canner Gauge Pressure (PSI) at Altitudes of:			
			0–2,000 ft	2,001–4,000 ft	4,001–6,000 ft	6,001–8,000 ft
Hot	Pints or Quarts	20 minutes	6 lbs	7 lbs	8 lbs	9 lbs
		15 minutes	11 lbs	12 lbs	13 lbs	14 lbs
*After the canner is completely depressurized, remove the weight from the vent port or open the petcock. Wait ten minutes; then unfasten the lid and remove it carefully. Lift the lid with the underside away from you so that the steam coming out of the canner does not burn your face.						

PROCESS TIMES FOR TOMATO SAUCE IN A WEIGHTED-GAUGE PRESSURE CANNER*

Style of Pack	Jar Size	Process Time	Canner Gauge Pressure (PSI) at Altitudes of:	
			0–1,000 ft	Above 1,000 ft
Hot	Pints or Quarts	20 minutes	5 lbs	10 lbs
		15 minutes	10 lbs	15 lbs
*After the canner is completely depressurized, remove the weight from the vent port or open the petcock. Wait ten minutes; then unfasten the lid and remove it carefully. Lift the lid with the underside away from you so that the steam coming out of the canner does not burn your face.				

2. Add bottled lemon juice or citric acid to jars (see page 170). Add 1 teaspoon of salt per quart to the jars, if desired. Fill jars, leaving ¼-inch headspace. Adjust lids and process.

QUANTITY

- An average of 21 pounds is needed per canner load of seven quarts.
- An average of 13 pounds is needed per canner load of nine pints.
- A bushel weighs 53 pounds and yields 15 to 21 quarts—an average of three pounds per quart.

Tomatoes, Whole or Halved, Packed in Water

Whole or halved tomatoes are used for scalloped tomatoes, savory pies (baked in a pastry crust with parmesan cheese, mayonnaise, and seasonings), or stewed tomatoes.

Directions

1. Wash tomatoes. Dip in boiling water for 30 to 60 seconds or until skins split; then dip in cold water. Slip off skins and remove cores. Leave whole or halve.
2. Add bottled lemon juice or citric acid to jars (see page 170). Add 1 teaspoon

PROCESS TIMES FOR WATER-PACKED WHOLE TOMATOES IN A BOILING-WATER CANNER*

Style of Pack	Jar Size	Process Time at Altitudes of:			
		0–1,000 ft	1,001–3,000 ft	3,001–6,000 ft	Above 6,000 ft
Hot or Raw	Pints	40 minutes	45 minutes	50 minutes	55 minutes
	Quarts	45 minutes	50 minutes	55 minutes	60 minutes
*After the process is complete, turn off the heat and remove the canner lid. Wait five minutes before removing jars.					

PROCESS TIMES FOR WATER-PACKED WHOLE TOMATOES IN A DIAL-GAUGE PRESSURE CANNER*

Style of Pack	Jar Size	Process Time	Canner Gauge Pressure (PSI) at Altitudes of:			
			0–2,000 ft	2,001–4,000 ft	4,001–6,000 ft	6,001–8,000 ft
Hot or Raw	Pints or Quarts	15 minutes	6 lbs	7 lbs	8 lbs	9 lbs
		10 minutes	11 lbs	12 lbs	13 lbs	14 lbs

*After the canner is completely depressurized, remove the weight from the vent port or open the petcock. Wait ten minutes; then unfasten the lid and remove it carefully. Lift the lid with the underside away from you so that the steam coming out of the canner does not burn your face.

PROCESS TIMES FOR WATER-PACKED WHOLE TOMATOES IN A WEIGHTED-GAUGE PRESSURE CANNER*

Style of Pack	Jar Size	Process Time	Canner Gauge Pressure (PSI) at Altitudes of:	
			0–1,000 ft	Above 1,000 ft
Hot or Raw	Pints or Quarts	15 minutes	5 lbs	10 lbs
		10 minutes	10 lbs	15 lbs

*After the canner is completely depressurized, remove the weight from the vent port or open the petcock. Wait ten minutes; then unfasten the lid and remove it carefully. Lift the lid with the underside away from you so that the steam coming out of the canner does not burn your face.

Spaghetti Sauce Without Meat

Homemade spaghetti sauce is like a completely different food than store-bought varieties—it tastes fresher, is more flavorful, and is far more nutritious. Adjust spices to taste, but do not increase proportions of onions, peppers, or mushrooms. This recipe yields about nine pints.

Ingredients

- 30 lbs. tomatoes
- 1 cup chopped onions
- 5 cloves garlic, minced
- 1 cup chopped celery or green pepper
- 1 lb. fresh mushrooms, sliced (optional)
- 4½ tsp salt
- 2 tbsps oregano
- 4 tbsps minced parsley
- 2 tsps black pepper
- ¼ cup brown sugar
- ¼ cup vegetable oil

Directions

1. Wash tomatoes and dip in boiling water for 30 to 60 seconds or until skins split. Dip in cold water and slip off skins. Remove cores and quarter tomatoes. Boil 20 minutes, uncovered, in large saucepan. Put through food mill or sieve.

of salt per quart to the jars, if desired. For hot pack products, add enough water to cover the tomatoes and boil them gently for 5 minutes.

3. Fill jars with hot tomatoes or with raw peeled tomatoes. Add the hot cooking liquid to the hot pack, or hot water for raw pack to cover, leaving ½-inch headspace. Adjust lids and process.

PROCESS TIMES FOR SPAGHETTI SAUCE WITHOUT MEAT IN A DIAL-GAUGE PRESSURE CANNER*

Style of Pack	Jar Size	Process Time	Canner Gauge Pressure (PSI) at Altitudes of:			
			0–2,000 ft	2,001–4,000 ft	4,001–6,000 ft	6,001–8,000 ft
Hot	Pints	20 minutes	11 lbs	12 lbs	13 lbs	14 lbs
	Quarts	25 minutes	11 lbs	12 lbs	13 lbs	14 lbs

*After the canner is completely depressurized, remove the weight from the vent port or open the petcock. Wait ten minutes; then unfasten the lid and remove it carefully. Lift the lid with the underside away from you so that the steam coming out of the canner does not burn your face.

PROCESS TIMES FOR SPAGHETTI SAUCE WITHOUT MEAT IN A WEIGHTED-GAUGE PRESSURE CANNER*

Style of Pack	Jar Size	Process Time	Canner Gauge Pressure (PSI) at Altitudes of:	
			0–1,000 ft	Above 1,000 ft
Hot	Pints	20 minutes	10 lbs	15 lbs
	Quarts	25 minutes	10 lbs	15 lbs

*After the canner is completely depressurized, remove the weight from the vent port or open the petcock. Wait ten minutes; then unfasten the lid and remove it carefully. Lift the lid with the underside away from you so that the steam coming out of the canner does not burn your face.

2. Sauté onions, garlic, celery or peppers, and mushrooms (if desired) in vegetable oil until tender. Combine sautéed vegetables and tomatoes and add spices, salt, and sugar. Bring to a boil.

3. Simmer uncovered, until thick enough for serving. Stir frequently to avoid burning. Fill jars, leaving 1-inch headspace. Adjust lids and process.

Tomato Ketchup

Ketchup forms the base of several condiments, including Thousand Island dressing, fry sauce, and barbecue sauce. And, of course, it's an American favorite in its own right. This recipe yields six to seven pints.

Ingredients

- 24 lbs. ripe tomatoes
- 3 cups chopped onions
- ¾ tsp ground red pepper (cayenne)
- 4 tsps whole cloves
- 3 sticks cinnamon, crushed
- 1½ tsp whole allspice
- 3 tbsps celery seeds
- 3 cups cider vinegar (5 percent acetic acid)
- 1½ cups sugar
- ¼ cup salt

Directions

1. Wash tomatoes. Dip in boiling water for 30 to 60 seconds or until skins split. Dip in cold water. Slip off skins and remove cores. Quarter tomatoes into 4-gallon stockpot or a large kettle. Add onions and red pepper. Bring to boil and simmer 20 minutes, uncovered.

2. Combine remaining spices in a spice bag and add to vinegar in a 2-quart saucepan. Bring to boil. Turn off heat and let stand until tomato mixture has been cooked 20 minutes. Then, remove spice bag and combine vinegar and tomato mixture. Boil about 30 minutes.

3. Put boiled mixture through a food mill or sieve. Return to pot. Add sugar and salt, boil gently, and stir frequently until volume is reduced by one-half or until mixture rounds up on spoon without separation. Fill pint jars, leaving ⅛-inch headspace. Adjust lids and process.

Chile Salsa (Hot Tomato-Pepper Sauce)

For fantastic nachos, cover corn chips with chile salsa, add shredded Monterey jack or cheddar cheese, bake under broiler for about five minutes, and serve with guacamole and sour cream. Be sure to wear rubber gloves while handling chiles or wash hands thoroughly with soap and water before touching your face. This recipe yields six to eight pints.

Ingredients

- 5 lbs. tomatoes
- 2 lbs. chile peppers
- 1 lb onions
- 1 cup vinegar (5 percent)
- 3 tsp salt
- ½ tsp pepper

Directions

1. Wash and dry chiles. Slit each pepper on its side to allow steam to escape. Peel peppers using one of the following methods:

   Oven or broiler method: Place chiles in oven (400°F) or broiler for 6 to 8 minutes until skins blister. Cool and slip off skins.

   Range-top method: Cover hot burner, either gas or electric, with heavy wire mesh. Place chiles on burner for several minutes until skins blister. Allow peppers to cool. Place in a pan and cover with a damp cloth. This will make peeling the peppers easier. After several minutes, peel each pepper.

2. Discard seeds and chop peppers. Wash tomatoes and dip in boiling water for 30 to 60 seconds or until skins split. Dip in cold water, slip off skins, and remove cores.

3. Coarsely chop tomatoes and combine chopped peppers, onions, and remaining ingredients in a large saucepan. Heat to boil, and simmer ten minutes. Fill jars, leaving ½-inch headspace. Adjust lids and process.

PROCESS TIMES FOR TOMATO KETCHUP IN A BOILING-WATER CANNER*

Style of Pack	Jar Size	Process Time at Altitudes of:		
		0–1,000 ft	1,001–6,000 ft	Above 6,000 ft
Hot	Pints	15 minutes	20 minutes	25 minutes

*After the process is complete, turn off the heat and remove the canner lid. Wait five minutes before removing jars.

PROCESS TIMES FOR CHILE SALSA IN A BOILING-WATER CANNER*

Style of Pack	Jar Size	Process Time at Altitudes of:		
		0–1,000 ft	1,001–6,000 ft	Above 6,000 ft
Hot	Pints	15 minutes	20 minutes	25 minutes

*After the process is complete, turn off the heat and remove the canner lid. Wait five minutes before removing jars.

Drying and Freezing

Drying

Drying fruits, vegetables, herbs, and even meat is a great way to preserve foods for longer-term storage, especially if your pantry or freezer space is limited. Dried foods take up much less space than their fresh, frozen, or canned counterparts. Drying requires relatively little preparation time, and is simple enough that kids will enjoy helping. Drying with a food dehydrator will ensure the fastest, safest, and best quality results. However, you can also dry produce in the sunshine, in your oven, or strung up over a woodstove.

For more information on food drying, check out *So Easy to Preserve, 5th ed.* from the Cooperative Extension Service, the University of Georgia. Much of the information that follows is adapted from this excellent source.

Drying with a Food Dehydrator

Food dehydrators use electricity to produce heat and have a fan and vents for air circulation. Dehydrators are efficiently designed to dry foods fast at around 140°F. Look for food dehydrators in discount department stores, mail-order catalogs, the small appliance section of a department store, natural food stores, and seed or garden supply catalogs. Costs vary depending on

features. Some models are expandable and additional trays can be purchased later. Twelve square feet of drying space dries about a half-bushel of produce.

Dehydrator Features to Look For

- Double-wall construction of metal or high-grade plastic. Wood is not recommended, because it is a fire hazard and is difficult to clean.
- Enclosed heating elements
- Countertop design
- An enclosed thermostat from 85 to 160°F
- Fan or blower
- Four to ten open mesh trays made of sturdy, lightweight plastic for easy washing
- Underwriters Laboratory (UL) seal of approval
- A one-year guarantee
- Convenient service
- A dial for regulating temperature
- A timer. Often the completed drying time may occur during the night, and a timer turns the dehydrator off to prevent scorching.

Types of Dehydrators

There are two basic designs for dehydrators. One has horizontal air flow and the other has vertical air flow. In units with horizontal flow, the heating element and fan are located on the side of the unit. The major advantages of horizontal flow are: it reduces flavor mixture so several different foods can be dried at one time; all trays receive equal heat penetration; and juices or liquids do not drip down into the heating element. Vertical air flow dehydrators have the heating element and fan located at the base. If different foods are dried, flavors can mix and liquids can drip into the heating element.

Fruit Drying Procedures

Apples—Select mature, firm apples. Wash well. Pare, if desired, and core. Cut in rings or slices $1/8$ to $1/4$ inch thick or cut in quarters or eighths. Soak in ascorbic acid, vinegar, or lemon juice for ten minutes. Remove from solution and drain well. Arrange in single layer on trays, pit side up. Dry until soft, pliable, and leathery; there should be no moist area in center when cut.

Apricots—Select firm, fully ripe fruit. Wash well. Cut in half and remove pit. Do not peel. Soak in ascorbic acid, vinegar, or lemon juice for ten minutes. Remove from solution and drain well. Arrange in single layer on trays, pit side up with cavity popped up to expose more flesh to the air. Dry until

soft, pliable, and leathery; there should be no moist area in center when cut.

Bananas—Select firm, ripe fruit. Peel. Cut in $1/8$-inch slices. Soak in ascorbic acid, vinegar, or lemon juice for ten minutes. Remove and drain well. Arrange in single layer on trays. Dry until tough and leathery.

Berries—Select firm, ripe fruit. Wash well. Leave whole or cut in half. Dip in boiling water thirty seconds to crack skins. Arrange on drying trays not more than two berries deep. Dry until hard and berries rattle when shaken on trays.

Cherries—Select fully ripe fruit. Wash well. Remove stems and pits. Dip whole cherries in boiling water thirty seconds to crack skins. Arrange in single layer on trays. Dry until tough, leathery, and slightly sticky.

Citrus peel—Select thick-skinned oranges with no signs of mold or decay and no color added to skin. Scrub oranges well with brush under cool running water. Thinly peel outer $1/16$ to $1/8$ inch of the peel; avoid white bitter part. Soak in ascorbic acid, vinegar, or lemon juice for ten minutes. Remove from solution

and drain well. Arrange in single layers on trays. Dry at 130°F for one to two hours; then at 120°F until crisp.

Figs—Select fully ripe fruit. Wash or clean well with damp towel. Peel dark-skinned varieties if desired. Leave whole if small or partly dried on tree; cut large figs in halves or slices. If drying whole figs, crack skins by dipping in boiling water for thirty seconds. For cut figs, soak in ascorbic acid, vinegar, or lemon juice for ten minutes. Remove and drain well. Arrange in single layers on trays. Dry until leathery and pliable.

Grapes and black currants—Select seedless varieties. Wash, sort, and remove stems. Cut in half or leave whole. If drying whole, crack skins by dipping in boiling water for thirty seconds. If halved, dip in ascorbic acid or other antimicrobial solution for ten minutes. Remove and drain well. Dry until pliable and leathery with no moist center.

Melons—Select mature, firm fruits that are heavy for their size; cantaloupe dries better than watermelon. Scrub outer surface well with brush under cool running water. Remove outer skin, any fibrous tissue, and seeds. Cut into ¼- to ½-inch-thick slices. Soak in ascorbic acid, vinegar, or lemon juice for ten minutes. Remove and drain well. Arrange in single layer on trays. Dry until leathery and pliable with no pockets of moisture.

Nectarines and peaches—Select ripe, firm fruit. Wash and peel. Cut in half and remove pit. Cut in quarters or slices if desired.

Soak in ascorbic acid, vinegar, or lemon juice for ten minutes. Remove and drain well. Arrange in single layer on trays, pit side up. Turn halves over when visible juice disappears. Dry until leathery and somewhat pliable.

Pears—Select ripe, firm fruit. Bartlett variety is recommended. Wash fruit well. Pare, if desired. Cut in half lengthwise and core. Cut in quarters, eighths, or slices 1/8 to ¼ inch thick. Soak in ascorbic acid, vinegar, or lemon juice for ten minutes. Remove and drain. Arrange in single layer on trays, pit side up. Dry until springy and suede-like with no pockets of moisture.

Plums and prunes—Wash well. Leave whole if small; cut large fruit into halves (pit removed) or slices. If left whole, crack skins in boiling water one to two minutes. If cut in half, dip in ascorbic acid or other antimicrobial solution for ten minutes. Remove and drain. Arrange in single layer on trays, pit side up, cavity popped out. Dry until pliable and leathery; in whole prunes, pit should not slip when squeezed.

Fruit Leathers

Fruit leathers are a tasty and nutritious alternative to store-bought candies that are full of artificial sweeteners and preservatives. Blend the leftover fruit pulp from making jelly or use fresh, frozen, or drained canned fruit. Ripe or slightly overripe fruit works best.

Drying and Freezing

Chances are the fruit leather will get eaten before it makes it into the cupboard, but it can keep up to one month at room temperature. For storage up to one year, place tightly wrapped rolls in the freezer.

Ingredients

- 2 cups fruit
- 2 tsp lemon juice or $1/8$ tsp ascorbic acid (optional)
- $1/4$ to $1/2$ cup sugar, corn syrup, or honey (optional)

Directions

1. Wash fresh fruit or berries in cool water. Remove peel, seeds, and stem.
2. Cut fruit into chunks. Use 2 cups of fruit for each 13 x 15-inch inch fruit leather. Purée fruit until smooth.
3. Add 2 teaspoons of lemon juice or $1/8$ teaspoon ascorbic acid (375 mg) for each 2 cups light-colored fruit to prevent darkening.
4. Optional: To sweeten, add corn syrup, honey, or sugar. Corn syrup or honey is best for longer storage because

SPICES, FLAVORS, AND GARNISHES

To add interest to your fruit leathers, include spices, flavorings, or garnishes.

- **Spices to try**—Allspice, cinnamon, cloves, coriander, ginger, mace, mint, nutmeg, or pumpkin pie spice. Use sparingly; start with $1/8$ teaspoon for each 2 cups of purée.
- **Flavorings to try**—Almond extract, lemon juice, lemon peel, lime juice, lime peel, orange extract, orange juice, orange peel, or vanilla extract. Use sparingly; try $1/8$ to $1/4$ teaspoon for each 2 cups of purée.
- **Delicious additions to try**—Shredded coconut, chopped dates, other dried chopped fruits, granola, miniature marshmallows, chopped nuts, chopped raisins, poppy seeds, sesame seeds, or sunflower seeds.
- **Fillings to try**—Melted chocolate, softened cream cheese, cheese spreads, jam, preserves, marmalade, marshmallow cream, or peanut butter. Spread one or more of these on the leather after it is dried and then roll. Store in refrigerator.

these sweeteners prevent crystals. Sugar is fine for immediate use or short storage. Use $1/4$ to $1/2$ cup sugar, corn syrup, or honey for each 2 cups of fruit. Avoid aspartame sweeteners as they may lose sweetness during drying.

5. Pour the leather. Fruit leathers can be poured into a single large sheet (13 x 15 inches) or into several smaller sizes. Spread purée evenly, about $1/8$ inch thick, onto drying tray. Avoid pouring purée too close to the edge of the cookie sheet.
6. Dry the Leather. Dry fruit leathers at 140°F. Leather dries from the outside edge toward the center. Larger fruit leathers take longer to dry. Approximate drying times are 6 to 8 hours in a

HINTS

- Applesauce can be dried alone or added to any fresh fruit purée as an extender. It decreases tartness and makes the leather smoother and more pliable.
- To dry fruit in the oven, a 13 x 15-inch cookie pan with *edges* works well. Line pan with plastic wrap, being careful to smooth out wrinkles. Do not use waxed paper or aluminum foil.

dehydrator, up to 18 hours in an oven, and 1 to 2 days in the sun. Test for dryness by touching center of leather; no indentation should be evident. While warm, peel from plastic and roll, allow to cool, and rewrap the roll in plastic. Cookie cutters can be used to cut out shapes that children will enjoy. Roll, and wrap in plastic.

Vegetable Leathers

Pumpkin, mixed vegetables, and tomatoes make great leathers. Just purée cooked vegetables, strain, spread on a tray lined with plastic wrap, and dry. Spices can be added for flavoring.

Mixed Vegetable Leather

- 2 cups cored, cut-up tomatoes
- 1 small onion, chopped
- ¼ cup chopped celery
- Salt to taste

Combine all ingredients in a covered saucepan and cook over low heat 15 to 20 minutes. Purée or force through a sieve or colander. Return to saucepan and cook until thickened. Spread on a cookie sheet or tray lined with plastic wrap. Dry at 140°F.

Pumpkin Leather

- 2 cups canned pumpkin or 2 cups fresh pumpkin, cooked and puréed
- ½ cup honey
- ¼ tsp cinnamon

- ⅛ tsp nutmeg
- ⅛ tsp powdered cloves
- Blend ingredients well. Spread on tray or cookie sheet lined with plastic wrap. Dry at 140°F.

Tomato Leather

Core ripe tomatoes and cut into quarters. Cook over low heat in a covered saucepan, fifteen to twenty minutes. Purée or force through a sieve or colander and pour into electric fry pan or shallow pan. Add salt to taste and cook over low heat until thickened. Spread on a cookie sheet or tray lined with plastic wrap. Dry at 140°F.

How to Make a Woodstove Food Dehydrator

1. Collect pliable wire mesh or screens (available at hardware stores) and use wire cutters to trim to squares 12 to 16 inches on each side. The trays should be of the same size and shape. Bend up the edges of each square to create a half-inch lip.
2. Attach one S hook from the hardware store or a large paperclip to each side of each square (four clips per tray) to attach the trays together.
3. Cut four equal lengths of chain or twine that will reach from the ceiling to the level of the top tray. Use a wire or metal loop to attach the four pieces together at the top and secure to a hook in the ceiling above the wood-

stove. Attach the chain or twine to the hooks on the top tray.
4. To use, fill trays with food to dry, starting with the top tray. Link trays together using the S hooks or strong paperclips. When the foods are dried, remove the entire stack and disassemble. Remove the dried food and store.

Herbs

Drying is the easiest method of preserving herbs. Simply expose the leaves, flowers, or seeds to warm, dry air. Leave the herbs in a well-ventilated area until the moisture evaporates. Sun drying is not recommended because the herbs can lose flavor and color.

VINE DRYING

One method of drying outdoors is vine drying. To dry beans (navy, kidney, butter, great northern, lima, lentils, and soybeans) leave bean pods on the vine in the garden until the beans inside rattle. When the vines and pods are dry and shriveled, pick the beans and shell them. No pretreatment is necessary. If beans are still moist, the drying process is not complete and the beans will mold if not more thoroughly dried. If needed, drying can be completed in the sun, an oven, or a dehydrator.

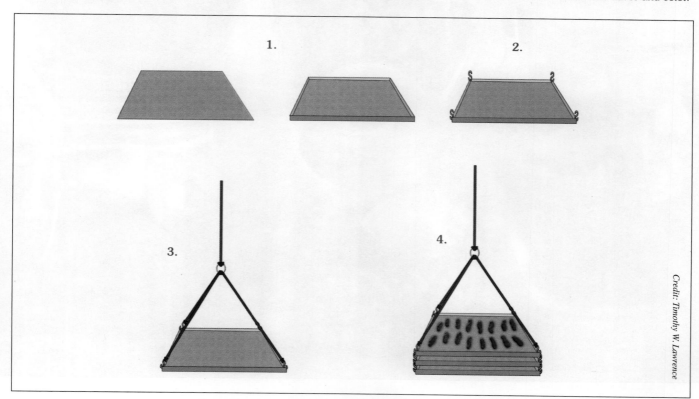

Credit: Timothy W. Lawrence

The best time to harvest most herbs for drying is just before the flowers first open when they are in the bursting bud stage. Gather the herbs in the early morning after the dew has evaporated to minimize wilting. Avoid bruising the leaves. They should not lie in the sun or unattended after harvesting. Rinse herbs in cool water and gently shake to remove excess moisture. Discard all bruised, soiled, or imperfect leaves and stems.

Dehydrator drying is another fast and easy way to dry high-quality herbs because temperature and air circulation can be controlled. Preheat dehydrator with the thermostat set to 95 to 115°F. In areas with higher humidity, temperatures as high as 125°F may be needed. After rinsing under

cool, running water and shaking to remove excess moisture, place the herbs in a single layer on dehydrator trays. Drying times may vary from one to four hours. Check periodically. Herbs are dry when they crumble, and stems break when bent. Check your dehydrator instruction booklet for specific details.

Less tender herbs—The more sturdy herbs, such as rosemary, sage, thyme, summer savory, and parsley, are the easiest to dry without a dehydrator. Tie them into small bundles and hang them to air dry. Air drying outdoors is often possible; however, better color and flavor retention usually results from drying indoors.

Tender-leaf herbs—Basil, oregano, tarragon, lemon balm, and the mints have a high moisture content and will mold if not dried quickly. Try hanging the tender-leaf herbs or those with seeds inside paper bags to dry. Tear or punch holes in the sides of the bag. Suspend a small bunch (large amounts will mold) of herbs in a bag and close the top with a rubber band. Place where air currents will circulate through the bag. Any leaves and seeds that fall off will be caught in the bottom of the bag.

Another method, especially nice for mint, sage, or bay leaf, is to dry the leaves separately. In areas of high humidity, it will work better than air drying whole stems. Remove the best leaves from the stems. Lay the leaves on a paper towel, without allowing leaves to touch. Cover with another towel and layer of leaves. Five layers may be dried at one time using this method. Dry in a very cool oven. The oven light of an electric range or the pilot light of a gas range furnishes enough heat for overnight drying. Leaves dry flat and retain a good color.

Microwave ovens are a fast way to dry herbs when only small quantities are to be prepared. Follow the directions that come with your microwave oven.

When the leaves are crispy, dry, and crumble easily between the fingers, they are ready to be packaged and stored. Dried leaves may be left whole and crumbled as used, or coarsely crumbled before storage. Husks can be removed from seeds by rubbing the seeds between the hands and blowing away the chaff. Place herbs in airtight containers and store in a cool, dry, dark area to protect color and fragrance.

Dried herbs are usually three to four times stronger than the fresh herbs. To substitute dried herbs in a recipe that calls for fresh herbs, use ¼ to ⅓ of the amount listed in the recipe.

Jerky

Jerky is great for hiking or camping because it supplies protein in a very lightweight form—plus it can be very tasty. A pound of meat or poultry weighs about four ounces after being made into jerky. In addition, because most of the moisture is

removed, it can be stored for one to two months without refrigeration.

Jerky has been around since the ancient Egyptians began drying animal meat that was too big to eat all at once. Native Americans mixed ground dried meat with dried fruit or suet to make pemmican. *Biltong* is dried meat or game used in many African countries. The English word *jerky* came from the Spanish word *charque*, which means, "dried salted meat."

Drying is the world's oldest and most common method of food preservation. Enzymes require moisture in order to react with food. By removing the moisture, you prevent this biological action.

Jerky can be made from ground meat, which is often less expensive than strips of meat and allows you to combine different kinds of meat if desired. You can also make it into any shape you want! As with strips of meat, an internal temperature of 160°F is necessary to eliminate disease-causing bacteria such as *E. coli*, if present.

Food Safety

The USDA Meat and Poultry Hotline's current recommendation for making jerky safely is to heat meat to 160°F and poultry to 165°F before the dehydrating process. This ensures that any bacteria present are destroyed by heat. If your food dehydrator doesn't heat up to 160°F, it's important to cook meat slightly in the oven or by steaming before drying. After heating, maintain a constant dehydrator temperature of 130 to 140°F during the drying process.

According to the USDA, you should always:

- Wash hands thoroughly with soap and water before and after working with meat products.
- Use clean equipment and utensils.
- Keep meat and poultry refrigerated at 40°F or slightly below; use or freeze ground beef and poultry within two days, and whole red meats within three to five days.
- Defrost frozen meat in the refrigerator, not on the kitchen counter.
- Marinate meat in the refrigerator. Don't save marinade to re-use. Marinades are used to tenderize and flavor the jerky before dehydrating it.
- If your food dehydrator doesn't heat up to 160°F (or 165°F for poultry), steam or roast meat before dehydrating it.
- Dry meats in a food dehydrator that has an adjustable temperature dial and will maintain a temperature of at least 130 to 140°F throughout the drying process.

Preparing the Meat

1. Partially freeze meat to make slicing easier. Slice meat across the grain $1/8$ to $1/4$ inch thick. Trim and discard all fat, gristle, and membranes or connective tissue.
2. Marinate the meat in a combination of oil, salt, spices, vinegar, lemon juice, teriyaki, soy sauce, beer, or wine.

Marinated Jerky

- $1/4$ cup soy sauce
- 1 tbsp Worcestershire sauce
- 1 tsp brown sugar
- $1/4$ tsp black pepper
- $1/2$ tsp fresh ginger, finely grated
- 1 tsp salt
- $1\frac{1}{2}$ to 2 lbs. of lean meat strips (beef, pork, or venison)

Drying and Freezing

1. Combine all ingredients except the strips, and blend. Add meat, stir, cover, and refrigerate at least one hour.
2. If your food dehydrator doesn't heat up to 160°F, bring strips and marinade to a boil and cook for 5 minutes.
3. Drain meat in a colander and absorb extra moisture with clean, absorbent paper towels. Arrange strips in a single layer on dehydrator trays, or on cake racks placed on baking sheets for oven drying.
4. Place the racks in a dehydrator or oven preheated to 140ºF, or 160°F if the meat wasn't precooked. Dry until a test piece cracks but does not break when it is bent (10 to 24 hours for samples not heated in marinade, 3 to 6 hours for preheated meat). Use paper towel to pat off any excess oil from strips, and pack in sealed jars, plastic bags, or plastic containers.

Freezing Foods

Many foods preserve well in the freezer and can make preparing meals easy when you are short on time. If you make a big pot of soup, serve it for dinner, put a small container in the refrigerator for lunch the next day, and then stick the rest in the freezer. A few weeks later you'll be ready to eat it again and it will only take a few minutes to thaw out and serve. Many fruits also freeze well and are perfect for use in smoothies and desserts, or served with yogurt for breakfast

or dessert. Vegetables frozen shortly after harvesting keep many of the nutrients found in fresh vegetables and will taste delicious when cooked.

Containers for Freezing

The best packaging materials for freezing include rigid containers such as jars, bottles, or Tupperware, and freezer bags or aluminum foil. Sturdy containers with rigid sides are especially good for liquids such as soup or juice because they make the frozen contents much easier to get out. They are also generally reusable and make it easier to stack foods in the refrigerator. When using rigid containers, be sure to leave headspace so that the container won't explode when the contents expand with freezing. Covers for rigid containers should fit tightly. If they do not, reinforce the seal with freezer tape. Freezer tape is specially designed to stick at freezing temperatures. Freezer bags or aluminum foil are good for meats, breads and baked goods, or fruits and vegetables that don't contain much liquid. Be sure to remove as much air as possible from bags before closing.

Type of Pack	Container with Wide Opening		Container with Narrow Opening	
	Pint	Quart	Pint	Quart
Liquid pack*	½ inch	1 inch	¾ inch	1½ inch
Dry pack**	½ inch	½ inch	½ inch	½ inch
Juices	½ inch	1 inch	1½ inch	1½ inch
*Fruit packed in juice, sugar syrup, or water; crushed or puréed fruit.				
**Fruit or vegetable packed without added sugar or liquid.				

Headspace to Allow Between Packed Food and Closure

Headspace is the amount of empty air left between the food and the lid. Headspace is necessary because foods expand when frozen.

Effect of Freezing on Spices and Seasonings

- Pepper, cloves, garlic, green pepper, imitation vanilla and some herbs tend to get strong and bitter.
- Onion and paprika change flavor during freezing.
- Celery seasonings become stronger.
- Curry develops a musty off-flavor.

FOODS THAT DO NOT FREEZE WELL

Food	Usual Use	Condition After Thawing
Cabbage*, celery, cress, cucumbers*, endive, lettuce, parsley, radishes	As raw salad	Limp, waterlogged; quickly develops oxidized color, aroma, and flavor
Irish potatoes, baked or boiled	In soups, salads, sauces, or with butter	Soft, crumbly, waterlogged, mealy
Cooked macaroni, spaghetti, or rice	When frozen alone for later use	Mushy, tastes warmed over
Egg whites, cooked	In salads, creamed foods, sandwiches, sauces, gravy, or desserts	Soft, tough, rubbery, spongy
Meringue	In desserts	Soft, tough, rubbery, spongy
Icings made from egg whites	Cakes, cookies	Frothy, weeps
Cream or custard fillings	Pies, baked goods	Separates, watery, lumpy
Milk sauces	For casseroles or gravies	May curdle or separate
Sour cream	As topping, in salads	Separates, watery
Cheese or crumb toppings	On casseroles	Soggy
Mayonnaise or salad dressing	On sandwiches (not in salads)	Separates
Gelatin	In salads or desserts	Weeps
Fruit jelly	Sandwiches	May soak bread
Fried foods	All except French fried potatoes and onion rings	Lose crispness, become soggy

* Cucumbers and cabbage can be frozen as marinated products such as "freezer slaw" or "freezer pickles." These do not have the same texture as regular slaw or pickles.

- Salt loses flavor and has the tendency to increase rancidity of any item containing fat.
- When using seasonings and spices, season lightly before freezing, and add additional seasonings when reheating or serving.

How to Freeze Vegetables

Because many vegetables contain enzymes that will cause them to lose color when frozen, you may want to blanche your vegetables before putting them in the freezer. To do this, first wash the vegetables thoroughly, peel if desired, and chop them into bite-size pieces. Then pour them into boiling water for a couple of minutes (or cook longer for very dense vegetables, such as beets), drain, and immediately dunk the vegetables in ice water to stop them from cooking further. Use a paper towel or cloth to absorb excess water from the vegetables, and then pack in resealable airtight bags or plastic containers.

How to Freeze Fruits

Many fruits freeze easily and are perfect for use in baking, smoothies, or sauces.

Blanching Times for Vegetables	
Artichokes	3–6 minutes
Asparagus	2–3 minutes
Beans	2–3 minutes
Beets	30–40 minutes
Broccoli	3 minutes
Brussels sprouts	4–5 minutes
Cabbage	3–4 minutes
Carrots	2–5 minutes
Cauliflower	6 minutes
Celery	3 minutes
Corn (off the cob)	2–3 minutes
Eggplant	4 minutes
Okra	3–4 minutes
Peas	1–2 minutes
Peppers	2–3 minutes
Squash	2–3 minutes
Turnips or Parsnips	2 minutes

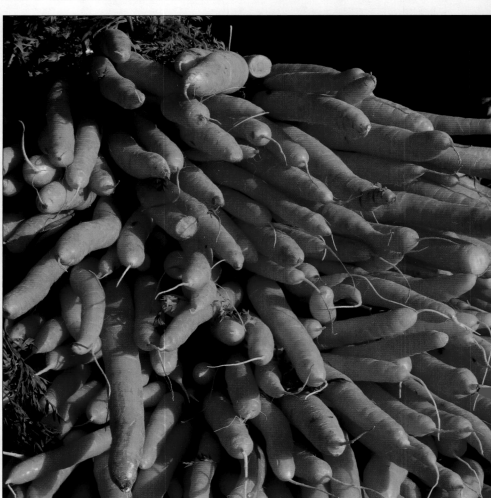

Wash, peel, and core fruit before freezing. To easily peel peaches, nectarines, or apricots, dip them in boiling water for fifteen to twenty seconds to loosen the skins. Then chill and remove the skins and stones.

Berries should be frozen immediately after harvesting and can be frozen in a single layer on a paper towel-lined tray or cookie sheet to keep them from clumping together. Allow them to freeze until hard (about three hours) and then pour them into a resealable plastic bag for long-term storage.

Some fruits have a tendency to turn brown when frozen. To prevent this, you can add ascorbic acid (crush a vitamin C in a little water), citrus juice, plain sugar, or a sweet syrup (one part sugar and two parts water) to the fruit before freezing. Apples, pears, and bananas are best frozen with ascorbic acid or citrus juice, while berries, peaches, nectarines, apricots, pineapple, melons, and berries are better frozen with a sugary syrup.

How to Freeze Meat

Be sure your meat is fresh before freezing. Trim off excess fats and remove bones, if desired. Separate the meat into portions that will be easy to use when preparing meals and wrap in foil or place in resealable plastic bags or plastic containers. Refer to the chart to determine how long your meat will last at best quality in your freezer.

Meat	Months
Bacon and sausage	1 to 2
Ham, hotdogs, and lunchmeats	1 to 2
Meat, uncooked roasts	4 to 12
Meat, uncooked steaks or chops	4 to 12
Meat, uncooked ground	3 to 4
Meat, cooked	2 to 3
Poultry, uncooked whole	12
Poultry, uncooked parts	9
Poultry, uncooked giblets	3 to 4
Poultry, cooked	4
Wild game, uncooked	8 to 12

Part 4

Country Crafts

"Happiness is not in the mere possession of money; it lies in the joy of achievement, in the thrill of the creative effort."

—President Franklin Delano Roosevelt

Spring	238
Summer	247
Autumn	261
Winter	275

Spring

Springtime lends itself to creativity. As new flowers push through the soil and birds begin to gather materials for their nests, you may find yourself eager to start new projects, too. It is a busy season; days are filled with preparing gardens and airing out the house, weekends are packed with weddings and graduations. But a rainy afternoon or a quiet Sunday may give you the opportunity for crafting you crave. Here is a smattering of ideas to give your inspiration some direction, whether you're creating a gift, an accent for your home, or a keepsake for a special celebration.

Springtime Wreath

Grapevines make an attractive and natural base for this welcoming wreath. Begin shaping the vines soon after cutting. If you do need to store them for more than a day before using them, remove the leaves and soak the vines for several hours before beginning the wreath to make them more pliable.

- Grapevines
- Decoration such as moss, baby's breath, etc.
- Florist tape

1. Cut several lengths of vine, keeping them as long as you can manage. Remove the leaves, but don't trim the tendrils.

2. Start with the thicker end of the vine and form a circle. Then begin coiling a second circle, winding and twisting it around the first. Continue until you are almost at the end of the vine and then wind the end around and around, using the vine tendrils or florist tape to secure the end. Add additional vines, winding them around the first circle and tucking the ends into the center.

3. Add moss, baby's breath, yarrow, ferns, feathers, decorative grasses, ribbon, etc. Tuck the ends into the vines and use florist tape as needed for added security.

Blown Eggs

If you want to keep your egg creations to display year after year, blow out the eggs before decorating them. Blown eggs are more fragile than hardboiled eggs, but they won't ever spoil.

- Eggs
- Needle
- Toothpick
- Tiny straw or a syringe

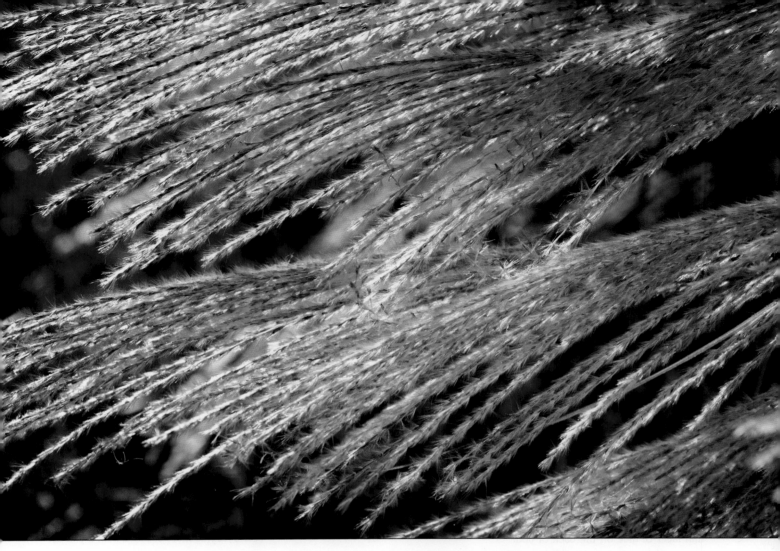

⌃ To dry flowers or grasses for use in wreaths or other decorations, cut them slightly before they reach their peak, leaving long stems. Tie the stems together and hang them upside down for several weeks in a dry, dark area.

1. Use a needle to poke one hole on each end of the egg. Insert a toothpick and wiggle it around to help widen the holes slightly. Stick the needle back into one hole and move it around inside the egg until the yolk breaks.

2. Insert the straw into one hole and blow through it until the insides of the egg drain out, or use a syringe to draw out the egg. Rinse thoroughly to wash away any remaining egg residue.

3. Bake the eggshell at 400°F for ten minutes.

PLANTS THAT DRY WELL FOR DECORATIVE USE

Flowers	Grasses	Herbs
Acrolinium, Baby's breath, Bachelor's button, Bells of Ireland, Cockscomb, Coneflower, Delphinium, Foxglove, Globe amaranth, Goldenrod, Heather, Hydrangea, Larkspur, Statice, Strawflower, Yarrow	Bristly foxtail, Cattails, Eulalia grass, Fountain grass, Hare's-tail grass, Indian grass, Northern sea oats, Pampas grass, Plume grass, Quaking grass, Spike grass, Squirrel-tail grass, Switch grass, Wheatgrass	Chamomile, Chives, Dill, Eucalyptus, Fennel, Lavender, Lemongrass, Rosemary, Sage, St. John's wort, Thyme

Leaf- or Flower- Stenciled Eggs

These eggs make a stunning centerpiece when displayed on a plate or in a basket. If you intend to eat the eggs later, use only food coloring to dye them. However, if you are using blown eggs or you do not intend to eat the eggs, fabric dye, or other natural dyes, and stencil paint will produce richer, more vibrant colors.

- Food coloring, fabric dye, or other natural dye (prepared accordingly)
- Large white eggs
- White glue
- Leaves, ferns, or small flowers
- Small paintbrush
- Wide, stiff paintbrush
- Stencil paint (if you do not intend to eat the egg later)

1. Fill a deep bowl or wide glass half full with the prepared dye solution.
2. Immerse an egg in the dye mixture and allow it to sit for a few minutes, or until the egg has reached the desired color. Rinse the egg in cold water and allow it to dry.
3. Using a small paintbrush, paint a thin layer of glue on the back of your leaf or fern. Stick the leaf or fern to the egg.
4. Dip a wide paintbrush in undiluted food coloring (if you intend to eat the eggs later) or stencil paint. Blot the brush on newspaper or paper towel to get rid of excess paint. Dab the ends of the bristles up and down over the leaf, allowing the first layer of color to show through to create a dappled effect.
5. When the egg is completely dry, peel the leaf away from the egg.

Natural Dyes for Easter Eggs

There are many ingredients from nature you can use to dye your Easter eggs. The colors may be more subdued than if you use

food coloring or paints, but you can achieve some beautiful pastels with berries, flowers, and other plants and foods. Mix dyestuff or the finished dyes to make more color variations. Note that liquid dyestuffs, such as grape juice, do not need to be simmered with water as described below. Simply add the vinegar and use!

- Dyestuff (see chart below)
- Water
- White vinegar
- Cooking oil or mineral oil

1. Place a handful or two of the dyestuff of your choice (or a couple of tablespoons if using herbs or spices) in a saucepan.
2. Add water until the dyestuff is fully submerged. Simmer on low for about fifteen minutes, or until the desired color is reached. Keep in mind that the eggs will turn out paler than the dye appears in the pan.
3. Strain dye into a liquid measuring cup. Add 2 tablespoons of white vinegar for every cup of dye. The dye is now ready for use.

4. After the dyed eggs are dry, rub the eggs with cooking oil or mineral oil to give them a glossier sheen.

Terrariums

Terrariums are miniature ecosystems that you can create and keep indoors. They're a wonderful way to learn about gardening on a small scale and can add interest to your home décor and oxygen to the air you breathe. Terrariums can contain only plants or can be homes for lizards, turtles, or other small animals. If you do wish to make your terrarium a home for a pet, be sure to include the proper shelter for the animal and a way to provide food and water. Some terrariums even include waterfalls so that animals can have a constant supply of fresh water! The size of the terrarium can be as large as a fish tank or as small as a thimble. Bowls, teapots, jars, and bottles have all been successfully transformed into miniature indoor gardens. Terrariums can be fully enclosed or can have an open top to allow fresh air to circulate. Because one of the benefits of a terrarium is the oxygen that the plants contribute to the air, these directions are for an open top terrarium.

- Container (preferably a clear glass container, so you can easily see your miniature garden)
- Coarse sand or pebbles
- Sphagnum moss

Color	Items to Dye With
Blue	Blueberries, red cabbage, purple grape juice
Brown or Beige	Coffee grounds, black walnut shells, black tea leaves
Brown Gold	Dill seeds
Brown	Chili powder
Green	Spinach leaves, liquid chlorophyll
Gray	Purple or red grape juice, beet juice
Lavender	Purple grape juice, violet blossoms plus a little lemon juice, Red Zinger tea
Orange	Yellow onion skins, carrots, paprika
Pink	Beets, crushed cranberries or cranberry juice, crushed raspberries, grape juice
Red	Lots of red onion skins, pomegranate juice, canned cherries, crushed raspberries
Violet or Purple	Violet blossoms, hibiscus tea, small quantity of red onion skins, red wine
Yellow	Orange or lemon peels, carrot tops, chamomile tea, celery seed, green tea, ground cumin, ground turmeric, saffron

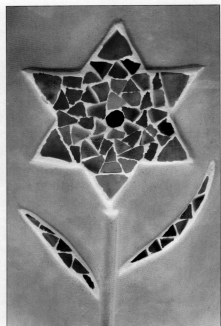

- Soil
- Seeds or seedlings
- Water
- Ornaments (optional)

1. Place a ½-inch to 1-inch layer of coarse sand or pebbles in the bottom of your container. This will help the soil to drain properly.
2. Add a layer of moss over the pebbles. The moss acts as a filter, allowing the soil to drain but not to seep down into the pebbles.
3. Pour the soil over the moss and spread evenly. How much soil you use will depend on how big your container is and how large the plants will grow. Pat the dirt down firmly.
4. Plant the seeds or seedlings. Think carefully about how you want the plants to be arranged. You may want taller plants in the center and shorter ones toward the outside so that you can see them all. Add pretty stones, pinecones, figurines, or other ornaments, if desired.
5. Place your terrarium in a sunny spot and water it regularly.

Mosaic Flowerpots

Spring is the time to start planting seeds so you'll have seedlings to transplant to the garden come summer. Make your pots unique by decorating them yourself with bits of beach glass, pottery, sea shells, or beads. This is a great project to do with kids, but be careful of sharp pieces of glass or pottery.

- Putty knife
- Ceramic tile grout
- Terra cotta flowerpot
- Pieces of beach glass, broken pottery or mirror, tile, beads, charms, etc.

1. Use the putty knife to spread a thick layer of grout around the outside of the flowerpot (at least ¼-inch thick).
2. Press the pieces into the grout and add more grout around each piece to cover any sharp or rough edges.
3. Allow pot to dry thoroughly. Then wipe away any grout residue from the pieces with a damp sponge.

TIP

Look for nontoxic grout online or at your local hardware store. Alternatively, you can make your own grout by mixing one part Portland cement to two parts sand in a tub. Add water slowly while stirring until the mixture is thick like mud. If desired, add natural iron oxide pigments to the grout for a more colorful background for your mosaic.

For an even simpler "faux grout," mix two parts white sand to one part white glue. Add acrylic paint or concentrated natural dyes as desired. The "faux grout" won't be as strong or as smooth as real grout, but it will work in a pinch.

Homemade Weddings

Weddings are wonderful, memorable, and very personal events. Every couple is unique and weddings are an opportunity to celebrate their distinct experiences of love, beauty, and joy. Couples who do some or all of the wedding preparations on their own or with the help of friends and family (rather than relying entirely on professionals) often find their day especially meaningful—and they're less likely to start off their married life in debt.

There are endless ways to add homemade touches to a wedding. The invitations, flowers, centerpieces, favors, food, and even the attire can be made or arranged without the help of professional florists, caterers, and other service providers. Couples should think about what aspects of the wedding matter most to them, what they would most enjoy doing on their own, and how much time they can realistically dedicate to wedding preparations. Delegation is also key—friends and family are often more than happy to be a part of the celebration by assisting with certain tasks.

Bouquets

Once you've chosen the flowers you want in your bouquets (often based on color and season), think about what shape the bouquets will be and how you will arrange them. Smaller bouquets can be secured with a wide ribbon tied around the stems a little below the flower heads. For larger bouquets or for flowers with rough or thorny stems, you may want to bind the stems together with a ribbon that covers the length of the stems.

Spring

- Flowers
- Floral tape
- Ribbon
- Corsage pins

1. Arrange the flowers in a bunch, cutting the stems evenly at the bottom. Wrap a piece of floral tape around the stems just below the blossoms. Cut a piece of ribbon about five feet long (or longer for very long stems).
2. Begin wrapping the ribbon around the stems just below the blossoms, leaving a tail of ribbon about two feet long.
3. Wrap the ribbon around and around the stems, working your way down and making sure the ribbon lies flat. When you reach the bottom, wrap over the bottom of the stems and work your way back up to the top.
4. Now you should have two ribbon tails that you can tie into a full bow. Use pearl-headed corsage pins to secure the ribbon in place. Trim the tails if desired.

Hanging Flower Pomander

Flower pomanders should be made as late as possible before the celebration, since the flowers will not be getting adequate water to stay fresh for as long as a bouquet in a vase would. If making the pomanders the day before the wedding, keep them refrigerated as long as possible. Choose flowers with stiff stems such as roses or carnations, as they'll be easier to insert into the ball.

- Ball of floristry foam
- 2½-inch-wide ribbon
- Straight pins or a glue gun
- Scissors
- Flowers

1. Soak floral foam in water until saturated and then set aside on a towel and allow to dry until just damp.
2. Wrap the ribbon once around the outside of the foam ball, pinning it at every inch or securing it to the ball with hot glue. Leave a foot of ribbon loose on each end.
3. Trim your flowers, leaving about an inch of stem for smaller blossoms and two to three inches for larger blossoms. Push the stems all the way into the ball, following the edge of the ribbon. Then do a line of flowers around the circumference so that the ball is divided into corners. Finally, fill in the four sectors, adding ferns or other greenery as desired to close up any spaces. If

desired, only cover the top two-thirds of the ball with flowers and line the bottom third with large green leaves, secured with pins.

4. Tie the two ends of the ribbon together in a tight knot and hang the pomander in a door or window or on a chandelier, or carry one down the aisle instead of a bouquet.

Wedding Cake

Your cake can be any flavor you like, and any shape, for that matter. But to make a traditional tiered wedding cake, follow the steps below. Be sure to do at least one test run well before the wedding.

- ½-inch-thick plywood or large cutting board
- Parchment paper
- Tape
- Frosting
- Cake tiers
- Corrugated cardboard
- Tinfoil

EDIBLE FLOWERS FOR CAKE DECORATING

Any nontoxic flowers can be used to decorate your cake, as long as you remove the flowers before serving. If you don't want to remove the flowers, choose edible flowers such as the following:

- Apple blossoms
- Orange blossoms
- Cornflowers
- Pansies
- Violets
- Nasturtiums
- Oregano
- Lavender
- Carnations

- Clover
- Calendula
- Jasmine
- Hollyhock
- Impatiens
- Lilacs
- Peonies
- Roses

- Knife or toothpick
- ¼-inch-thick dowels
- Heavy duty shears
- Cake decorations

1. Prepare the cake board. Cover a ½-inch-thick piece of plywood or a large cutting board with parchment paper, taping it to the underside of the board. The board should be at least 4 inches larger than the largest tier of your cake. This board will stay with your cake until it's eaten, so do a neat job of covering it.
2. Frost and fill the bottom, largest cake tier and place it in the center of the cake board.
3. Cut out two circles of corrugated cardboard that are the same size as the second tier, and two circles the same

size as the third tier. Tape the same-sized circles together so that they are double thickness. Cover each cardboard circle with tinfoil, making the foil as smooth as possible.

4. Gently place the medium-sized cardboard circle onto the center of the bottom tier and mark around the circle with a knife or toothpick. Now you can remove the cardboard and still see the circle.
5. Push a ¼-inch-thick dowel into the very center of the cake. Mark the height of the cake on the dowel and remove it. Use heavy duty shears to cut eight dowels at that length. Insert the dowels into the cake at even intervals around the circle marked on the bottom tier.
6. Place the second cake tier on the medium-sized cardboard circle and fill

≫ A few fresh blossoms floating in a bowl make a simple and elegant centerpiece.

≫ Refer to page 263 for directions for making sachets. Sachets filled with candy, a few homemade cookies, or handmade soap make lovely favors.

Spring

and frost it. Place it carefully on the center of the bottom tier.

7. Repeat steps 5 and 6 with the top layer. To help keep the layers from sliding, sharpen one long dowel with a knife and insert it through the center of the top layer straight down through all the layers. Trim the dowel so that it's even with the top of the cake.

8. Decorate the cake with piping around the bottom of each tier, marzipan fruit, or fresh fruit or flowers.

Programs for the Ceremony

A handmade program will give your guests a unique keepsake by which to remember your ceremony. To make your own paper, refer to page 294.

- Paper in various colors
- Pen
- Paste
- Decorations

1. Cut a piece of sturdy, attractive paper to 7 x 11 inches, or desired size.

2. Write or print the text on paper of a contrasting color or texture, making sure it fits on a rectangle that is slightly smaller than the original cut rectangle (about 6 x 10 inches).

3. Cut out the second rectangle and paste it to the first, centering it carefully. Allow to dry and then fold the program in half.

4. If desired, paste ribbon, pressed flowers, or other accents to the outside of the program.

« Slip flowers into folded napkins to add color to simple table settings.

Summer

Ripe berries, gardens full of flowers, warm breezes—summer is the season of simple pleasures. What better way to enjoy it than by relaxing in a handmade hammock, flying a kite you made yourself, or concocting a potpourri recipe that will remind you of the sweet smells of summer all year round?

Hammocks

Hammocks are wonderful for relaxing outside with a good book or for taking an afternoon nap in the shade of two trees.

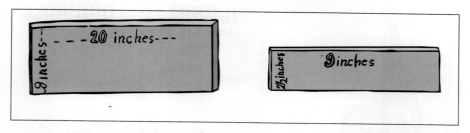

≈ Mesh sticks are used to help keep your meshes even. You can easily make them yourself, following the dimensions shown here. The one on the left is 20" × 8" and beveled on both edges. The one on the right is 9" × 2 ½" and beveled on the long edge.

Making your own hammock is quite simple if you follow these directions:

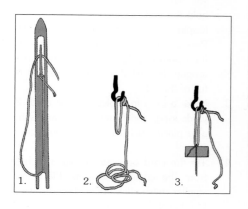

« Figure 1 (left) shows how to thread the needle. Make a loop and fasten it to a doorknob or hook, as shown in figure 2 (center). Figure 3 (right) shows the smaller mesh stick under the cord, beveled edge close to the loop.

» Figure 4 (bottom) shows the first half of the knot. Figure 5 (center) shows the loop as it is being formed. Continue the sequence to create a number of knots, as in figure 6 (top). "A" shows a loosened knot. "B" shows the cord running to the needle. "C" is the cord and "D" is the mesh stick.

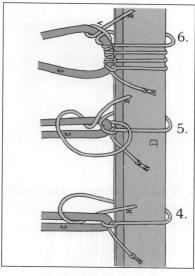

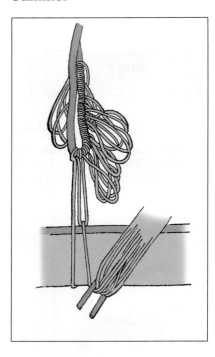

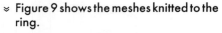

« Figure 7. Once 30 meshes are complete, shove them off the mesh stick.

⩘ Fisherman's Knot.

⩘ Figure 8 shows the start of the second row.

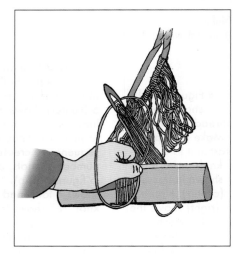

⩘ Figure 9 shows the meshes knitted to the ring.

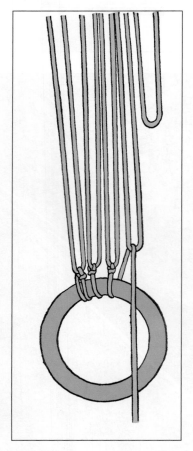

Figure 10 shows the first needleful to the loop. »

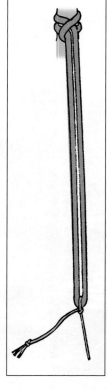

Materials

1 lb macrame cord, No. 24, or hammock twine

1 hammock needle roughly 9 inches long (you can find this at a craft store)

2 iron rings (each should be roughly 2½ inches in diameter)

2 mesh sticks (one 20 inches long and 8 inches wide; one 9 inches long and 2½ inches wide), edges beveled

Directions

1. Wind the cord into balls so it is easier to handle.
2. Thread the needle by taking it in your left hand and using your thumb to hold the end of the cord in place, and looping it over the tongue (figure 1). Pass the cord down under the needle to the opposite side and catch it over the tongue. Repeat this until the needle is full of thread.
3. Make a loop in a piece of the cord, 2 yards long, and fasten this to any suitable place, such as a doorknob (figure 2).
4. Tie the cord on your needle 3 inches from the end of the loop. Place the smaller mesh stick under the cord with the beveled edge close to the loop (figure 3).
5. With your thumb on the cord, holding it in place, pass the needle around the stick and, with the needle point toward you, pass it through the loop from the top, and bring it over the stick. This will form the first half of the knot (figure 4). Pull this taut and throw the cord over your hand, forming a loop (figure 5).
6. Pass the needle from under through the loops, drawing it tight to fasten the knot. Hold it in place with your thumb and repeat this for the next knot (figure 6).
7. When 30 meshes are finished, push them off the short stick (figure 7). Now your hammock will be sufficiently wide.
8. Begin the next row by placing the stick under the cord and taking up the first mesh and drawing it close to the stick. Hold it in place with your thumb while throwing the cord over your hand (figure 8). Pass the needle on the left-hand side of the mesh from under and through the loop thrown over your hand. Pull this tight and you'll have tied the common knitting knot.

9. Proceed to carry out the steps in number 8 until the row is finished. When your needle needs to be rethreaded, tie the ends of the cord with a fisherman's knot and then wrap each end of the cord from the knot securely to the main cord with strong thread to give it a neat appearance.

10. Continue knitting until there are 30 rows.

11. Using the larger stick, knit one row on the short side first, and then knit a row on the long side. After this is complete, knit the meshes to the ring by passing the needle through it from the top, knitting them to the ring in rotation as if they were on a stick (figure 9). When finished, tie the string securely to the ring and one end of your hammock is complete.

12. Cut the loop where the first row was knitted and pass it through the knots. Tie the end of the cord onto your needle to the same piece used in fastening the end of the first needleful to the loop (figure 10). Knit the long meshes to the other ring as described above. This completes the hammock.

In order to use your hammock, attach two pieces of strong rope to the rings of your hammock and secure them tightly between two trees or other sturdy poles. The two trees or poles should be about 12 feet apart and at least 10 feet high, to allow for your hammock to swing freely.

Berry Ink and Feather Pens

During the Civil War, soldiers made ink out of berry juice and used feathers or corn stalks to write important letters. Use berry ink and a quill pen for special invitations, place cards, or just for fun. The vinegar helps the ink to retain its color and the salt acts as a natural preservative. Store extra ink in an airtight jar.

- Strainer
- Bowl
- Spoon
- ¼ cup berries (raspberries, strawberries, currants, or any other brightly colored berry will work)
- ¼ teaspoon vinegar

- ¼ teaspoon salt
- Bird feather
- Scissors
- Paper

1. Place the strainer over the bowl and use the back of a spoon to squish the berries so that the juice runs into the bowl. Once all the juice is extracted, discard

the berry remains (or add them to your compost pile).

2. Add the vinegar and salt to the berry juice. Mix thoroughly.

3. Cut the sharp tip of the feather off at an angle. Dip the quill into the ink and begin writing. Practice on scrap paper before attempting an important project.

Potpourri from Your Garden

Rose petals and sweet geranium leaves are the primary ingredients in most potpourri. You can also add lavender, sweet verbena leaves, bay leaves, rosemary, dried orange peels, or pine needles. Orrisroot powder will help preserve the scent of the other ingredients.

The best time to collect flowers, seeds, or roots is in the morning, just after the dew has evaporated. Be careful not to bruise flower petals when gathering them, as damaged flowers will lose their scent.

- Flower petals, flower heads, stems, roots, or fruit (see chart below)
- Screen
- String
- Food dehydrator, oven, or solar oven
- Spices and orrisroot
- Essential oil

1. To dry individual petals or flower heads, lay them on a screen and leave them in a dry place, out of direct sunlight, for about two weeks.
2. For stems or roots, bunch them together, tie with string, and hang upside down for one to two weeks.
3. Flowers or fruit (such as citrus slices or peels) can also be dried in a food dehydrator set at its lowest temperature, or in the oven set to 180°F. Drying in the oven takes several hours. Leave the oven door open slightly to allow moisture to escape. Using a solar oven will take longer but is much more energy efficient!
4. To enhance the fragrance of your potpourri, add spices and orrisroot (available online or from many florists) to your final mixture. Gather violet powder, ground allspice, ground cloves, ground mixed spice, ground mace, whole mace, and/or whole cloves.

Fragrance Category	Ingredients
Spicy	Allspice Berry, Bay, Cardamom, Caraway, Cinnamon, Clove, Clary Sage, Clove, Coriander, Cumin, Fir, Frankincense, Hyssop, Myrrh, Neroli, Nutmeg, Rosewood,
Sweet	Anise, Clary sage, Coriander, Frankincense, Geranium, Jasmine, Lavender, Rose, Sandalwood, Vanilla
Fruity/Citrusy	Bergamot, Berries, Lemon, Orange, Lemongrass, Ylang ylang, Wintergreen,
Earthy	Cedarwood, Chamomile, Eucalyptus, Fir, Juniper, Myrrh, Patchouli, Pine, Rosemary, Spruce

« Mint and otherherbs can be dried and included in your potpourri.

5. Once all the components are thoroughly dried, you can mix them together. To make the potpourri smell stronger, you can add a few drops of essential oil and store the mixture in a crock with a tight-fitting lid or a lidded jar for a few weeks or even months. As the mixture sits, the fragrance tends to become much richer.

This chart shows the main types of fragrances and the ingredients associated with them. When experimenting with creating a recipe, first decide whether you want your potpourri to be primarily spicy, sweet, fruity/citrusy, or earthy. Use mostly ingredients from that category and then add smaller amounts from the other categories, if desired.

"You Choose" Potpourri Recipe

Potpourri can be comprised of a mix of flowers, leaves, and herbs—depending on which of these are available to you. The ingredients in this recipe are interchangeable and can all be used together or you can pick and choose which you want to use in making your own potpourri.

Essential Ingredients

   1 ounce orrisroot
   1 ounce allspice
   1 ounce bay salt
   1 ounce cloves

Assorted Ingredients (to add as you like)

   Rose petals
   Lavender
   Lemon plant
   Verbena
   Myrtle
   Rosemary
   Bay leaf
   Violets
   Thyme

   Mint
   Essence of lemon
   Essence of lavender

Mix the orrisroot, allspice, bay salt, and cloves together. Combine this with about twelve handfuls of the dried petals and leaves and store in an airtight jar or bowl. A small quantity of essence of lemon and/or lavender may be added but these are not necessary. Let the mixture stand for a few weeks. If it becomes too moist, add additional powdered orrisroot. Once the potpourri is dry and very fragrant, parcel it into bowls to set around the house or give as gifts in jars or sachets.

Rose Potpourri

Ingredients

   1 lb rose petals (already pressed and from a jar)

   Dried lavender (any proportion you like)

   Lemon verbena leaves (any proportion you like)

   A dash of orange thyme

   A dash of bergamot

   1 dozen young bay leaves (dried and broken up)

   A pinch of musk

   2 ounces orrisroot, crushed

   1 ounce cloves, crushed

   1 ounce allspice, crushed

   ½ ounce nutmeg, crushed

   ½ ounce cinnamon, crushed

Combine all the ingredients together in a large crock, jar, or bowl with a lid. Seal and allow it to sit for a few weeks, until the aroma is to your liking. Parcel into small bowls to fragrance rooms or give as gifts.

A Simple Recipe for Sachet Potpourri

If you are unable to procure large amounts of petals and leaves, here is a simple recipe, using oils as substitutes, that makes fine potpourri sachets.

Ingredients

    2 drams alcohol
    10 drops bergamot
    20 drops eucalyptus oil
    4 drops oil of roses
    ½ tsp cloves
    1 ounce orrisroot
    ¼ tsp cinnamon
    ½ tsp mace
    1 ounce rose sachet powder

Mix these ingredients together in a large stone crock or in a large glass bowl. When the ingredients are thoroughly mixed, store the potpourri in small wooden boxes or sachets and place them around the house. The potpourri gives off a pleasing fragrance to any room or drawer.

Sachet Bags

Sachets are small bags filled with sweet-smelling potpourri. Hang them in a closet or tuck them in a drawer to lend your clothes a gentle fragrance.

- Fabric
- Scissors
- Needle and thread or sewing machine
- Ribbon or string

1. Cut two 3 x 5-inch rectangles of fabric. Hem one of the 3-inch sides of both rectangles. Stack the pieces one on top of the other, placing the hemmed edges together and lining up all sides exactly. The right sides of the fabric should be facing toward each other.
2. Sew along three sides of the fabric, about ¼-inch from the edges. Leave

one of the 3-inch sides unsewn so that the bag has an opening.
3. Turn the bag inside out so that the right side of the fabric is visible and the seams are hidden. Fill the bag half-full with potpourri and tie a ribbon or string around the top.

Preserving Flowers

Pressed Flowers and Leaves

Pressed flowers can be used to decorate stationery, handmade boxes, bookmarks, scrapbooks, or picture frames. Kids can glue pressed wildflowers to a blank book and add species names and descriptions to make their own field guides.

- Large book or newspapers
- Blotting paper

2. Use the newspapers for leaves and ferns. Blotting paper is best for the flowers. Both the flowers and leaves should be fresh and without moisture. Place them as nearly as possible in their natural positions in the book or papers, and press, allowing several thicknesses of paper between each layer.

3. Remove the flowers and leaves onto dry papers each day until they are perfectly dried.

Some flowers, like orchids, must be immersed—all but the flower head—in boiling water for a few minutes before pressing, to prevent them from turning black.

In order to preserve your flowers forever, get a blank book or pieces of stiff, white paper on which to mount your preserved flowers and leaves. You can glue them down to the paper with hot glue or regular Elmer's glue. The sooner you mount the specimens, the better. Place them carefully on the paper and, beneath each flower or leaf, write the name of the plant, where it was found, and the date.

Natural Wax Flowers

- Paraffin
- Saucepan
- Fresh flowers or leaves and ferns
- Wax paper
- Iron

1. To make wax flowers, dip the fresh buds and blossoms in paraffin that is just hot enough to be liquefied. First dip the stems of the flowers. When

these have cooled and hardened, then dip the flowers or sprays. Be sure to hold them by the stalks and move them gently.

2. When they are completely covered, remove the flowers from the wax and shake them lightly in order to throw off the excess wax. Allow the flowers to dry completely. The flowers will keep their beautiful coloring and natural forms, and even their fragrance for a short while.

3. For leaves, ferns, or flat flowers, you can place the plant between two sheets of wax paper and run a hot iron over the paper. Allow the paper to cool slightly and then carefully remove it.

- Weights
- Leaves, ferns, or flowers

1. Have a large book or a quantity of old newspapers and blotting paper, and several weights ready.

QUALITIES OF A GOOD KNOT

1. It can be tied quickly.
2. It will hold tightly.
3. It can be untied easily.

THREE PARTS OF A ROPE

1. **The standing part:** this is the long, unused part of the rope.
2. **The bight:** this is the loop formed whenever the rope is turned back.
3. **The end:** this is the part used in leading.

Pine Cone Bird Feeder

To attract wild birds to your yard, all you need is a pine cone, a bunch of seeds, and a little peanut butter. Keep in mind that wild animals also enjoy backyard treats, so you may end up attracting bears, raccoons, or other unwanted fauna. But in the meantime, the birds will appreciate your generosity!

- Cord or string (at least two feet long)
- Large pine cone
- Peanut butter
- Birdseed

1. Loop the cord around the top petals of the pine cone and tie it tightly. Then spoon a little peanut butter between each layer of pine cone petals.
2. Roll the pine cone in the birdseed (you can spread the seeds in a pie dish or on a sheet of waxed paper first) and hang the feeder in a tree.

Tying Knots

Knowing how to tie a variety of knots is invaluable, especially if you are involved in boating, rock climbing, fishing, or other outdoor activities.

Strong knots are typically those that are neat in appearance and are not bulky. If a knot is tied properly, it will almost never loosen and will still be easy to untie when necessary.

The best way to learn how to tie knots effectively is to sit down and practice with a piece of cord or rope. Listed below are a few common knots that are useful to know:

- **Bowline knot:** Fasten one end of the line to some object. After the loop is made, hold it in position with your left hand and pass the end of the line up through the loop, behind and over the line above, and through the loop once again. Pull it tightly and the knot is now complete.

- **Clove hitch:** This knot is particularly useful if you need the length of the running end to be adjustable.

- **Halter:** If you need to create a halter to lead a horse or pony, try this knot.

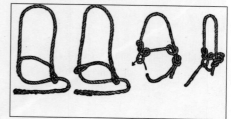

- **Sheepshank knot:** This is used for shortening ropes. Gather up the amount to be shortened and then make a half hitch around each of the bends.

- **Slip knot:** Slip knots are adjustable, you can tighten them around an object after they're tied.

Summer

- **Square/reef knot:** This is the most common knot for tying two ropes together.

- **Timber hitch:** If you need to secure a rope to a tree, this is the knot to use. It is easy to untie, too.

- **Two half hitches:** Use this knot to secure a rope to a pole, boat mooring, washer, tire, or similar object.

Kites

Flying kites is a wonderful way to spend a breezy summer day. Making your own kites is easy and fun and adds to the enjoyment and satisfaction of kite flying. Here are a few examples of kites that can easily be made and are particularly fun for children.

Frog Kite

Materials

Two 2-foot-long sticks (with the thinner part bent to form knees)

1-foot, 7-inch-long stick (for the spine)

2-foot, 5-inch-long piece of rattan (to make the body of the frog)

1. Place the two leg sticks one above the other and then place the spine stick on top of them. The tops of all three sticks should be perfectly even. Eight inches from the top, drive a pin through all three sticks and carefully clamp it on the other side where the point comes through.

2. For the body, bend the piece of rattan to form a circle (allow the ends to overlap about an inch or so—this will help in the binding of the ends to make a joint).

HOW TO PASTE TISSUE PAPER TO YOUR KITE FRAME

Make a good paste out of flour and water by boiling it until it reaches the consistency of starch. Put the paste on with a bristle brush, make the seams hardly more than ¼ inch wide, and press them together with a soft rag.

To adhere tissue paper to your frame, place the tissue paper on the ground and lay the frame over it, holding the frame down with heavy books. Cut the paper around the frame, leaving a ½-inch edge, and make a slit in the edge every 6 or 7 inches and at the angles. With your brush, coat the edge with paste, one section at a time, turn the sections over, and press down with the rag.

The circle, when complete, should be about 8 inches in diameter.

3. Take the three sticks that are pinned together, lay them on the floor, and spread them apart to form an irregular star. The top of the spine should be just about halfway between the tops of the legs (5 inches from each). Place the rattan circle over the sticks, with the intersection of the sticks located in the middle of the circle.

4. With pins and thread, fasten the frame together. The lower limbs will be spread wide apart and they must be carefully drawn together and held in position by a string that is tied near the termination of each leg stick.

5. Cross-sticks for hands and feet may now be added and the strings attached to various points. Cover the kite with green tissue paper and decorate it to look like a frog.

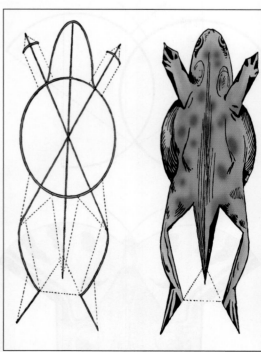

⌃ **Layout of Frog Kite**

Summer

Butterfly Kite

1. Make a thin, straight stick out of a piece of elastic wood or split rattan. At the top of this, attach a piece of thread or string.
2. Bend the stick as you would a bow until it forms an arc or half circle. Then, holding the stick in this position, tie the other end of the string to a point a few inches above the bottom end of the stick. At a point on the stick about one-quarter the distance from the top, tie another string, draw it taut, and fasten it to the bottom end of the bow.
3. Take another stick of exactly the same length and thickness as the first and go through the same process, making a frame that is exactly the same as the first. Then, fasten the two frames together (Figure 13), allowing the arcs to overlap several inches, and bind the joints securely with thread.
4. Make the head of the insect by attaching two pipe cleaners to the top part of the wings where they are joined together. The straws must be crossed and the projecting ends can serve as the antennae.
5. Select a piece of yellow or blue tissue paper, place your frame over it, cut it to the correct measurements, and paste. After the kite is dry, draw some markings on the wings with black paint or cut

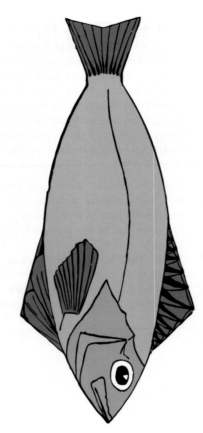

out markings in dark-colored paper and paste them on.

Fish Kite

1. Cut two straight pine sticks and shave them down until they are thin enough to bend easily. They should be exactly the same length and roughly the same weight. Fasten the top ends together by driving a pin through them.
2. Bend each stick to form a bow and hold them in this position until you have secured a third stick across them at right angles about one-third the way down from the top. The kite should now be half as broad as it is long.
3. Let the lower ends of the side, or bow, sticks cross each other far enough up to form a tail for the fish and fasten the sticks together at their intersection.
4. Before stringing the frame, see that the cross-stick protrudes an equal distance from each side of the fish.
5. To make the tail, tie a string across the bottom from the end of one cross-stick to the end of the other and tie another string to this string in the middle. Pass the string up to the base of the tail, draw it taut, and fasten it there at the intersection of the side-

sticks. This will make a natural look for the caudal fin.

6. The remainder of the strings can be put on—take care that the dorsal and back fin are made exactly the same size. Choose yellow, red, or green tissue paper to cover the kite and decorate as you see fit.
7. Tie the strings of the breast-band to the side-sticks near the head and tail, and let them cross each other as in a common kite. Attach the tail-band to the tail of the fish.

Shield Kite

1. Make the frame of four sticks: two straight cross-sticks and two bent side-sticks.
2. Cover it with red, white, and blue tissue paper (making it look like a flag) or use any color of your choosing. Make sure to cut the paper so it looks like a shield.

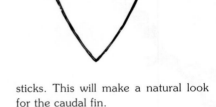

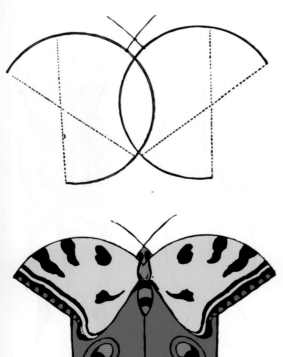

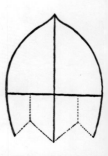

Japanese Square Kite

1. This kite is not actually perfectly square. It is rectangular and made with a framework of very thin bamboo or cane sticks, bound together.

2. The frame should be covered with Japanese paper and all sides of the paper should be glued down well.

3. The kite should be bent backwards, making it slightly convex in the front. To hold the kite in this position, use strings that are tied from end to end of the cross-sticks at the back. The breastband may be attached like any other six-sided kite.

4. Instead of a tail band, with a single tail attached, this kite carries two tails, one tied at each side to the protruding ends of he diagonal sticks at the bottom of the kite.

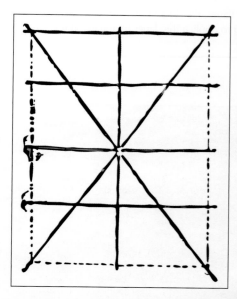

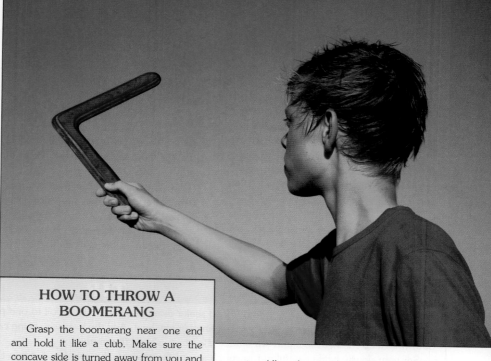

HOW TO THROW A BOOMERANG

Grasp the boomerang near one end and hold it like a club. Make sure the concave side is turned away from you and the convex side is toward you. Find something to take aim at and then throw the boomerang at the object. If the boomerang is well made, it should return to you after its flight. Be careful not to throw the boomerang when others are close by—it may end up hitting them and it can leave a bad welt. It is best to throw your boomerang in a large, open field by yourself.

Boomerangs

In order to make a boomerang, scald a piece of well-seasoned elm, ash, or hickory plank (free from any knots) in a pot of boiling water. Allow the wood to remain in the water until it becomes pliable enough to bend into a slight V-shaped form. When the wood has assumed the proper shape, nail on the side pieces to hold the wood in position until it is thoroughly dry. After the plank is completely dry, the side pieces can be removed—the wood will keep the curved shape.

Saw the wood into as many pieces as it will allow and each piece will become a boomerang. If the edges are very rough, trim them with a pocketknife and scrape them smooth. You can use a large file to help shape the boomerang. The efficiency of your boomerang (how well it soars and returns to you) will vary in each piece, depending on the curvature.

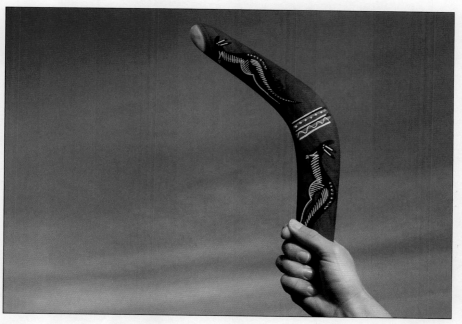

Sundial

Sundials have been used for well over 5,000 years as a means of telling time based on the sun's position. The vertical axis, or gnomon (in this case a chopstick), casts a shadow over the horizontal axis (the wooden disk). The shadow moves as the sun travels across the sky. You can tell what time it is by seeing where the shadow falls.

- One wooden disk or rectangle
- One chopstick or wooden dowel
- Drill
- Air-dry clay
- Paint (optional)
- Permanent marker
- Spray acrylic sealer

1. Drill a hole in the wood. The hole should be just large enough for the wider end of the chopstick to fit.

2. Press the clay into the hole and stick the wide end of the chopstick firmly into the clay. Be sure the chopstick is completely vertical—not leaning one way or the other. Allow the clay to harden. If desired, paint the disk and gnomon (chopstick).

3. Place the sundial in a sunny spot outdoors early in the day. Every hour, on the hour, make a mark where the gnomon's shadow falls and place a number near the mark to indicate the time. You may want to do this in pencil first and then outline it with permanent marker.

4. When all twelve hours are marked, spray the entire sundial with a clear acrylic sealer. Allow to dry and then apply a second coat.

TIP

Adults can make their own non-toxic wood sealer using five parts mineral oil to one part beeswax. Children should not attempt this on their own. Heat the mineral oil in a saucepan over low heat. Add the beeswax, being careful not to splash the oil, and allow it to melt. *Caution: Keep the heat low and stir slowly. Beeswax is very flammable.*

Stir very slowly, being careful not to splash any of the mixture. If it splashes out into the flame, it will ignite. Once all the beeswax is melted, use a funnel to pour the mixture into a mason jar. Once the mixture is cool, use a rag to apply a thin layer to any wood you want to protect from the elements.

⥤ A simple, temporary sundial can be made in snow or sand.

Autumn

Fall days may be filled with harvesting the gardens and savoring the fleeting rays of sunshine, but a cool autumn evening is perfect for cozy craft projects. Try dipping candles, making fragrant soaps, or weaving baskets reeds or rushes. The kids will have fun making dolls from the husks leftover from late summer corn.

Handmade Candles

Making candles is a great activity for a fall afternoon. Simple beeswax candles can be completed in a few minutes, but give yourself several hours to make dipped candles. The process is fun, creative, and productive. Give your handmade candles to friends or family or burn them at home to create atmosphere and save on your electricity bill.

TIP

Rather than pouring leftover wax down the drain (which will clog your drain and is bad for the environment), dump it into a jar and set it aside. You can melt it again later for another project.

TIP

When making candles, keep a box of baking soda nearby. If wax lights on fire, it reacts similarly to a grease fire, which is aggravated by water. Douse a wax fire with baking soda and it will extinguish quickly.

Autumn

Rolled Beeswax Candle

Beeswax candles are cheap, eco-friendly, non-allergenic, dripless, non-toxic, and they burn cleanly and beautifully. They're also very simple and quick to make—perfect for a short afternoon project.

Materials

Sheets of beeswax (you can find these at your local arts and crafts store or from a local beekeeper)

Wick (you can purchase candle wicks at your local arts and crafts store)

Supplies

Scissors

Hair dryer (optional)

Directions

1. Fold one sheet of beeswax in half. Cut along the crease to make two separate pieces.
2. Cut your wick to about 2 inches longer than the length of the beeswax sheet.
3. Lay the wick on the edge of the beeswax sheet, closest to you. Make sure the wick hangs off of each end of the sheet.
4. Start rolling the beeswax over the wick. Apply slight pressure as you roll to keep the wax tightly bound. The tighter you roll the beeswax, the sturdier your candle will be and the better it will burn.
5. When you reach the end, seal off the candle by gently pressing the edge of the sheet into the rolled candle, letting your body heat melt the wax.
6. Trim the wick on the bottom (you may also want to slice off the bottom slightly to make it even so it will stand up straight) and then cut the wick to about ½ inch at the top.

Taper Candles

Taper candles are perfect for candlesticks, and they can be made in a variety of sizes and colors.

TIP

If you are having trouble using the beeswax and want to facilitate the adhering process, you can use a hair dryer to soften the wax and to help you roll it. Start at the end with the wick and, moving the hair dryer over the wax, heat it up. Keep rolling until you reach a section that is not as warm, heat that up, and continue all the way to the end.

TIP

Old crayons can be melted and used instead of paraffin for candle-making.

Materials

Wick (be sure to find a spool of wick that is made specifically for taper candles)

Wax (paraffin is best)

Candle fragrances and dyes (optional)

Supplies

Pencil or chopstick (to wind the wick around to facilitate dipping and drying)

Weight (such as a fishing lure, bolt, or washer)

Dipping container (this should be tall and skinny. You can find these containers at your local arts and craft store, or you can substitute a spaghetti pot)

Stove

Large pot for boiling water

Small trivet or rack

Glass or candle thermometer

Newspaper

Drying rack

Directions

1. Cut the wick to the desired length of your candle, leaving about 5 additional inches that will be tied onto the pencil or chopstick for dipping and drying purposes. Attach a weight (a fishing lure, bolt, or heavy metal washer) to the dipping end of the wick to help with the first few dips into the wax.

2. Ready your dipping container. Put the wax (preferably in smaller chunks—this will speed up the melting process) into the container and set aside.

3. In a large pot, start to boil water. Before putting the dipping container full of wax into the larger pot, place a small trivet, rack, or other elevating device into the bottom of the larger pot. This will keep the dipping container from touching the bottom of the larger pot and will prevent the wax from burning and possibly combusting.

4. Put the dipping container into the pot and start to melt the wax, keeping a thermometer in the wax at all times. The wax should be heated and melted between 150 and 165°F. Stir frequently in order to keep the chunks of paraffin from burning and to make sure all the wax is thoroughly melted. (If you want to add fragrance or dye, do so when the wax is completely melted and stir until the additives are dissolved.)

5. Once your wax is completely melted, it's time to start the dipping process. Removing the container from the stove, take your wick that's tied onto a stick and dip it into the wax, leaving it there for a few minutes. Continue to lower the wick in and out of the dipping container, and by the eighth or ninth dip, cut off the weight from the bottom of the wick—the candle should be heavy enough now to dip well on its own.

6. To speed up the cooling process—and to help the wax to continue to adhere and build up on the wick—blow on the hot wax each time you lift the candle out of the dipping pot.

7. When the candle is at the desired length and thickness, you may want to lay it down on a very smooth surface (such as a countertop) and gently roll it into shape.

8. On a drying rack (which can be made from a box long enough so the candles do not touch the bottom or from another device), carefully hang your taper candle to dry for a good twenty-four hours.

9. Once the candle is completely hardened, trim the wick to just above the wax.

LAYERED TAPER CANDLES

For a more ornate candle, add different shades of food coloring to three or four separate pots of melted wax (or melt down old crayons). Alternate between the different colors of wax as you dip the wick, creating different layers of color. Once the candle is the desired thickness and is mostly cooled, use a paring knife to carefully peel away strips of the wax around the outside of the candle. Allow the wax strips to curl downward as you peel, revealing a rainbow of colors.

GOURD VOTIVES

Small gourds make perfect votive candleholders. Carve a circle out of the top of the gourd, making it the same size as the circumference of the candle you intend to place in it. Gently pry off the top and set the candle in the indentation. If necessary, cut the hole slightly larger, but keep it small enough that the candle fits snugly.

⌃ Stubs of taper candles can be melted into empty walnut shells to make unique floating candles.

Jarred Soy Candles

Soy candles are environmentally friendly and easy to make. You can find most of the ingredients and materials needed to make soy candles at your local arts and crafts store—or even in your own kitchen!

Materials

1 lb soy wax (either in bars or flakes)
1 ounce essential oil (for fragrance)

Natural dye (try using dried and powdered beets for red, turmeric for yellow, or blueberries for blue)

Supplies

Stove
Pan to heat wax (a double boiler is best)
Spoon
Glass thermometer
Candle wick (you can find this at your local arts and crafts store)
Metal washers

TIP

To make floating candles, pour hot wax into a muffin tin until each muffin cup is about one-third full. Allow the wax to cool until a film forms over the tops of the candles. Insert a piece of wick into the center of each candle (use a toothpick to help poke the hole if necessary). Allow candles to finish hardening and then pop them out of the tin. Trim wicks to about ¼ inch.

Pencils or chopsticks

Heatproof cup to pour your melted wax into the jar(s)

Jar to hold the candle (jelly jars or other glass jars work well)

Directions

1. Put the wax in a pan or a double boiler and heat it slowly over medium heat. Heat the wax to 130 to 140°F or until it's completely melted.

2. Remove the wax from the heat. Add the essential oil and dye (optional) and stir into the melted wax until completely dissolved.

3. Allow the wax to cool slightly, until it becomes cloudy.

4. While the wax is cooling, prepare your wick in the glass container. It is best to have a wick with a metal disk on the end—this will help stabilize it while the candle is hardening. If your wick does not already have a metal disk at the end, you can easily attach a thin metal washer to the end of the wick. Position the wick in the glass container and wrap the excess wick around the middle of a pen or chopstick. Lay the pencil or chopstick on the rim of the container and position the wick so it falls in the center.

5. Using a heat proof cup or the container from the double boiler, carefully pour the wax into the glass container, being careful not to disturb the wick from the center.

6. Allow the candle to dry for at least twenty-four hours before cutting off the excess wick and using.

TIP

Add citronella essential oil and a few drops of any of the following other essential oils to make your candle a mosquito repellant:

- Catnip
- Cloves
- Cedarwood
- Lavender
- Lemongrass
- Eucalyptus
- Peppermint
- Rosemary
- Rose geranium
- Thyme

Soap Making

When you make your own soap, you get to choose how you want it to look, feel, and smell. Adding dyes, essential oils, texture (with oatmeal, seeds, etc.), or pouring it into molds will make your soap unique. Making soap requires time, patience, and caution, as you'll be using some caustic and potentially dangerous ingredients—especially lye (sodium hydroxide). Avoid coming into direct contact with the lye; wear goggles, rubber gloves, and long sleeves, and work in a well-ventilated area. Be careful not to breathe in the fumes produced by the lye and water mixture.

Soap is made up of three main ingredients: water, lye, and fats or oils. While lard and tallow were once used exclusively for making soaps, it is perfectly acceptable to use a combination of pure oils for the "fat" needed to make soap. Saponification is the process in which the mixture becomes completely blended and the chemical reactions between the lye and the oils, over time, turn the mixture into a hardened bar of usable soap.

Cold-Pressed Soap

Ingredients

6.9 ounces lye (sodium hydroxide)

2 cups distilled water, cold (from the refrigerator is the best)

2 cups canola oil

2 cups coconut oil

2 cups palm oil

the mixture to cool to around 110°F (the chemical reaction of the lye mixing with the water will cause it to heat up quickly at first).

4. While the lye is cooling, combine the oils in a pot on medium heat and stir well until they are melted together. Place a thermometer into the pot and allow the mixture to cool to 110°F.

5. Carefully pour the lye mixture into the oil mixture in a small, consistent stream, stirring continuously to make sure the lye and oils mix properly. Continue stirring, either by hand (which can take a very long time) or with a handheld stick blender, until the mixture traces (has the consistency of thin pudding). This may take anywhere from thirty to sixty minutes or more, so be patient. It is well worth the time invested to make sure your mixture traces. If it doesn't trace all the way, it will not saponify correctly and your soap will be ruined.

6. Once your mixture has traced, pour carefully into the mold(s) and let sit for a few hours. Then, when the mixture is still soft but congealed enough not to melt back into itself, cut the soap with a table knife into bars. Let sit for

Supplies

Goggles, gloves, and mask (optional) to wear while making the soap

Mold for the soap (a cake or bread loaf pan will work just fine; you can also find flexible plastic molds at your local arts and crafts store)

Plastic wrap or wax paper to line the molds

Glass bowl to mix the lye and water

Wooden spoon for mixing

2 thermometers (one for the lye and water mixture and one for the oil mixture)

Stainless steel or cast iron pot for heating oils and mixing in lye mixture

Handheld stick blender (optional)

Directions

1. Put on the goggles and gloves and make sure you are working in a well-ventilated room.

2. Ready your mold(s) by lining with plastic wrap or wax paper. Set them aside.

3. Slowly add the lye to the cold, distilled water in a glass bowl (*never* add the water to the lye) and stir continually for at least a minute, or until the lye is completely dissolved. Place one thermometer into the glass bowl and allow

a few days, then take the bars out of the mold(s) and place on brown paper (grocery bags are perfect) in a dark area. Allow the bars to cure for another four weeks or so before using.

If you want your soap to be colored, add special soap-coloring dyes (you can find these at the local arts and crafts store) after the mixture has traced, stirring them in. Or try making your own dyes using herbs, flowers, or spices.

To make a yummy-smelling bar of soap, add a few drops of your favorite essential oils (such as lavender, lemon, or rose) after the tracing of the mixture and stir in. You can also add aloe and vitamin E at this point to make your soap softer and more moisturizing.

To add texture and exfoliating properties to your soap, you can stir some oats into the traced mixture, along with some almond essential oil or a dab of honey. This will not only give your soap a nice, pumice-like quality but it will also smell wonderful. Try adding bits of lavender, rose petals, or citrus peel to your soap for variety.

To make soap in different shapes, pour your mixture into molds instead of making them into bars. If you are looking to have round soaps, you can take a few bars of soap you've just made, place them into a resealable plastic bag, and warm them by putting the bag into hot water (120°F) for thirty minutes. Then, cut the bars up and roll them into balls. These soaps should set in about one hour or so.

SOAP OILS

Oil	Qualities
Almond Butter	Conditioning. Creamy Lather. Moderate Iodine.
Almond Oil, sweet	Conditioning. Fragrant. High Iiodine.
Apricot Kernel Oil	Conditioning. Fragrant. High Iodine.
Avocado Oil	Conditioning. Creamy Lather. High Iodine.
Babassu Oil	Cleansing. Bubbly. Very Low Iodine.
Canola Oil	Conditioning. Inexpensive. High Iodine.
Cocoa Butter	Creamy Lather. Low Iodine.
Coconut Oil	Bubbly Lather. Cleansing. Low Iodine.
Emu Oil	Conditioning. Creamy Lather. Moderate Iodine.
Evening Primrose Oil	Conditioning. Very High Iodine.
Flax Oil, Linseed	Conditioning. Very High Iodine.
Ghee	Cleansing. Bubbly Lather. Very Low Iodine.
Grapseed Oil	Conditioning. Very High Iodine.
Hemp Oil	Conditioning. Very High Iodine.
Lanolin liquid wax	Low Iodine.
Neem Tree Oil	Conditioning. Creamy Lather. High Iodine.
Olive Oil	Conditioning. Creamy Lather. High Iodine.
Palm Oil	Conditioning. Creamy Lather. Moderate Iodine.
Rapeseed Oil	High Iodine.
Safflower Oil	Conditioning. Very High Iodine.
Sesame Oil	Conditioning. High Iodine.
Shea Butter	Conditioning. Creamy Lather. Moderate Iodine.
Ucuuba Butter	Conditioning. Creamy Lather. Low Iodine.

THE JUNIOR HOME-STEADER

How Soap Works

Teach kids how soap cleans with this simple experiment.

1. Half-fill two Mason jars with water and add a few drops of food coloring. Pour several tablespoons of oil into each jar (corn oil, olive oil, or whatever you have on hand will be fine). You will see that the oil and water form separate layers. This is because the molecules in oil are hydrophobic, meaning that they repel water.

2. Add a few drops of liquid soap to one of the jars. Close both jars securely and shake for about thirty seconds. The oil and water should be thoroughly mixed.

3. Let both jars rest undisturbed. The jar with the soap in it will stay mixed, whereas the jar without the soap will separate back into two distinct layers. Why? Soap is made up of long molecules, each with a hydrophobic end and a hydrophilic (water-loving) end. The water bonds with the hydrophilic end and the oil bonds with the hydrophobic end. The soap serves as a glue that sticks the oil and water together. When you rinse off the soap, it sticks to the water, and the oil sticks to the soap, pulling all the oil down the drain.

NATURAL BATH SALTS AND SCRUBS

Follow these recipes to make your own luxurious bath products.

Lavender Bath Salt

Pour several tablespoons of this into your bath as it fills for an extra-soothing, relaxing, and cleansing experience. You can also add powdered milk or finely ground old-fashioned oatmeal to make your skin especially soft. Toss in a few lavender buds if you have them.

Ingredients

2 cups coarse sea salt
½ cup Epsom salts
½ cup baking soda
4 to 6 drops lavender essential oil
Red and blue food coloring, if desired (use more red than blue to achieve a lavender color)

Mix all ingredients thoroughly and store in a glass jar or other airtight container.

Citrus Scrub

Use this invigorating scrub to wake up your senses in the morning. The vitamin C in oranges serves as an astringent, making it especially good for oily skin.

Ingredients

½ orange or grapefruit
3 tbsps cornmeal
2 tbsps Epsom salts or coarse sea salt

Squeeze citrus juice and pulp into a bowl and add cornmeal and salts to form a paste. Rub gently over entire body and then rinse.

Healing Bath Soak

This bath soak will relax tired muscles, help to calm nerves, and leave skin soft and fragrant. You may also wish to add blackberry, raspberry, or violet leaves. Dried or fresh herbs can be used.

2 tbsps comfrey leaves
1 tbsp lavender
1 tbsp evening primrose flowers
1 tsp orange peel, thinly sliced or grated
2 tbsps oatmeal

Combine herbs and tie up in a small muslin or cheesecloth sack. Leave under faucet as the tub fills with hot water. If desired, empty herbs into the bath water once the tub is full.

Rosemary Peppermint Foot Scrub

Use this foot rub to remove calluses, soften skin, and leave your feet feeling and smelling wonderful.

Ingredients

1 cup coarse sea salt
¼ cup sweet almond or olive oil
2 to 3 drops peppermint essential oil
1 to 2 drops rosemary essential oil
2 sprigs fresh rosemary, crushed, or ½ tsp dried rosemary

Combine all ingredients and massage into feet and ankles. Rinse with warm water and follow with a moisturizer.

NATURAL DYES FOR SOAP OR CANDLES	
Light/Dark Brown	Cinnamon, ground cloves, allspice, nutmeg, coffee
Yellow	Turmeric, saffron, calendula petals
Green	Liquid chlorophyll, alfalfa, cucumber, sage, nettles
Red	Annatto extract, beets, grapeskin extract
Blue	Red cabbage
Purple	Alkanet root

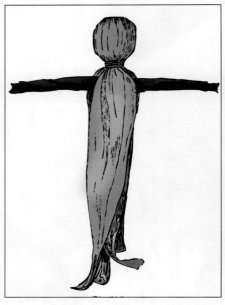

Almost any oil can be used to make soap, but different oils have different qualities; some oils create a creamier lather, some create a bubbly lather. Oils that are high in iodine will produce a softer soap, so be sure to mix with oils that are lower in iodine. Online soap calculators are very helpful when creating your own recipes.

Cornhusk Dolls

This old-fashioned doll makes a wonderful gift for young children and also a unique, decorative, homemade item for your home or for sale at a craft fair. Cornhusk dolls are quite easy to make if you just follow these simple steps:

- Corn husks
- Thread or string
- Pen or marker
- Natural materials for decorations

1. Gather husks from several large ears of corn (you may have these from your garden, if you grow corn, or you might find them at a garden center or a farmers' market in the fall). Select the soft, white husks that grow closest to the ear.
2. Place the stiff ends of two husks together, fold one of the long, soft husks in a strip lengthwise, and wrap it around the ends.
3. Choose the softest and widest husk you can find, fold it across the center, and place a piece of strong thread or string round it and tie it tightly in a knot.
4. Bring this down over the already-wound husks and tie it with a thread underneath. This will form the head and neck of the doll.

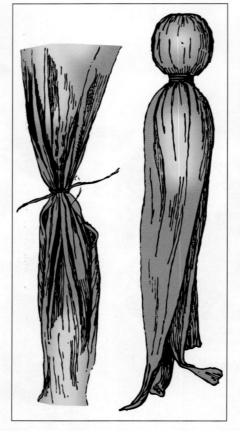

5. To make the arms, divide the husk below the neck into two equal parts. Fold two or three husks together and insert them in the space you've made by the division. Hold the arms in place with one hand and use your other hand to fold several layers of husk over each shoulder, allowing them to extend down the back of the figure.
6. When the figure seems substantial enough, use your best husks for the topmost layers and wrap the waist with strong thread, tying it tightly.
7. Divide the husks below the waist band and make the legs by neatly wrapping each portion with thread. Trim the husks off evenly for the feet.

Once you have the basics down, try adding skirts, costumes, or other details to your dolls.»

8. Twist the arms once or twice, tie them, and trim them evenly for the hands.

9. You can draw a face on the doll with a pen or marker or you can glue tiny natural items on the head to make a face.

10. If you want your doll to have clothing or a specific "costume," you can make these from any kind of material and in any way you wish.

Once you have the basics down, try adding skirts, costumes, or other details to your dolls.»

For a variation on a traditional corn husk doll, try the following:

1. Gather a young ear of corn (whose silk has not yet browned), a crab apple for the head, and a leaf from the cob for a dress.

2. Cut off the bottom of the ear of corn where the husks are puckered and carefully take the silk from the other end, making sure not to disturb the closely wrapped husks remaining.

3. Roll part of the leaf (Fig. 120) for the arms and fasten the crab apple to the leaf arms with a small stick. Stick the other end of the twig into the small end of the corncob.

4. Now you can dress the doll. The hat for the doll can be made from a leaf (just where it joins the stalk. This can be fastened to the doll's head with a small twig or thorn. Make sure the silk is placed on the head to form hair before securing the hat (Fig. 121).

5. Make a scarf by folding a leaf around the shoulders and securing it with small pins or thorns (Fig. 122).

6. Stick tiny thorns into the crab apple to make eyes and a nose.

Basketweaving

Basketweaving is one of the oldest, most common, and useful crafts. The materials used in making baskets are primarily reed or rattan, raffia, corn husks, splints, and natural grasses. Rattan grows in tropical forests, where it twines about the trees in great lengths. It is numbered according to its thickness, and numbers 2, 3, and 4 are the best sizes for small baskets. For scrap baskets, 3, 5, and 6 are the best sizes. Rattan should be thoroughly soaked before using. Raffia is the outer cuticle of a palm, and comes from Madagascar. Cattail reeds can also be excellent for baskets and may be more readily available, as they frequently grow near ponds or swampy areas. Most basket making materials can also be found at local craft stores.

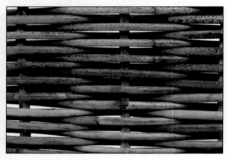

Small Reed Basket

Most reed baskets have at least sixteen spokes, and for small baskets and where small reeds are used these spokes are often woven in pairs. You can vary the look of your reed basket by combining and interweaving two different colored reeds.

Materials Needed

Sixteen 16-inch spokes, No. 2 reed
Five weavers of No. 2 brown reed

Directions

Separate the sixteen spokes into groups of four each. Mark the centers and lay the first group on the table in a vertical position. Across the center of this group place the second group horizontally. Place the third group diagonally across these, having the upper ends at the right of the vertical spokes. Lay the fourth group diagonally with the upper ends at the left of the vertical spokes.

Soak the reeds well and then start the basket by laying the weaver's end over the group to the left of the vertical group, just above the center; then bring it under the vertical group, over the horizontal and then under, and so on until it reaches the vertical group again. Repeat this weave three or four times. Then separate the spokes into twos and bring the weaver over the pair at the left of the upper vertical group, and so on, over and under until it comes around again, when it is necessary to pass under two groups of spokes and then continue weaving over and under alternate spokes. At the beginning of each new row the weaver passes under two groups of spokes, always under the last of the two under which it went before and the group at the right of it.

Weave the bottom until it is 4 inches in diameter; then wet and turn the spokes gradually up and weave 1 inch. After that, turn the spokes in sharply and draw them in with three rows of weaving. Now weave four rows, going over and under the same spokes, making an ornamental band; then weave three rows of over- and under-weaving, followed by four rows without changing the weave. Continue to draw the side in with four rows of over- and under-weaving, and then bind it all off. Finish with the following border:

Always wet the spokes till they are pliable before starting the border. Bring each group under the first group at the right and over the next and inside the basket. Finally, cut the reeds long enough to allow them to rest on the group ahead.

Note: Leave the first two groups a little loose so that the last ones can be easily woven into them.

Basket with Triple Twist

Materials

Sixteen 24-inch spokes, No. 1 gray-green reed

8 weavers of No. 1 natural-colored reed

4 weavers of No. 1 gray-green reed

Directions

Weave the center as in the small reed basket until it measures 2 inches in diameter. Then, separate the pairs of reeds and weave over and under each spoke separately until the bottom measures 3½ inches. Now turn the reeds up sharply and weave six rows of under- and over-weaving. By this time the spokes should stand straight. Begin the triple-weave by inserting two new weavers in addition to the one already started.

Insert a green weaver between the two spokes to the right of the one already in use. Place another natural-colored reed between the two spokes to the right of the green reed. Pass the first weaver in front of the first two spokes to the right, behind the third spoke and out. Now pass the colored weaver in front of the next two spokes behind the third and out. Do the same with the third weaver, and then with the first one again and continue until the basket is 5 inches tall. Soak the mat and finish as follows:

1. Pass each spoke in succession behind the one to its left and out. Press down close to the basket and put the last spoke left standing under the first one and turn it downwards.

2. Pass each spoke in succession in front of the one at its left and turn it inward. Put the last one under the first one that is turned in.

3. Pass each spoke in succession behind the one at the left and then out. Press down sharply between the outside weave of the border and the basket and cut off the excess reed with wire cutters close to the border.

Coiled Basket

Sweet grass, corn husks, or any pliable grasses can be used for this type of basket, and with a contrasting color for sewing, the basket can be very attractive.

Materials

A bunch of grasses

A bunch of raffia

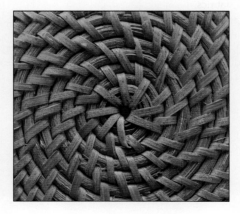

Directions

1. Cut off the hard ends of the grasses and take only a small bunch for the center to start. Split the raffia very fine and use a sharp needle for extra help.

2. Hold the grasses and the end of raffia in your left hand, about 2 inches from the end of the coil, and wind the raffia around the coil to the end of the grasses.

3. Bend the end of the coil into a small round center and sew over and under, binding the first two coils very firmly together. The next time around, leave a very small space between each stitch, and take the stitch only through the upper portion of the coil below. It is necessary that the spaces between the stitches be very small in the first few rows—this will determine the regularity of the spirals to come.

4. In sewing through the coil, place the needle diagonally from the right of the stitch through the coil to the left of the stitch.

5. When the bottom measures 4 inches across, begin shaping the sides by raising the coil up slightly on the coil below and continue to bind the coils together as before. When the basket measures about 6 inches across, begin shaping the sides by pushing the coil slightly in toward the center.

BIRCH BARK BASKET

Birch bark baskets are a wonderful way to display dried wildflowers and make nice gifts. Making a basket in the shape of a canoe works very well with the bark. Gather bark strips (do not string directly from the tree; try to find these either on cut wood or from a craft or lumber store) that are 6½ inches long and 4 inches wide. Sew the ends of the bark together with a thick thread (Fig. 34), leaving one side of each strip unstitched. Sew a ribbon on each end of the canoe—these will serve as handles. Now the basket may be filled with dried wildflowers or other things from nature (such as pine cones) and hung on the wall for display.

6. To finish, gradually decrease the size of the coil but do not increase the number of stitches. Fasten the raffia, after the last stitch, by running it through the coil and cut it off close to the border. If necessary, bind the ends more closely by sewing over and over with a thin thread of very fine raffia the same color as the grasses.

Lampshades

Making your own lampshades will enable you to match or complement the décor of your room. Choose material that is fire-restistant; a 100 watt light bulb can heat up to more than 200 degrees. Keep in mind that lighter-colored, thin fabrics will let more light through than dark, heavy fabrics.

- Newspaper
- Lampshade frame
- Fabric
- Pins
- Scissors
- Masking tape
- Measuring tape or ruler
- Fabric glue

1. Measure the circumference of the top and bottom of your lampshade frame. To facilitate this, you can place a strip of masking tape around one circle and then peel it off and use a ruler or measuring tape to determine the exact length of the tape. Next, measure the distance from the bottom, wide circle of the frame to the top, narrow circle. Add 1 ½ inches to all measurements.

2. Use your measurements to draw a pattern on newspaper. The pattern will be in the shape of a trapezoid. The bottom of the trapezoid is the length of the bottom circle of the frame; the top of the trapezoid is the length of the top circle of the frame; and the distance between them is the height of your frame.

3. Cut out the pattern and lay it on the backside of your fabric. Pin the pattern to the fabric and carefully cut out the fabric. Remove the pattern.

4. Cover the backside of the fabric with fabric glue or spray adhesive. Place the frame onto the fabric and begin rolling it, pressing the fabric to the frame as you go. The fabric should hang over the top and bottom of the frame slightly.

5. Once the fabric is wrapped all the way around the lampshade, tuck in the raw edge of the fabric to create a seam and glue it down. Then glue and tuck the fabric over the top circle and under the bottom circle of the frame. Use clothespins to hold the fabric in place until it dries completely.

6. If desired, add ribbons or tassels to the top or bottom of the shade.

TIP

Fabric with a higher percentage of cotton will adhere to the frame better than synthetic fabrics. Also keep in mind that the darker the fabric, the less light will shine through.

Winter

Winter is perhaps the ideal season for crafting. The colder weather lends itself to long afternoons by the fire spent piecing together a quilt or organizing photos and keepsakes into a scrapbook. Or try your hand at making pottery or designing jewelry!

Quilting

Crazy quilts first became popular in the 1800s and were often hung as decorative pieces or displayed as keepsakes, but they can also be warm and practical. They can be made out of scraps of fabric that are too small for almost any other use, and there is endless room for variation in colors, patterns, and texture.

Quilts are generally made up of many small squares that are sewn together into a large rectangle and layered with batting (a thick layer of fabric to add warmth—usually wool or cotton) and backing (the material that will show on the underside of the quilt).

- Non-woven interfacing or lightweight muslin (prewashed)
- Fabric pieces in a variety of colors and patterns
- Scissors
- Ruler
- Pencil
- Needle or sewing machine
- Pins
- Thread
- Iron

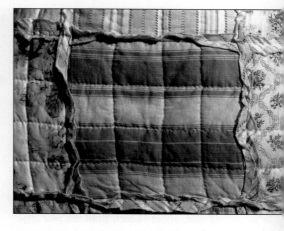

1. Make the foundation squares. If your foundation fabric (interfacing or muslin) is wrinkled, iron it carefully until it is completely flat. Then use a ruler to measure and draw a 13 x 13-inch square in one corner (your final square will be 12 x 12 inches, but it's a good idea to leave yourself a little extra fabric to work with). Repeat until all of the foundation fabric is cut into squares.

2. Cut a small piece of patterned fabric into a shape with three or five straight edges. Pin it right-side up on the center of one foundation square. Cut another small piece of fabric with straight edges and lay it right-side down on top of the first piece. Sew a ¼-inch seam along the edge where the two fabrics overlap. If the second piece is longer than the first piece, don't sew beyond the edge of the first piece. Turn the second piece of fabric over so that it's facing up and iron it. Trim the second piece to align with the first, so you have one larger shape with straight edges.

3. Continue with a third piece of fabric, making sure it is large enough to extend the length of the first two patches combined. Sew the seam, flip the fabric upright, iron, trim, and proceed with a fourth piece. Work clockwise, each piece getting larger as you move toward the edge of the foundation square. Once the square is filled, trim off any overhanging fabric so that you have one neat square. Repeat steps 2 and 3 until all foundation pieces are filled.

4. Sew all foundation squares together with a ¼- to ½-inch seam.

5. Sandwich the quilt by placing the backing face down, the batting on top of it, and then the foundation on top, with the patterned squares facing up. Baste around the quilt to hold the three layers together, using long stitches and

staying about ¼-inch from the outside edge of the patterned fabric.

6. Binding your quilt covers the rough edges and creates an attractive border around the edges of the quilt. To make the binding, cut strips of fabric 2 ½ inches wide and as long as one side of your quilt plus 2 inches. Fold the fabric in half lengthwise and press.

7. Lay the strip along one edge of the quilt. The raw edges of the quilt and the binding should be stacked together. Leave a ½ inch extra hanging off the first corner. Sew along the length of the quilt, about ¼ inch from the raw edge. Trim the binding, leaving a ½ inch extra. Fold the fabric over the rough edges to the back of the quilt and slip-stitch the binding to the backside. Fold the loose ends of the binding over the edge of the quilt and stitch to the backside. Repeat with all sides of the quilt.

8. Finish your crazy quilt by adding decorative stitching between small pieces of fabric, sewing on buttons, tassels, or ribbons, or using stitching or fabric markers to record important names or dates.

Pottery Basics

Clay is the basic ingredient for making pottery. Clay is decomposed rock containing water (both in liquid and chemical forms). Water in its liquid form can be separated from the clay by heating the mass to a boiling point—a process that restores the clay to its original condition once dried. The water in the clay that is found in chemical forms can also be removed by ignition—a process commonly referred to as "firing." After being fired, clay cannot be restored to any state of plasticity—this is called "pottery." Some clay requires greater heat in order to be fired, and these are known as "hard clays." These types of clay must be subjected to a "hard-firing" process. However, in the making of simple pottery, soft clay is generally used and is fired in an over-glaze (soft glaze) kiln.

Pottery clays can either be made by hand (by finding clay in certain soils) or bought from craft stores. If you have clay soil available on your property, the process of separating the clay from the other soil materials is simple. Put the earthen clay into a large bucket of water to wash the soil away. Any rocks or other heavy matter will sink to the bottom of the bucket. The milky fluid that remains—which is essentially water mixed with clay—may then be drawn off and allowed to settle in a separate container, the clear water eventually collecting on the top. Remove the excess water by using a siphon. A repetition of this process will refine the clay and make it ready for use.

THE JUNIOR HOMESTEADER

Pine Cone Birds

Pine cones lend themselves to all sorts of fanciful creations. With a little string, some construction paper, and a marker you can transform pine cones of varying shapes and sizes into a flock of birds!

Things You'll Need:

- Pine cones (any size or shape)
- Pencil
- Scissors
- Construction paper
- Glue
- Colored marker
- Strong thread or dental floss

The pine cones will be the bodies of the birds. To add feathers, cut out leaf-shaped pieces of construction paper, draw vein patterns on the pieces, and glue them to the sides or one end of the cone. For a turkey, tilt a round pinecone on its side and glue the feathers upright to the top petals of the cone. Add a neck and head with more construction paper, drawing on the eyes. For an owl, a round cone can be stood upright on its bottom petals—just add big round eyes and an oval beak cut out of construction paper and you have a wise old owl! For a blue bird, use an elongated cone and glue the construction paper wings to either side. Glue a head to the bottom petals of the cone. Don't forget to draw the beak and eyes!

To hang your bird ornaments, secure the strong thread or dental floss around a pine cone petal and tie a knot. Hang them on your Christmas tree or in the window!

You can also purchase clay at your local craft store. Usually, clay sold in these stores will be in a dry form (a grayish or yellowish powder), so you will need to prepare it in order to use it in your pottery. To prepare it for use, you must mix the powder with water. If there are directions on your clay packet, then follow those closely to make your clay. In general, though, you can make your clay by mixing equal parts of clay powder and water in a bowl and allowing the mixture to soak for ten to twelve hours. After it has soaked, you must knead the mixture thoroughly to disperse the water evenly throughout the clay and pop any air bubbles. Air bubbles, if left in the clay, could be detrimental to your pottery once kilned, as the bubbles would generate steam and possibly crack your creation. However, be careful not to knead your clay mixture too much, or you may increase the chance of air bubbles becoming trapped in the mixture.

If, after kneading, you find that the clay is too wet to work with (test the wetness of the clay on your hands and if it tends to slip around your palm very easily, it is probably too wet), the excess water can be removed by squeezing or blotting out with a dry towel or dry-board.

The main tools needed for making pottery are simply your fingers. There are wooden tools that can be used for adding finer detail or decoration, but typically, all you really need are your own two hands. A loop tool (a piece of fine, curved wire) may also be used for scraping off excess clay where it is too thick. Another tool has ragged edges and this can be used to help regulate the contour of the pottery. Remember that homemade pottery will not always be symmetrical, and that is what makes it so special.

inside the structure and the other outside. If you are building a vase, you can extract one finger at a time as you reach closer and closer to the top of the model.

5. Make sure the clay is moist throughout the entire molding process. If you need to stop molding for an extended period of time, cover the item with a moist cloth to keep it from drying out.

6. When your model has reached the size you want, you may turn it upside down and smooth and refine the contours of the object. You can also make the base much more detailed and shaped to a more pleasing design.

7. Allow your model to air dry.

Embellishing Your Clay Models

You may eventually want to make something that requires a handle or a spout, such as a cup or teapot. Adding handles and

Basic Vase or Urn

Try making this simple vase or urn to get used to working with clay.

1. Take a lump of clay. The clay should be about the size of a small orange and should be rather elastic feeling. Then, begin to mold the base of your object—let's say it is either a bowl or a vase.

2. Continue molding your base. By now, you'll have a rather heavy and thick model, hollowed to look a little like a bird's nest. Now, using this base as support, start adding pieces of clay in a spiral shape. Press the clay together firmly with your fingers. Make sure that your model has a uniform thickness all around.

3. Continue molding your clay and making it grow. As you work with the clay, your hands will become more accustomed to its texture and the way it molds, and you will have less difficulty making it do what you want. As you start to elongate and shape the model, remember to keep the walls of the piece substantial and not too thin—it is easier to remove extra thickness than it is to add it.

4. Don't become frustrated if your first model fails. Even if you are being extra careful to make your bowl or vase sturdy, there is always the instance when a nearly complete vase will fall over. This usually happens when one side of the structure becomes too thin or the clay is too wet. To keep this from happening, it is sometimes helpful to keep one hand

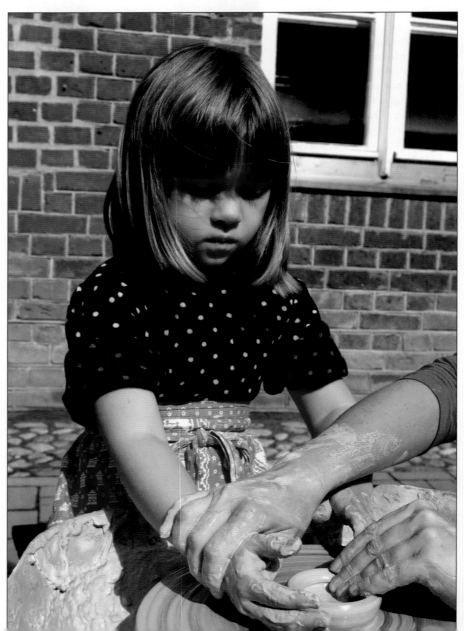

278

spouts can be tricky, but only if you don't remember some simple rules. Spouts can be modeled around a straw or any other material that is stiff enough to support the clay and light enough to burn out in the firing. In the designing of spouts and handles, it is still important to keep them solid and thick. Also, keeping them closer to the body of your model is more practical, as handles and spouts that are elongated are harder to keep firm and can also break off easily. Although more time consuming and difficult to manage, handles and spouts can add a nice aesthetic to your finished pottery.

The simplest way to decorate your pottery is by making line incisions. Line incision designs are best made with wooden, finger-shaped tools. It is completely up to you as to how deep the lines are and into what pattern they are made.

Wheel-working and Firing Pottery

If you want to take your pottery-making one step further, you can experiment with using a potters' wheel and also glazing and firing your model to create beautiful pottery. Look online or at your local craft store for potters' wheels. Firing can leave your pottery looking two different ways, depending on whether you decide to leave the clay natural (so it maintains a dull and porous look) or to give it a color glaze.

Colored glazes come in the form of powder and are generally metallic oxides, such as iron oxides, cobalt oxide, chromium oxide, copper oxide, and copper carbonate.

The colors these compounds become will vary depending on the atmosphere and temperature of the kiln. Glazes often come in the form of powder and need to be combined with water in order to be applied to the clay. Only apply glaze to dried pottery, as it won't adhere well to wet clay. Use a brush, sponge, or putty knife to apply the glaze. Your pottery is then ready to be fired.

There are various different kinds of kilns in which to fire your pottery. An over-glaze kiln is sufficient for all processes discussed here, and you can probably find a kiln in your surrounding area (check online and in your telephone book for places that have kilns open to the public). Schools that have pottery classes may have over-glaze kilns. It is important, whenever you are using a kiln, that you are with a skilled pottery maker who knows how to properly operate a kiln.

After the pottery has been colored and fired, a simple design may be made on the pottery by scraping off the surface color so as to expose the original or creamy-white tint of the clay.

Unglazed pottery may be worked with after firing by rubbing floor wax on the outer surface. This fills up the pores and gives a more uniform quality to the whole piece.

Pottery offers so many opportunities for personal experimentation and enjoyment; there are no set rules as to how to make a piece of pottery. Keep a journal about the different things you try while making pottery, so you can remember what works best and what should be avoided in the future. Note the kind of clay you used and its consistency, the types of colors that have worked well, and the temperature and positioning within the kiln, if you use firing. Above all, enjoy making unique pieces of pottery!

Making Jars, Candlesticks, Bowls, and Vases

Making pottery at home is simple and easy, and is a great way for you to make personalized, unique gifts for family and friends. Clay can be purchased at local arts and crafts stores. Clay must always be kneaded before you model with it, because it contains air that, if left in the clay, would form air bubbles in your pottery and spoil it. Work out this air by kneading it the same way that you knead bread. Also guard against making the clay too moist, because that causes the pottery to sag, which spoils the shape.

In order to make your own pottery, you need modeling clay, a board on which you can work, a pie tin on which to build, a knife, a short stick (one side should be pointed), and a ruler.

Jars

To start a jar, put a handful of clay on the board, pat it out with your hand until it is an inch thick, and smooth off the surface. Then, take a coffee cup, invert it upon the base, and, with your stick, trim the clay outside the rim.

To build up the walls, put a handful of clay on the board and use a knife to smooth it out into a long piece, ¼ inch thick. With the knife and a ruler, trim off one edge of the piece and cut a number of strips $^3/_4$ inch wide. Take one strip, stand it on top of the base, and rub its edge into the base on both sides of the strip; then, take another strip and add it to the top of the first one, and continue building in this way, placing one strip on another, joining each to the one beneath it, and smoothing over the joints as you build. Keep doing this until the walls are as high as you want them to be. Remember to keep one hand inside the jar while you build, for extra support. Fill uneven places with bits of clay and smooth out rough spots with your fingers, having moistened your fingers with water first. When you are finished, you may also add decoration, or ornament, to your jar.

Candlestick

Making a pottery candlestick requires a round base ½ inch thick and 4 inches in diameter. After preparing the base, put a lump of clay in the center, work it into the base, place another lump on top, work it into the piece, and continue in this way until the candlestick has been built as high as you want it. Then, force a candle into the moist clay, twisting it around until it has made a socket deep enough to place a candle into.

A cardboard "templet," with one edge trimmed to the proper shape, will help make it easy to make the walls of the candlestick symmetrical and the projecting cap on the top equal on all sides. Run the edge of the templet around the walls as you work, and it will show you exactly where and how much to fill out, trim, and straighten the clay.

If you want to make a candlestick with a handle, make a base just as stated above. Then cut strips of clay and build up the wall as if building a jar, leaving a center hole just large enough to hold a candle. When the desired height for the wall has been reached, cut a strip of clay ½ inch wide and ½ inch thick, and lay it around the top of the wall with a projection of ¼ inch over the wall. Smooth this piece on top, inside, and outside with your modeling stick and fingers. For the handle, prepare a strip 1 inch wide and ½ inch thick, and join one end to the top band and the other end to the base. Use a small lump of clay for filling around where you join the piece, and smooth off the piece on all sides.

When the candlestick is finished, run a round stick the same size as the candle down into the hole, and let it stay put until the clay is dry, to keep the candlestick straight.

Bowls

Bowls are simple to make. Starting with a base, lay strips of clay around the base, building upon each strip as you did when making a jar. Once the bowl is at the desired height and width, allow it to dry.

Vases

Vases can be made as you'd make a jar, only bringing the walls up higher, like a candlestick. Experiment with different shapes and sizes and always remember to keep one hand inside the vase to keep everything even and to prevent your vase from caving in on itself.

Decorating Your Pottery

Pottery may be ornamented by scratching a design upon it with the end of your modeling stick. You can do a simple, straight-line design by using a ruler to guide the stick in drawing the lines. Ornamentation on vases and candlesticks can be done by hand-modeling details and applying them to your item.

Glazing and Firing

Pottery that you buy is generally glazed and then fired in a pottery kiln, but firing is not necessary to make beautiful, sturdy pottery. The clay will dry hard enough, naturally, to keep its shape, and the only thing you must provide for is waterproofing (if the pottery will be holding liquids). To do this, you can take bathtub enamel and apply it to the inside (and outside, if desired) of the pottery to seal off any cracks and to keep the item from leaking.

If you do want to try glazing and firing your own pottery, you will need a kiln. Below are instructions for making your own.

Sawdust Kiln

This small, homemade kiln can be used to bake and fire most small pottery projects. It will only get up to about 1,200°F, which is not hot enough to fire porcelain or stonewear. However, it will suffice for clay pinch pots and other decorative pieces.

You will need:

- 20 to 30 red or orange bricks
- Chicken wire
- Sawdust
- Newspaper and kindling
- Sheet metal

1. Choose a spot outdoors that is protected from strong winds. Clear away any dried branches or other flammables from the immediate area. A concrete patio or paved area makes an ideal base, but you can also place bricks or stones on the ground.
2. Stack bricks in a square shape, building each wall up at least four bricks high. Fill the kiln with sawdust.
3. Place the chicken wire on top of the bricks and add another layer or two of bricks. Carefully place your pottery in the center of the mesh, spacing the pieces at least ½ inch apart. Cover the pottery with sawdust.
4. Add another piece of chicken wire, add bricks and pottery, and cover with sawdust. Repeat until your kiln is the desired height.
5. Light the top layer of sawdust on fire, using kindling and newspaper if needed. Cover with the sheet metal, using another layer of bricks to hold it in place.
6. Once the kiln stops smoking, leave it alone until it completely cools down. Then carefully remove the sheet metal lid.

Hemp Jewelry

Hemp jewelry is easy to make, the materials required are very affordable, and it is durable and recyclable. Most hemp jewelry is created by making a series of knots and by incorporating beads or charms if you want. By repeating one of two knot patterns, you will make a spiral piece of jewelry; by incorporating two different knot patterns, you will be able to make a flat pattern. It is quite simple to make hemp necklaces and bracelets by following these simple steps.

- A ball of hemp twine (about 1 mm thick and the smoother the better)
- Beads or charms (optional—but if you do use these, make sure the stringing holes are large enough for two pieces of hemp to fit through)
- Clasp, cord tip, or large bead for securing the jewelry (optional)
- Tape (to hold down the hemp while you are knotting)
- Scissors
- Ruler or measuring tape
- Glue (optional—used to attach the securing bead at the end)

1. To start, you will need four lengths of hemp: two for the base strings (these should be a little longer than your overall desired length) and two for knotting (these should be five times the length of the base strings). If you want to make a looped closure, you can also cut two lengths of hemp string roughly 3 yards long, fold these strings in half to create four strings, and then simply make a loop at the end where the strings are folded. Whichever method you choose, be sure to tape down the folded end of the string so you have greater ease in knotting your jewelry.

2. Then begin knotting. You may use just one of these knots (spiral pattern) or alternate both (flat pattern) when making your jewelry. Just remember that you are never knotting the two inner (base) strings.

a. First knot: Take the right string, pass it over the base strands, and pull it under the left strand (this should make a shape that looks like a backwards 4). Then, take the left strand, pass it under the base strings, and bring it up and over the right string. Pull both ends and tighten the knot.

b. Second knot: Take the left string, pass it over the base strings, and bring it under the right string (this should make a shape that looks like a regular 4). Then, take the right string, pass it under the base strings, and bring it up and over the left string. Pull both ends and tighten the knot.

c. If you want a flat pattern, continue alternating the first and second knots.

3. If you want to add beads or charms to your hemp, simply slide the bead or charm onto the base strings and make a knot around it. It is a good idea to try to space the beads or charms equally on your necklace or bracelet. You can easily do this by counting the number of knots in between each bead or charm and then adding another item after the designated amount of knots.

4. Continue knotting the hemp until the desired length is reached.

5. To finish the jewelry, you may either tie both ends together (the easiest way); knot a larger bead to the end so it will pass through the top loop, securing the jewelry (place a dab of glue onto the bed to help it stay fast); or use metal clasps, which you can buy at your local arts and crafts store.

Handcrafted Paper

Instead of throwing away your old newspapers, office paper, or wrapping paper, use it to make your own unique paper! The paper will be much thicker and rougher than regular paper, but it makes great stationery, gift cards, and gift wrap.

Materials

Newspaper (without any color pictures or ads if at all possible), scrap paper, or wrapping paper (non-shiny paper is preferable)

2 cups hot water for every ½ cup shredded paper

2 tsps instant starch (optional)

Supplies

Blender or egg beater

Mixing bowl

Flat dish or pan (a 9 x 13-inch or larger pan will do nicely)

Rolling pin

8 x 12-inch piece of non-rust screen

4 pieces of cloth or felt to use as blotting paper, or at least 1 sheet of Formica

10 pieces of newspaper for blotting

Directions

1. Tear the newspaper, scrap paper, or wrapping paper into small scraps. Add hot water to the scraps in a blender or large mixing bowl.
2. Beat the paper and water in a blender or with an egg beater in a large bowl. If you want, mix in the instant starch (this will make the paper ready for ink). The paper pulp should be the consistency of a creamy soup when it is complete.
3. Pour the pulp into the flat pan or dish. Slide the screen into the bottom of the pan. Move the screen around in the pulp until it is evenly covered.
4. Carefully lift the screen out of the pan. Hold it level and let the excess water drip out of the pulp for a minute or two.
5. With the pulp side up, put the screen on a blotter (felt) that is situated on top of some newspaper. Put another blotter on the top of the pulp and put more newspaper on top of that.
6. Using the rolling pin, gently roll the pin over the blotters to squeeze out the excess water. If you find that the newspaper on the top and bottom is becoming completely saturated, add more (carefully) and keep rolling.
7. Remove the top level of newspaper. Gently flip the blotter and the screen over. Very carefully, pull the screen off of the paper. Leave the paper to dry on the blotter for at least twelve to twenty-four hours. Once dry, peel the paper off the blotter.

To add variety to your homemade paper:

- To make colored paper, add a little bit of food coloring or natural dye to the pulp while you are mixing in the blender or with the egg beater.
- You can also try adding dried flowers (the smoother and flatter, the better) and leaves or glitter to the pulp.
- To make unique bookmarks, add some small seeds to your pulp (hardy plant seeds are ideal), make the paper as in the directions, and then dry your paper quickly using a hairdryer. When the paper is completely dry, cut out bookmark shapes and give to your friends and family. After they are finished using the bookmarks, they can plant them and watch the seeds sprout.

Bookbinding

Simple Homemade Book Covers

If you have loose papers that you want bound together, there is an easy way to make a book cover:

1. Take two pieces of heavy cardboard that are slightly larger than the pages of the book you wish to assemble. Make three holes near the edges of each cardboard piece with a hole-punch. Then, punch three holes (at the same distance from each other as in the cardboard pieces) in the papers that are to be in the interior of the book. Be sure that your book is not too thick or the cover won't be as effective.

2. String a narrow ribbon through these holes and tie the ribbons in knots or bows. If the leaves of your book are thin, you can punch more than three holes into the paper and cardboard, and then lace strong string or cord between the holes, like shoelaces.

3. To decorate your covers, you can paint them with watercolors or you can simply use colored cardboard. You can also take some fabric, cut it so the fabric just folds over the inside edges of the cardboard pieces, and then hot glue (or use a special paste—see below) the fabric to the cardboard. Or, you can glue on photographs or cutouts from magazines and protect the cover with laminating paper.

Bind Your Own Book

Making your own blank book or journal takes careful measuring, folding, and gluing, but the end product will be something unique that you'll treasure for ages. You can also rebind old, worn out books that you'd like to preserve for more years to come.

Before binding anything very special, practice by binding a "dummy" book, or a book full of blank pages—you could use this blank book as a journal if it comes out fairly well, so your efforts will not be wasted. In order to make a blank book, you need to plan out what you'll need—how thick the book will be, what the dimensions are, the quality of the paper being used (at least a medium-grade paper in white or cream), and so on. Carefully fold and cut the paper to the appropriate size.

- White or cream-colored paper (at least 32 sheets) or the pages you wish to rebind
- Stapler or needle and thread
- Binder clips or vise
- Decorative paper (2 sheets)
- Stiff cardboard
- Cloth or leather
- Silk or cloth cord
- Glue
- Scissors (or a metal ruler and craft knife)

1. Make four stacks of eight sheets of paper. These stacks, once folded, are called "folios," Four stacks will make a sixty-four-page book. If you wish to make it longer or shorter you can do more or fewer stacks of eight sheets. Carefully fold each stack in half.

2. Unfold the stacks and staple or sew along the crease. If stapling, only use two staples: one at the top of the fold and one at the bottom.

3. Refold all the stacks and pile them on top of each other. Use binder clips or a vise to hold them together. Cut a rectangular piece of fabric that is the same length as the spine of your book and about five times as wide. So if your stack of folded papers is 8 ½ inches long and 1 inch high, your fabric should be 8 ½ inches long and 5 inches wide.

4. Using a hot glue gun or regular white glue, cover the spine with glue and stick on the fabric. The fabric should hang off either side of the spine.

5. For the cover, cut two pieces of sturdy cardboard that are the same size as the pages of your book. Using a metal ruler and a craft knife will help you make the cuts straight and smooth. Place one piece of cardboard at the bottom of your stack of papers and another on the top. Cut another piece of cardboard that is the same height and width as the book's spine, including the pages and both covers.

6. Select a piece of fabric (or leather) to cover your book and lay it flat on a table. Place the three pieces of cardboard on the fabric with the spine between the two cover pieces. Use a ruler to measure and mark a rectangle on the fabric that is 1 inch larger on all sides than the combined pieces of cardboard. Remove the cardboard pieces and cut out the rectangle.

7. Lay the fabric on the table face down. Cover one side of the cardboard pieces with white glue or rubber cement. If using white glue, use a stiff brush, putty knife, or scrap piece of cardboard to spread the glue in an even thin layer so there are no lumps. Place the cardboard glue-side-down on the fabric so that all three pieces are aligned with the spine between the two covers. Leave a gap of about two thicknesses of the cardboard between the spine and the two covers.

8. Smear glue on the top and bottom edges of the cardboard pieces and fold over the fabric. Then repeat with the outside edges.

9. Smear glue on the inside edges of the cover boards. Don't glue the spine. Place the stack of pages

1.

2.

3. + 4.

5.

6.

7. + 8.

9.

10. + 11.

12.

Credit: Timothy W. Lawrence

TO MAKE FLOUR PASTE FOR YOUR BOOK COVERS

Mix ½ cup of flour with enough cold water to make a very thin batter. This must be smooth and free of lumps. Put the batter on top of the stove in a tin saucepan and stir it continually until it boils. Remove the pan from the stove, add three drops of clove oil, and pour the paste into a cup or tumbler and cover.

spine-side-down on top of the boards. The extra material that is hanging off of the spine should adhere to the glue on the cover boards. Place two solid bookends, rocks, or jars of food on either side of the papers to hold them upright until they dry thoroughly.

10. Select a decorative piece of paper to use for endpapers. This will cover the inside front and back covers so that you won't see the folded material and cardboard. It can be a solid color or patterned, according to your preference. Cut it to be slightly wider than the pages you started with (before being folded) and not quite as tall.

11. Open your book and cover the inside front cover and first page with glue or rubber cement. Fold your endpaper in half to create a crease, open it back up, and then stick it to the inside cover and first page, making sure the crease slides into the space between the spine and front cover slightly. Allow to dry and then repeat at the back of the book.

12. Cut two pieces of thin cord for the head and tail band, which will cover the top and bottom of the spine. They should be the same length as the width of the spine. Use a hot glue gun or white glue to adhere them to the top and bottom of the spine, where the pages are gathered together.

Scrapbooking

Creating a scrapbook is a great way to process and preserve memories. Keep travel brochures from special trips, ticket stubs, event programs, photographs, and so on in a box or file throughout the year. Then, when cold or rainy weather sets in, you can pull out your book and begin arranging your keepsakes in a meaningful way.

Depending on the nature of your keepsakes and your own preferences, you can choose to base your scrapbook on a theme (musical or sporting events, family vacations, your wedding, home renovations, etc.) or to organize it chronologically. Of course, really you can do anything you want, including creating a collage-like book of scattered memories, keepsakes, magazine clippings, or even journal entries. Scrapbooking is a personal project and can be done with as much thoughtful organization or creative chaos as you wish.

There are many ways to create a unified feel to your scrapbook, whether or not the subject matter itself maintains any continuity. Here are a few suggestions:

- Layout. Design each page in a similar way, so that (for example) there is always a favorite quote on the top of the page, a photo on the side, text below, and an embellishment across from it. Or create a pattern of layouts, so that every third or fourth page has a similar design.
- Colors. Choose two or three colors or a single color palette to use throughout the book. Incorporate these colors in borders, frames for photos, backgrounds, with colored inks, and so on.
- Fonts or Script. Use the same style of writing for photo captions or other text throughout the book.
- Backgrounds. You may want a clean white background if you want to showcase high-quality photographs without any distractions to the eye. Or you can use scrap pieces of wallpaper all in the same style or color, colored card stock, or even thin fabric as the backdrop for your keepsakes.

Once you have a general plan for your scrapbook, you're ready to begin:

1. Create a background. Unless you want a plain white background, select the cardstock or other colored papers or fabrics, cut them to size, and use rubber cement to glue them to the page. You may choose to use several coordinated pieces of paper to create a patterned background.

2. Matt your photos. Cut squares of paper that are slightly larger than your photos and glue the photos onto them. This will create an attractive border around your photos. Later you can add frames or leave them as they are.

3. Add text or other embellishments. Use neat handwriting or printed text to caption your photos, add favorite quotes, etc. Pieces of ribbon, stickers, or magazine clippings can also be added as accents. Avoid any thick materials that will make your page bumpy and keep the album from closing all the way.

Rag Rugs

The first rag rugs were made by homesteaders over two centuries ago who couldn't afford to waste a scrap of fabric. Torn garments or scraps of leftover material could easily be turned into a sturdy rug to cover dirt floors or stave off the cold of a bare wooden floor in winter. Any material can be used for these rugs, but cotton or wool fabrics are traditional.

- Rags or strips of fabric
- Darning needle
- Heavy thread

1. Cut long strips of material about 1 inch wide. Sew strips together end to end to make three very long strips (or you can start with shorter strips and sew on more pieces later). To make a clean seam between strips, hold the two pieces together at right angles to form a square corner. Sew diagonally across the square and trim off excess fabric.

2. Braid the three strips together tightly, just as you would braid hair.
3. Start with one end of the braid and begin coiling it around itself, sewing each coil to the one before it with circular stitches. Keep the coil flat on the floor or on a table to keep it from bunching up.
4. When the rug is as large as you want it to be, tack the end of the braid firmly to the edge of the rug.

TIP

Use thinner strips of fabric toward the end of your rug to make it easier to tack to the edge of the rug.

≫ **Cut strips along the bias to keep them from unraveling**

Toboggans

Toboggans can fit several people on them, making the sledding experience all the more fun. Toboggans are suitable for deep snow and heavy drifts, due to their broad, smooth bottoms, enabling the sled to glide well over crusted snow.

To make a toboggan, you will need two pieces of quarter-inch pine lumber that are either 8 or 10 feet long and 1 foot wide. Place the two boards side by side and join them together by means of round cross-sticks. Bind the cross-sticks to the lumber via thongs that pass through holes in the bottom boards on each side of the cross-stick and are held tightly by hammock hitches. Where the thongs pass underneath, cut grooves in the bottom board deep enough to prevent the cord from sticking out. These grooves are necessary as any cords that would be sticking out from the toboggan would not only slow down the sled but would eventually wear out, causing the toboggan to fall apart.

On top of the cross-sticks, lash together two side bars. Then, curl over the front ends of the boards and hold them in place by two thongs that are tied tightly onto the ends. Then you can take your handmade toboggan out during the first big snow and give it a test ride.

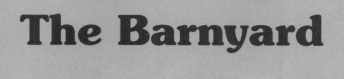
The Barnyard

"Agriculture is our wisest pursuit, because it will in the end contribute most to real wealth, good morals, and happiness."

—Letter from Thomas Jefferson to George Washington (1787)

Chickens	292
Ducks	297
Turkeys	301
Rabbits	304
Horses	307
Beekeeping	312
Goats	316
Sheep	320
Llamas	324
Cows	327
Pigs	332
Butchering	335

Part 5

Chickens

Raising chickens in your yard will give you access to fresh eggs and meat, and since chickens are some of the easiest creatures to keep, even families in urban areas are able to raise a few in a small backyard. Four or five chickens will supply your whole family with eggs on a regular basis.

Housing Your Chickens

You will need to have a structure for your chickens to live in—to protect them from predators and inclement weather, and to allow the hens a safe place to lay their eggs. See page 382 for several types of structures you can make for housing your chickens and other poultry.

≈ Building a chicken coop close to your house will make it easier to tend the chickens and gather eggs in inclement weather.

Placing your henhouse close enough to your own home will remind you to visit it frequently to feed the chickens and to gather eggs. It is best to establish the house and yard in dry soil and to stay away from areas in your yard that are frequently damp or moist, as this is the perfect breeding ground for poultry diseases. The henhouse should be well-ventilated, warm, protected from the cold and rain, have a few windows that allow sunlight to shine in (especially if you live in a colder climate), and have a sound roof.

The perches in your henhouse should not be more than 2½ feet above the floor, and you should place a smooth platform under the perches to catch the droppings so they can easily be cleaned. Nesting boxes should be kept in a darker part of the house and should have ample space around them.

The perches in your henhouse can be relatively narrow and shouldn't be more than a few feet from the floor.

A simple, movable chicken coop can be constructed out of two-by-fours and two wheels. The floor of the coop should have open slats so that the manure will fall onto the ground and fertilize the soil. An even simpler method is to construct a pen that sits directly on the ground, making sure that it has a roof to offer the chickens suitable shade. The pen can be moved once the area is well fertilized.

Selecting the Right Breed of Chicken

Take the time to select chickens that are well suited for your needs. If you want chickens solely for their eggs, look for chickens that are good egg-layers. Mediterranean poultry are good for first-time chicken owners as they are easy to care for and only need the proper food in order to lay many eggs. If you are looking to slaughter and eat your chickens, you will want to have heavy-bodied fowl (Asiatic poultry) in order to get the most meat from them. If you are looking to have chickens that lay a good amount of eggs and that can also be used for meat, invest in the Wyandottes or Plymouth Rock breeds. These chickens are not incredibly bulky but they are good sources of both eggs and meat.

Wyandottes have seven distinct varieties: Silver, White, Buff, Golden, and Black are the most common. These varieties are hardy and they are very popular in the United States. They are compactly built and lay excellent dark brown eggs. They are good sitters and their meat is perfect for broiling or roasting.

Plymouth Rock chickens have three distinct varieties: Barred, White, and Buff. They are the most popular varieties in the United States and are hardy birds that grow to a medium size. These chickens are good for laying eggs, roost well, and also provide good meat.

Feeding Your Chickens

Chickens, like most creatures, need a balanced diet of protein, carbohydrates, vitamins, fats, minerals, and water. Chickens with plenty of access to grassy areas will find most of what they need on their own. However, if you don't have the space to allow your chickens to roam free, commercial chicken feed is readily available in the form of mash, crumbles, pellets, or scratch. Or you can make your own feed out of a combination of grains, seeds, meat scraps or protein-rich legumes, and a gritty substance such as bone meal, limestone, oyster shell, or granite (to aid digestion, especially in winter). The correct ratio of food for a warm, secure chicken should be 1 part protein to 4 parts carbohydrates. Do not rely too heavily on

⌄ Chickens that are allowed to roam freely ("free-range" chickens) will be able to scavenge most of the food they need, as long as there is plenty of grass or other vegetation available.

much room they have to exercise. Often it's easiest and best for the chickens to leave feed available at all times in several locations within the chickens' range. This will ensure that even the lowest chickens in the pecking order get the feed they need.

Hatching Chicks

If you are looking to increase the number of chickens you have, or if you plan to sell some chickens at the market, you may want some of your hens to lay eggs and hatch chicks. In order to hatch a chick, an egg must be incubated for a sufficient amount of time with the proper heat, moisture, and position. The period for incubation varies based on the species of chicken. The average incubation period is around twenty-one days for most common breeds.

If you are only housing a few chickens in your backyard, natural incubation is the easiest method with which to hatch chicks. Natural incubation is dependent upon the instinct of the mother hen and the breed of hen. Plymouth Rocks and Wyandottes are good hens to raise chicks. It is important to separate the setting hen from the other chickens while she is nesting and to also keep the hen clean and free from lice. The nest should also be kept clean and the hens should be fed grain food, grit, and clean, fresh water.

It is important, when you are considering hatching chicks, to make sure your hens are healthy, have plenty of exercise, and are fed a balanced diet. They need materials on

---

Chicken Feed

- 4 parts corn (or more in cold months)
- 3 parts oat groats
- 2 parts wheat
- 2 parts alfalfa meal or chopped hay
- 1 part meat scraps, fish meal, or soybean meal
- 2 to 3 parts dried split peas, lentils, or soybean meal
- 2 to 3 parts bone meal, crushed oyster shell, granite grit, or limestone
- ½ part cod-liver oil

You may also wish to add sunflower seeds, hulled barley, millet, kamut, amaranth seeds, quinoa, sesame seeds, flax seeds, or kelp granules. If you find that your eggs are thin-shelled, try adding more calcium to the feed (in the form of limestone or oystershell). Store feed in a covered bucket, barrel, or other container that will not allow rodents to get into it. A plastic or galvanized bucket is good, as it will also keep mold-causing moisture out of the feed.

corn as it can be too fattening for hens; combine corn with wheat or oats for the carbohydrate portion of the feed. Clover and other green foods are also beneficial to feed your chickens.

How much food your chickens need will depend on breed, age, the season, and how

BACTERIA ASSOCIATED WITH CHICKEN MEAT

- *Salmonella*—This is primarily found in the intestinal tract of poultry and can be found in raw meat and eggs.

- *Campylobacter jejuni*—This is one of the most common causes of diarrheal illness in humans, and is spread by improper handling of raw chicken meat and not cooking the meat thoroughly.

- *Listeria monocytogenes*—This causes illness in humans and can be destroyed by keeping the meat refrigerated and by cooking it thoroughly.

STORING EGGS

Eggs are among the most nutritious foods on earth and can be part of a healthy diet. Hens typically lay eggs every twenty-five hours, so you can be sure to have a fresh supply on a daily basis, in many cases. But eggs, like any other animal by-product, need to be handled safely and carefully to avoid rotting and spreading disease. Here are a few tips on how to best preserve your farm-fresh eggs:

1. Make sure your eggs come from hens that have not been running with roosters. Infertile eggs last longer than those that have been fertilized.
2. Keep the fresh eggs together.
3. Choose eggs that are perfectly clean.
4. Make sure not to crack the shells, as this will taint the taste and make the egg rot much quicker.
5. Place your eggs directly in the refrigerator where they will keep for several weeks.

which to scratch and should not be infested with lice and other parasites. Free range chickens, which eat primarily natural foods and get lots of exercise, lay more fertile eggs than do tightly confined hens. The eggs selected for hatching should not be more than twelve days old and should be clean.

You'll need to construct a nesting box for the roosting hen and the incubated eggs. The box should be roomy and deep enough to retain the nesting material. Treat the box with a disinfectant before use to keep out lice, mice, and other creatures that could infect the hen or the eggs. Make the nest of damp soil a few inches deep placed in the bottom of the box, and then lay sweet hay or clean straw on top of that.

Place the nesting box in a quiet and secluded place away from the other chickens. If space permits, you can construct a smaller shed in which to house your nesting hen.

A hen can generally sit on anywhere between nine and fifteen eggs. The hen should only be allowed to leave the nest to feed, drink water, and take a dust bath. When the hen does leave her box, check the eggs and dispose of any damaged ones. An older hen will generally be more careful and apt to roost than a younger female.

Once the chicks are hatched, they will need to stay warm and clean, have lots of exercise, and have access to food regularly. Make sure the feed is ground finely enough that the chicks can easily eat and digest it. They should also have clean, fresh water.

Ducks

Ducks tend to be somewhat more difficult than chicks to raise, but they do provide wonderful eggs and meat. Ducks tend to have pleasanter personalities than chickens and are often prolific layers. The eggs taste similar to chicken eggs, but are usually larger and have a slightly richer flavor. Ducks are happiest and healthiest when they have access to a pool or pond to paddle around in and when they have several other ducks to keep them company.

Breeds of Ducks

There are six common breeds of ducks: White Pekin, White Aylesbury, Colored Rouen, Black Cayuga, Colored Muscovy, and White Muscovy. Each breed is unique and has its own advantages and disadvantages.

1. White Pekin—The most popular breed of duck, these are also the easiest to raise. These ducks are hardy and do well in close confinement. They are timid and must be handled carefully. Their large frame gives them lots of meat and they are also prolific egg layers.
2. White Aylesbury—This breed is similar to the Pekin but the plumage is much whiter and they are a bit heavier than the former. They are not as popular in the United States as the White Pekin duck.
3. Colored Rouens—These darkly plumed ducks are also quite popular and fatten easily for meat purposes.
4. Black Cayuga and Muscovy breeds— These are American breeds that are easily raised but are not as productive as the White Pekin.

≫ **White Pekins were originally bred from the Mallard in China and came to the United States in 1873.**

⩘ According to Mrs. Beeton in her *Book of Household Management*, published in 1861, "[Aylesbury ducks'] snowy plumage and comfortable comportment make it a credit to the poultry-yard, while its broad and deep breast, and its ample back, convey the assurance that your satisfaction will not cease at its death."

Housing Ducks

You don't need a lot of space in which to raise ducks—nor do you need water to raise them successfully, though they will be happier if you can provide at least a small pool of water for them to bathe and paddle around in. Housing for ducks is relatively simple. The houses do not have to be as warm or dry as for chickens but the ducks cannot be confined for long periods as chickens can. They need more exercise out of doors in order to be healthy and to produce more eggs. A house that is protected from dampness or excess rain water and that has straw or hay covering the floor is adequate for ducks. If you want to keep your ducks somewhat confined, a small fence about 2½ feet high will do the trick. Ducks don't require nesting boxes, as they lay their eggs on the floor of the house or in the yard around the house.

⩗ A Black Cayuga (bottom) stands with two Saxony ducks.

together. After this time, ducks should be fed on a mixture of 2 parts cornmeal, 1 part wheat bran, 1 part flour, some coarse sand, and green foods.

Hatching Ducklings

The natural process of incubation (hatching ducklings underneath a hen) is the preferred method of hatching ducklings. It is important to take good care of the setting hen. Feed her whole corn mixed with green food, grit, and fresh water. Placing the feed and water just in front of the nest for the first few days will encourage the hen to eat and drink without leaving the nest. Hens will typically lay their eggs on the ground, in straw or hay that is provided for them. Make sure to clean the houses and pens often so the laying ducks have clean areas in which to incubate their eggs.

Caring for Ducklings

Young ducklings are very susceptible to atmospheric changes. They must be kept warm and free from getting chilled. The ducklings are most vulnerable during the first three weeks of life; after that time, they are more likely to thrive to adulthood. Construct brooders for the young ducklings and keep them very warm by hanging strips of cloth over the door cracks. After three weeks in

Feeding and Watering Ducks

Ducks require plenty of fresh water to drink, as they have to drink regularly while eating. Ducks eat both vegetable and animal foods. If allowed to roam free and to find their own food stuff, ducks will eat grasses, small fish, and water insects (if streams or ponds are provided).

Ducks need their food to be soft and mushy in order for them to digest it. Ducklings should be fed equal parts corn meal, wheat bran, and flour for the first week of life. Then, for the next fifty days or so, the ducklings should be fed the above mixture in addition to a little grit or sand and some green foods (green rye, oats, clover) all mixed

the warm brooder, move the ducklings to a cold brooder as they can now withstand fluctuating temperatures.

Common Diseases

On a whole, ducks are not as prone to the typical poultry diseases, and many of the diseases they do contract can be prevented by making sure the ducks have a clean environment in which to live (by cleaning out their houses, providing fresh drinking water, and so on).

Two common diseases found in ducks are botulism and maggots. Botulism causes the duck's neck to go limp, making it difficult or even impossible for the duck to swallow. Maggots infest the ducks if they do not have any clean water in which to bathe, and are typically contracted in the hot summer months. Both of these diseases (as well as worms and mites) can be cured with the proper care, medications, and veterinary assistance.

Turkeys

Turkeys are generally raised for their meat (especially for holiday roasts) though their eggs can also be eaten. Turkeys are incredibly easy to manage and raise as they primarily subsist on bugs, grasshoppers, and wasted grain that they find while wandering around the yard. They are, in a sense, self-sustaining foragers.

If you are looking to raise a turkey for Thanksgiving dinner, it is best to hatch the turkey chick in early spring, so that by November, it will be about 14 to 20 pounds.

Breeds of Turkeys

The largest breeds of turkeys found in the United States are the Bronze and Narragansett. Other breeds, though not as popular, include the White Holland, Black turkey, Slate turkey, and Bourbon Red.

Bronze breeds are most likely a cross between a wild North American turkey and domestic turkey, and they have beautiful rich plumage. This is the most common type of turkey to raise, as it is the largest, is very hardy, and is the most profitable. The White Holland and Bourbon Red, however, are said to be the most "domesticated" in their habits and are easier to keep in a smaller roaming area.

Housing Turkeys

Turkeys flourish when they can roost in the open. They thrive in the shelter of trees, though this can become problematic as they are more vulnerable to predators than if they are confined in a house. If you do build a house for them, it should be airy, roomy, and very clean.

It is important to allow turkeys freedom to roam; if you live in a more suburban or neighborhood area, raising turkeys may not be the best option for you, as your turkeys

≈ **Bronze turkeys like this one are some of the most common in the United States.**

may wander into a neighboring yard, upsetting your neighbors. Turkeys need lots of exercise to be healthy and vigorous. When turkeys are confined for large periods of time, it is more difficult to regulate their feeding (turkeys are natural foragers and thrive best on natural foods), and they are more likely to contract disease than if they are allowed to range freely.

What Do Turkeys Eat?

Turkeys gain most of their sustenance from foraging, either in lawns or in pastures. They typically eat green vegetation, berries, weed seeds, waste grain, nuts, and various kinds of acorns. In the summer months,

turkeys especially like to get grasshoppers. Due to their love of eating insects that can damage crops and gardens, turkeys are quite useful in keeping your growing produce free from harmful insects and parasites.

Turkeys may be fed grain (similar to a mixture given to chickens) if they are going to be slaughtered, in order to make them larger.

Hatching Turkey Chicks

Turkey hens lay eggs in the middle of March to the first of April. If you are looking to hatch and raise turkey chicks, it is vital to watch the hen closely for when she lays the eggs, and then gather them and keep the eggs warm until the weather is more stable. Turkey hens generally aim to hide their nests from predators. It is best, for the hen's sake, to provide her with a coop of some sort, which she can freely enter and leave. Or, if no coop is available, encourage the hen to lay her eggs in a nest close to your house (putting a large barrel on its side and heaping up brush near the house may entice the hen to nest there). This way, you can keep an eye on the eggs and hatchlings.

Hens are well adapted to hatch all of the eggs that they lay. It takes twenty-seven to twenty-nne days for turkey eggs to hatch. While the hens are incubating the eggs, they should be given adequate food and water, placed close to their nest. Wheat and corn are the best food during the laying and incubation period.

Raising the Poults

Turkey chicks, also known as "poults," can be difficult to raise and require lots of care and attention for their first few weeks of life. In this sense, a turkey raiser must be "on call" to come to the aid of the hen and her poults at any time during the day for the first month or so. Many times, the hens can raise the poults quite well, but it is important that they receive enough food and warmth in the early weeks to allow them to grow healthy and strong. The poults should stay dry, as they become chilled easily. If you are able, encouraging the poults and their mother into a coop until the poults are stronger will aid their growth to adulthood.

Poults should be fed soft and easily digestible foods. Stale bread, dipped in milk and then dried until it crumbles, is an excellent source of food for the young turkeys.

Diseases

Turkeys are hardy birds but they are susceptible to a few debilitating or fatal diseases. It is a fact that the mortality rate among young turkeys, even if they are given all the care and exercise and food needed, is relatively high (usually due to environmental and predatory factors).

The most common disease in turkeys is blackhead. Blackhead typically infects young turkeys between six weeks and four months old. This disease will turn the head darker colored or even black and the bird will become very weak, will stop eating, and will have an insatiable thirst. Blackhead is usually fatal.

Another disease that turkeys occasionally contract is roup. Roup generally occurs when a turkey has been exposed to extreme dampness or cold drafts for long periods of time. Roup causes the turkey's head to swell around the eyes and is highly contagious to other turkeys. Nutritional roup is caused by a vitamin A deficiency, which can be alleviated by adding vitamin A to the turkeys' drinking water. It is best to consult a veterinarian if your turkey seems to have this disease.

SLAUGHTERING POULTRY

If you are raising your own poultry, you may decide that you'd like to use them for consumption as well. Slaughtering your own poultry enables you to know exactly what is in the meat you and your family are consuming, and to ensure that the poultry is kept humanely before being slaughtered. Here are some guidelines for slaughtering poultry:

1. To prepare a fowl for slaughter, make sure the bird is secured well so it is unable to move (either hanging down from a pole or laid on a block that is used for chopping wood).

2. Killing the fowl can be done in two ways: one way is to hang the bird upside down and to cut the jugular vein with a sharp knife. It is a good idea to have a funnel or vessel available to collect the draining blood so it does not make a mess and can be disposed of easily. The other option is to place the bird's head on a chopping block and then, in one clean movement, chop its head off at the middle of the neck. Then, hang the bird upside down and let the blood drain as described above.

3. Once the bird has been thoroughly drained of blood, you can begin to pluck it. Have a pot of hot water (around 140ºF) ready, into which to dip the bird. Holding the bird by the feet, dip it into the pot of hot water and leave it for about forty-five seconds—you do not want the bird to begin to cook! Then, remove the bird from the pot and begin plucking immediately. The feathers should come off fairly easily, but this process takes time, so be patient. Discard the feathers.

4. Once the bird has been completely rid of feathers, slip back the skin from the neck and cut the neck off close to the base of the body. Then, remove the crop, trachea, and esophagus from the bird by loosening them and pulling them out through the hole created from chopping off the neck. Cut off the vent to release the main entrails (being careful not to puncture the intestines or bacteria could be released into the meat) and make a horizontal slit about an inch above it so you can insert two fingers. Remove the entrails, liver (carefully cutting off the gallbladder), gizzard, and heart from the bird and set the last three aside if you want to eat them later or make them into stuffing. If you are going to save the heart, slip off the membrane enclosing it and cut off the veins and arteries. Make sure to clean out the gizzard as well if you will be using it later.

5. Wash the bird thoroughly, inside and out, and wipe it dry.

6. Cut off the feet below the joints and then carefully pull out the tendons from the drumsticks.

7. Once the carcass is thoroughly dry and clean, store it in the refrigerator if it will be used that same day or the next. If you want to save the bird for later use, place it in a moisture-proof bag and set it in the freezer (along with any innards that you may have saved).

8. Make sure you clean and disinfect any surface you were working on to avoid the spread of bacteria and other diseases.

Rabbits

Rabbits are very social and docile animals, and easy to maintain. They like to play, but because of their skittish nature, are not necessarily the best pets for young children. Larger rabbits, bred for eating, often make good pets because of their more relaxed personalities. Rabbits are easier to raise than chickens and can provide you with beautiful fur and lean meat. In fact, rabbits will take up less space and use less money than chickens.

Breeds

There are over forty breeds of domestic rabbits. Below are ten of the most commonly owned varieties, along with their traits and popular uses.

1. Californian: 6–10 lbs. Short fur. Relaxed personality. Choice for eating.
2. Dutch: 3–5 lbs. Short fur. Relaxed personality. Choice pet. Good for young children.
3. Flemish Giant: 9+ lbs. Medium-length fur. Calm personality. Choice for eating.
4. Holland Lop: 3–5 lbs. Medium-length fur. Curious personality. Choice pet. One of the lop-eared rabbits, its ears flop down next to its face. A similar popular breed is the American Fuzzy Lop.
5. Jersey Wooly: 2–4 lbs. Long fur. Relaxed personality. Choice pet.
6. Mini Lop: 4–7 lbs. Medium-length fur. Relaxed personality. Choice pet. Lop-eared. Some reports of higher biting tendencies.

7. Mini Rex: 3–5 lbs. Very short, velvety fur. Curious personality. Choice pet. Tend to have sharp toenails.
8. Netherland Dwarf: 2–4 lbs. Medium-length fur. Excitable personality. Choice pet.
9. New Zealand: 9+ lbs. Short fur. Curious personality. Choice for eating. Variable reputation for biting.
10. Satin: 9+ lbs. Medium-length fur. Relaxed personality. Fur is finer and denser than other furs.

Housing

Rabbits should be kept in clean, dry, spacious homes. You will need a hutch, similar to a henhouse, to house your rabbits. It is important to provide your rabbits with lots of air. The best hutch will have a wide, over-hanging roof and is elevated about six inches off the ground. This way, your rabbits

will not only have shade, but their homes will be prevented from getting damp.

Food

A rabbit's diet should be made up of three things: a small portion pellets (provided they are high in fiber), a continual source of hay, and vegetables. Rabbits love vegetables that are dark and leafy or root vegetables. Avoid feeding them beans or rhubarb, and limit the amount of spinach they eat. If you want to give rabbits a treat, try a small piece of fruit, such as a banana or apple. Remember that all of their food needs to be fresh (pellets should not be more than six weeks old), and like all other animals, be careful not to over-feed them. Also, to keep them from dehydrating, provide them with plenty of clean water every day.

Note: If you have a pregnant doe, allow her to eat a little more than usual.

Breeding

When you want to breed rabbits, put a male and female together in the morning or evening. After they have mated, you may separate them again. A female's gestation period is approximately a month in length, and litters range from six to ten babies. Baby rabbits' eyes will not open until two weeks after birth. Their mother will nurse them for a month, and for at least the first week, you must not touch any of the litter; you can alter their smell and the mother may stop feeding them. At two months, babies should be weaned from their mother,

and at four months, or approximately 4.5 pounds, they are old enough to sell, eat, or continue breeding. Larger rabbit varieties may take six to twelve months to sexually mature.

Health Concerns

The main issues that may arise in your rabbits' health are digestive problems and bacterial infections. Monitor your rabbit's droppings carefully. Diarrhea in rabbits can be fatal. Some diarrhea is easy to identify, but also be on the lookout for droppings that are misshapen, softer in consistency, a lack of droppings altogether, and loud tummy growling. Diarrhea requires antibiotics from your veterinarian. In bacterial infections, your rabbit may have a runny nose or eyes, a high temperature, or a rattling or coughing respiratory noise. This also requires medical attention and an antibiotic specific to the type of infection.

Hairballs are another issue you may encounter and also require some attention. Every three months, rabbits shed their hair, and these sheds will alter between light and heavy. Since rabbits will attempt to groom themselves as cats do, but cannot vomit hair as cats can, you must groom them addition-ally, to prevent too much hair ingestion. Brush and comb them when their shedding begins, and provide them with ample fresh hay and opportunity for exercise. The fiber in the hay will help the hair to pass through their digestive tracts, and the exercise will keep their metabolisms active.

If your rabbit has badly misaligned teeth, they may interfere with his or her ability to eat and will need to be trimmed by the veteri-narian.

Never give your rabbit amoxicillin or use cedar or pine shavings in their hutches. Penicillin-based drugs carry high risks for rabbits, and the shavings emit a carbon that can cause respiratory or liver damage to small animals like rabbits.

Horses

A horse can be a chauffeur (whether you're in a saddle or in a carriage trailing behind), a farmer's best friend when the fields need to be plowed, a spotlight-loving performer, and most importantly, an incredible pet and companion. The versatility of horses is what makes them such appealing animals. Yet with their great size, comes even greater responsibility. Horses have very specific needs: their living quarters, their diet, their exercise regimen, and their grooming, to name a few. They are complicated animals, but if taken care of the correct way, all the reaped rewards will absolutely be worth the work.

Breeds of Horses

There are hundreds of horse breeds, from the rare Abtenauer bred in a secluded valley just south of Salzburg, Austria, to the elegant Zweibrücken hailing from Germany. While it is fascinating to really explore all the differences in each breed, it is more practical to break the breeds up into three main categories based on their body type. The first two are "light horses" and "draft horses"; light horses are used for undemanding work and for their speed, whereas draft horses are able to complete more arduous tasks. You are likely to see light horses galloping around a racetrack or in a dressage show, while draft horses and their carriages are a popular method of transportation for princesses or, more likely, tourists in New York's Central Park. The final type of horse is a pony, sizing in at a mere 14.2 hands or under. Ponies have always been especially popular amongst children.

1. Light horses—An example of a light horse is the Thoroughbred. This breed's native home is England, where their human counterparts have fostered racing for over a thousand years. They have slender necks, deep, wide chests with long, slanting shoulders, and hind quarters that are high and muscular. Their legs are long and they end in rounded, well-shaped feet. Their lithe, angular bodies favor speed and are in complete opposition to a draft horse.
2. Draft horses—A Clydesdale is a well known draft breed. Bred in Scotland and later imported to America, this horse is famous for its intimidating size and its gigantic, hairy feet. Clydesdales are tall, and long limbed; they have medium-sized necks joining with slanting shoulders that meet strong legs, heavily fringed below the knee. This hair was often criticized by horsemen as being a fault of the breed; it was too hard to keep clean and free of disease, but many are fans of the defining characteristic.
3. Ponies—A Shetland pony is one of the more popular breeds. Shetlands resemble small draft horses, with their foretops, manes, and tails heavy and long. Being tinier than other horses,

a pony is a great option for a child learning how to ride—they are exceedingly intelligent and can be readily trained.

Figure out what you want to do with your horse before you select a breed; if you want your horse to be ridden, who will be the primary rider? Do you plan on leisurely trail rides or do you want your horse to be a racer? Once you figure these things out you can narrow down your choices. For many people, especially if they are new to having horses, the breed is not necessarily the deciding factor. Most care more about a horse's personality and how well-trained or how green the horse is. If selecting a horse to begin breeding, then the breed would obviously have to be taken into deep consideration.

Housing

Horses should have some kind of shelter; all horses need to be protected from inclement weather. Housing can be indoor or outdoor, or even a combination of the two. It is decided by the use of the horse on a day-to-day basis. Indoor housing is ideal for horses that are being ridden or used every day in the winter time. The size of the stall should be around 12 x 12, which is the industry standard, but if you are housing a draft horse then the dimensions should be increased to 16 x 16. Horses should be provided with bedding for their stall (different types of straw or shavings); they need a dry, soft bed. Their bedding should also be kept fresh and clean, which means cleaning the stall daily and removing soiled bedding and manure. If an owner neglects the horse's stall, it lessens the comfort and promotes disease. Outdoor housing has a myriad of advantages, for example lower construction costs and less labor where cleaning the building is concerned. The building should be a three-sided structure, like a run-in shed.

Grooming

Unlike other farm animals, horses must be groomed. Without grooming, a horse is susceptible to discomfort and diseases; this includes taking care of a horse's hooves because feet are the horse's most complicated structures, and the most integral to look after. A horse's hooves should be cleaned (with a pick) every day, before and after riding or being sent out to do work. The feet are the most liable to injury from the effects of hard work and mismanagement, and subsequently,

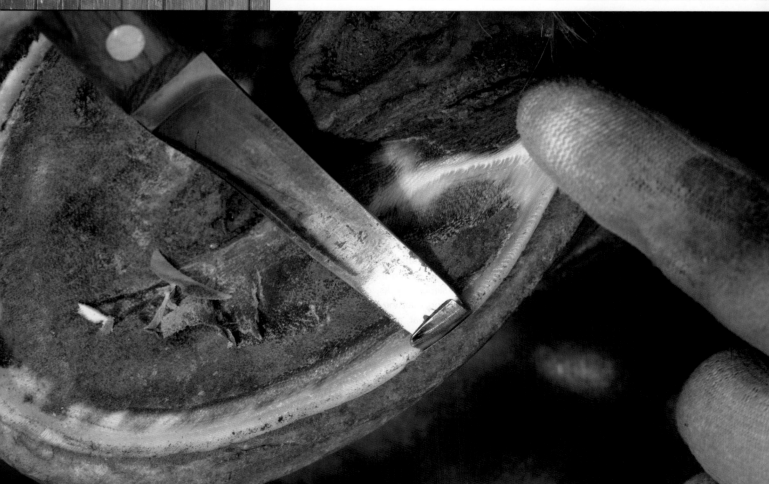

there is no body part that more requires care in both health and disease. Prevention, in the foot's case, is much better than cure. It is also a good idea to get the horse shoes, which prevent the hoof from wearing down—a hoof wears down faster than it grows back. A stable horse should be thoroughly groomed each day, before and after the horse's work; it is necessary to scrub away the "scurf" (the small shreds of epidermis that are continually exfoliated from the skin) that obstructs the pores. Doing so admits free perspiration and promotes circulation to the extremities. Cleaning and rubbing the skin is obviously important, but it is also imperative for the legs and feet because it preserves soundness.

Feeding and Watering

With horses, it is unnecessary to limit the amount of water; it should be left up to the horse's discretion. He will take only as much as he wants. Food, however, is a different story. Horses can be overfed with hay (horses will eat it just for the amusement) and with their oats or grain. They should be fed according to however much they work—a

work horse would be fed a more substantial amount than an idle horse, for example. If a horse is continually overfed, its appetite will eventually increase and, of course, the horse will become overweight. Horses should be fed regularly, at a scheduled time, as they anticipate the hour they will be fed and will become nervous it is too long delayed. As for what to feed horses—oats have been proven especially great for them; they favor speed and endurance more than any other food. If oats are too expensive, grains like barley, wheat, rye, and bran can be substituted. Horses are sometimes even known to eat corn. It should also be noted that if one is changing a horse's feed, it needs to be gradual, otherwise, there could be digestive problems.

Breeding

For horse owners, the idea of breeding from their own mare has much appeal. The prospect of producing a foal with qualities similar to its mother, or even better, has many attractions.

Before any breeding decisions are made, it is best to have knowledge about normal breeding behavior, what should happen at foaling, and how a newborn foal should

behave and develop is essential. For this reason, it is best for a novice horse breeder to seek professional help with mating and foaling from a stud.

Common Diseases

One of the most common, and most dangerous, diseases a horse can contract is colic. Colic is a broad term that covers any acute gastrointestinal problem: colic could be anything from a stomachache or cramp from changing food too fast to impaction of waste in a horse's intestines. Colic can be fatal; it is the leading cause of premature death among domesticated horses, and therefore if an owner suspects his or her horse has colic, it is absolutely necessary to call a vet. Some symptoms of colic are restlessness, lying down, kicking with the hind feet upward and toward the belly, jerky swishing of the tail, groaning, frequent position changes, and stretching as if to urinate, but with greater intensity, the movements become violent: the horse may throw himself down, roll, assume unnatural positions (for example, sitting on his haunches), and grunt loudly. With colic, the pain is not constant, so during the periods of peace, the horse may act completely normal. However, during the periods of pain, the horse will be sweating profusely.

As stated earlier, it is imperative that the owner takes care of their horses sensitive hooves. There are many diseases that result from neglected feet. One common disease is Laminitis, where the horse's digital laminae (attached to the hoof wall and coffin bone) become inflamed. It will eventually become impossible for the horse to walk without pain. It is impossible to discern that a horse has laminitis without radiographs (so if an owner suspects it, they should consult a vet); but the owner should be wary of laminitis if the horse has any of the following symptoms: sweating, increased vital signs, and a tendency to favor the afflicted foot.

Founder is not a disease, exactly, but a complication of many horse diseases. Founder occurs when a disease (like laminitis) goes untreated for a very long time. What will happen is that the coffin bone will sink through the frog of the hoof making moving, and even standing, impossible without extreme pain and discomfort. A horse that founders and refuses to stand could very likely end up with colic—and that means the horse is in a lot of danger of dying. Any horses that founder will need constant attention and possibly even hospitalization in an equine clinic.

Beekeeping

Beekeeping (also known as apiculture) is one of the oldest human industries. For thousands of years, honey has been considered a highly desirable food. Beekeeping is a science and can be a very profitable employment; it is also a wonderful hobby for many people in the United States. Keeping bees can be done almost anywhere—on a farm, in a rural or suburban area, and even, at times, in urban areas (even on rooftops!). Anywhere there are sufficient flowers from which to collect nectar, bees can thrive.

Apiculture relies heavily on the natural resources of a particular location and the knowledge of the beekeeper in order to be successful. Collecting and selling honey at your local farmers' market or just to family and friends can supply you with some extra cash if you are looking to make a profit from your apiary.

Why Raise Bees?

Bees are essential in the pollination and fertilization of many fruit and seed crops. If you have a garden with many flowers or fruit plants, having bees nearby will only help your garden flourish and grow year after year. Furthermore, nothing is more satisfying than extracting your own honey for everyday use.

How to Avoid Getting Stung

Though it takes some skill, you can learn how to avoid being stung by the bees you keep. Here are some ways you can keep your bee stings to a minimum:

1. Keep gentle bees. Having bees that, by sheer nature, are not as aggressive will reduce the number of stings you are likely to receive. Carniolan bees are one of the gentlest species, and so are the Caucasian bees introduced from Russia.
2. Obtain a good "smoker" and use it whenever you'll be handling your bees. Pumping smoke of any kind into and around the beehive will render your bees less aggressive and less likely to sting you.
3. Purchase and wear a veil. This should be made out of black bobbinet and worn over your face. Also, rubber gloves help protect your hands from stings.
4. Use a "bee escape." This device is fitted into a slot made in a board the same

size as the top of the hive. Slip the board into the hive before you open it to extract the honey, and it allows the worker bees to slip below it but not to return back up. So, by placing the "bee escape" into the hive the day before you want to gain access to the combs and honey, you will most likely trap all the bees under the board and leave you free to work with the honeycombs without fear of stings.

What Type of Hive Should I Build?

Most beekeepers would agree that the best hives have suspended, moveable frames where the bees make the honeycombs, which are easy to lift out. These frames, called Langstroth frames, are the most popular kind of frame used by apiculturists in the United States.

inhabit a cool, shaded hive than one that is baking in the hot summer sun.

Sometimes it is beneficial to try to prevent swarming, such as if you already have completely full hives. Removing the new honey frequently from the hive before swarming begins will deter the bees from swarming. Shading the hives on warm days will also help keep the bees from swarming.

Bee Pastures

Bees will fly a great distance to gather food but you should try to contain them, as well as possible, to an area within 2 miles of the beehive. Make sure they have access to many honey-producing plants, which you can grow in your garden. Alfalfa, asparagus, buckwheat, chestnut, clover, catnip, mustard, raspberry, roses, and sunflowers are some of the best honey-producing plants and trees. Also make sure that your bees always have access to pure, clean water.

Preparing Your Bees for Winter

If you live in a colder region of the United States, keeping your bees alive throughout the winter months is difficult. If your queen bee happens to die in the fall, before a young queen can be reared, your whole colony will die throughout the winter. However, the queen's death can be avoided by taking simple precautions and giving careful attention to your hive come autumn.

Whether you build your own beehive or purchase one, it should be built strongly and should contain accurate bee spaces and a close-fitting, rainproof roof. If you are looking to have honeycombs, you must have a hive that permits the insertion of up to eight combs.

Where Should the Hive Be Situated?

Hives and their stands should be placed in an enclosure where the bees will not be disturbed by other animals or humans and where it will be generally quiet. Hives should be placed on their own stands at least 3 feet from each other. Do not allow weeds to grow near the hives and keep the hives away from walls and fences. You, as the beekeeper, want to be able to easily access your hive without fear of obstacles.

Swarming

Swarming is simply the migration of honeybees to a new hive and is led by the queen bee. During swarming season (the

warm summer days), a beekeeper must remain very alert. If you see swarming above the hive, take great care and act calmly and quietly. You want to get the swarm into your hive, but this will be tricky. It they land on a nearby branch or in a basket, simply approach and then "pour" them into the hive. Keep in mind that bees will more likely

Colonies are usually lost in the winter months due to insufficient winter food storages, faulty hive construction, lack of protection from the cold and dampness, not enough or too much ventilation, or too many older bees and not enough young ones.

If you live in a region that gets a few weeks of severe weather, you may want to move your colony indoors, or at least to an area that is protected from the outside elements. But the essential components of having a colony survive through the winter season are to have a good queen; a fair ratio of healthy, young, and old bees; and a plentiful supply of food. The hive needs to retain a liberal supply of ripened honey and a thick syrup made from white cane sugar (you should feed this to your bees early enough so they have time to take the syrup and seal it over before winter).

To make this syrup, dissolve 3 pounds of granulated sugar in 1 quart of boiling water and add 1 pound of pure extracted honey to this. If you live in an extremely cold area, you may need up to 30 pounds of this syrup, depending on how many bees and hives you have. You can either use a top feeder or a frame feeder, which fits inside the hive in the place of a frame. Fill the frame with the syrup and place sticks or grass in it to keep the bees from drowning.

Extracting Honey

To obtain the extracted honey, you'll need to keep the honeycombs in one area of the hive or packed one above the other. Before removing the filled combs, you should allow the bees ample time to ripen and cap the honey. To uncap the comb cells, simply use a sharp knife (apiary suppliers sell knives specifically for this purpose). Then put the combs in a machine called a honey extractor to extract the honey. The honey extractor whips the honey out of the cells and allows you to replace the fairly undamaged comb into the hive to be repaired and refilled.

The extracted honey runs into open buckets or vats and is left, covered with a tea towel or larger cloth, to stand for a week. It should be in a warm, dry room where no ants can reach it. Skim the honey each day until it is perfectly clear. Then you can put it into cans, jars, or bottles for selling or for your own personal use.

Making Beeswax

Beeswax from the honeycomb can be used for making candles, can be added to lotions or lip balm, and can even be used in baking. Rendering wax in boiling water is especially simple when you only have a small apiary.

Collect the combs, break them into chunks, roll them into balls if you like, and put them in a muslin bag. Put the bag with the beeswax into a large stockpot and bring the water to a slow boil, making sure the bag doesn't rest on the bottom of the pot and burn. The muslin will act as a strainer for the wax. Use clean, sterilized tongs to occasionally squeeze the bag. After the wax is boiled out of the bag, remove the pot from the heat and allow it to cool. Then, remove the wax from the top of the water and then re-melt it in another pot on very low heat, so it doesn't burn.

Pour the melted wax into molds lined with wax paper or plastic wrap and then cool it before using it to make other items or selling it at your local farmers' market.

Extra Beekeeping Tips

General Tips

1. Clip the old queen's wings and go through the hives every ten days to destroy queen cells to prevent swarming.
2. Always act and move calmly and quietly when handling bees.
3. Keep the hives cool and shaded. Bees won't enter a hot hive.

When Opening the Hive

1. Have a smoker ready to use if you desire.
2. Do not stand in front of the hive while the bees are entering and exiting.
3. Do not drop any tools into the hive while it's open.
4. Do not run if you become frightened.
5. If you are attacked, move away slowly and smoke the bees off yourself as you retreat.
6. Apply ammonia or a paste of baking soda and water immediately to any bee sting to relieve the pain. You can also scrape the area of the bee sting with your fingernail or the dull edge of a knife immediately after the sting.

When Feeding Your Bees

1. Keep a close watch over your bees during the entire season, to see if they are feeding well or not.
2. Feed the bees during the evening.
3. Make sure the bees have ample water near their hive, especially in the spring.

Making a Beehive

The most important parts of constructing a beehive are to make it simple and sturdy. Just a plain box with a few frames and a couple of other loose parts will make a successful beehive that will be easy to use and manipulate. It is crucial that your beehive be well adapted to the nature of bees and also the climate where you live. Framed hives usually suffice for the beginning beekeeper. To the right is a diagram of a simple beehive that you can easily construct for your backyard beekeeping purposes.

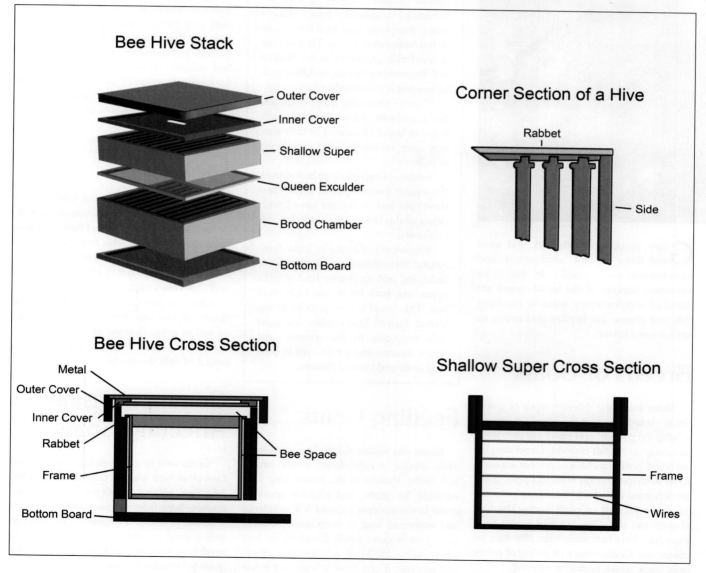

Bee Hive Stack
- Outer Cover
- Inner Cover
- Shallow Super
- Queen Exculder
- Brood Chamber
- Bottom Board

Corner Section of a Hive
- Rabbet
- Side

Bee Hive Cross Section
- Metal
- Outer Cover
- Inner Cover
- Rabbet
- Frame
- Bottom Board
- Bee Space

Shallow Super Cross Section
- Frame
- Wires

Credit: Timothy W. Lawrence

Goats

Goats provide us with milk and wool and thrive in arid, semitropical, and mountainous environments. In the more temperate regions of the world, goats are raised as supplementary animals, providing milk and cheese for families and acting as natural weed killers.

Breeds of Goats

There are many different types of goats. Some breeds are quite small (weighing roughly 20 pounds) and some are very large (weighing up to 250 pounds). Depending on the breed, goats may have horns that are corkscrew in shape, though many domestic goats are dehorned early on to lessen any potential injuries to humans or other goats. The hair of goats can also differ—various breeds have short hair, long hair, curly hair, silky hair, or coarse hair. Goats come in a variety of colors (solid black, white, brown, or spotted).

SIX MAJOR U.S. GOAT BREEDS

Alpine—Originally from Switzerland, these goats may have horns, are short haired, and are usually white and black in color. They are also good producers of milk.

Anglo-Nubian—A cross between native English goats and Indian and Nubian breeds, these goats have droopy ears, spiral horns, and short hair. They are quite tall and do best in warmer climates. They do not produce as much milk, though it is much higher in fat than other goats. They are the most popular breed of goat in the United States.

LaMancha—A cross between Spanish Murciana and Swiss and Nubian breeds, these goats are extremely adaptable, have straight noses, short hair, may have horns, and do not have external ears. They are not as good milk producers as the Saanen and Toggenburg breeds, and their milk fat content is much higher.

Pygmy—Originally from Africa and the Caribbean, these dwarfed goats thrive in hotter climates. For their size, they are relatively good producers of milk.

Saanen—Originally from Switzerland, these goats are completely white, have short hair, and sometimes have horns. Goats of this breed are wonderful milk producers.

Toggenburg—Originally from Switzerland, these goats are brown with white facial, ear, and leg stripes; have straight noses; may have horns; and have short hair. This breed is very popular in the United States. These goats are good milk producers in the summer and winter seasons and survive well in both temperate and tropical climates.

Feeding Goats

Goats can sustain themselves on bushes, trees, shrubs, woody plants, weeds, briars, and herbs. Pasture is the lowest cost feed available for goats, and allowing goats to graze in the summer months is a wonderful and economic way to keep goats, even if your yard is quite small. Goats thrive best when eating alfalfa or a mixture of clover and timothy. If you have a lawn and a few goats, you don't need a lawn mower if you plant these types of plants for your goats to eat. The one drawback to this is that your goats (depending on how many you own) may quickly deplete these natural resources, which can cause weed growth and erosion. Supplementing pasture feed with other food stuff, such as greenchop, root crops, and wet brewery grains will ensure that your yard does not become overgrazed and that your goats remain well-fed and healthy. It is also beneficial to supply your goats with unlimited access to hay while they are grazing. Make sure that your goats have easy access to shaded areas and fresh water, and offer a salt and mineral mix on occasion.

Dry forage is another good source of feed for your goats. It is relatively inexpensive to grow or buy and consists of good quality legume hay (alfalfa or clover). Legume hay is high in protein and has many essential minerals beneficial to your goats. To make sure your forages are highly nutritious, be sure that there are many leaves that provide protein and minerals and that the forage had an early cutting date, which will allow for easier digestion of the nutrients. If your forage is green in color, it most likely contains more vitamin A, which is good for promoting goat health.

Goat Milk

Goat milk is a wonderful substitute for those who are unable to tolerate cow's milk, or for the elderly, babies, and those suffering from stomach ulcers. Milk from goats is also high in vitamin A and niacin but does not have the same amount of vitamins B6, B12, and C as cow's milk.

Lactating goats do need to be fed the best quality legume hay or green forage possible, as well as grain. Give the grain to the doe at a rate that equals ½ pound grain for every pound of milk she produces.

Common Diseases Affecting Goats

Goats tend to get more internal parasites than other herd animals. Some goats develop infectious arthritis, pneumonia, coccidiosis, scabies, liver fluke disease, and mastitis. It is advisable that you establish a relationship with a good veterinarian who specializes in small farm animals to periodically check your goats for various diseases.

Milking a Goat

Milking a goat takes some practice and patience, especially when you first begin. However, once you establish a routine and rhythm to the milking, the whole process should run relatively smoothly. The main thing to remember is to keep calm and never pull on the teat, as this will hurt the goat and she might upset the milk bucket. The goat will pick up on any anxiousness or nervousness on your part and it could affect how cooperative she is during the milking.

Supplies

- A grain bucket and grain for feeding the goat while milking is taking place
- Milking stand
- Metal bucket to collect the milk
- A stool to sit on (optional)
- A warm sterilized wipe or cloth that has been boiled in water
- Teat dip solution (2 tbsps bleach, 1 quart water, one drop normal dish detergent mixed together)

Directions

1. Ready your milking stand by filling the grain bucket with enough grain to last throughout the entire milking. Then retrieve the goat, separating her from any other goats to avoid distractions and unsuccessful milking. Place the goat's head through the head hold of the milking stand so she can eat the grain and then close the lever so she cannot remove her head.
2. With the warm, sterilized wipe or cloth, clean the udder and teats to remove any dirt, manure, or bacteria that may be present. Then, place the metal bucket on the stand below the udder.
3. Wrap your thumb and forefinger around the base of one teat. This will help trap the milk in the teat so it can be squirted out. Then, starting with your middle finger, squeeze the three remaining fingers in one single, smooth motion to squirt the milk into the bucket. Be sure to keep a tight grip on the base of the teat so the milk stays there until extracted. Remember: the first squirt of milk from either teat should not be put into the bucket as it may contain dirt or bacteria that you don't want contaminating the milk.
4. Release the grip on the teat and allow it to refill with milk. While this is happening, you can repeat this process on the other teat and can alternate between teats to speed up the milking process.
5. When the teats begin to look empty (they will be somewhat flat in appearance), massage the udder just a little bit to see if any more milk remains. If so, squeeze it out in the same manner as above until you cannot extract much more.
6. Remove the milk bucket from the stand and then, with your teat dip mixture in a disposable cup, dip each teat into the solution and allow to air dry. This will keep bacteria and infection from going into the teat and udder.
7. Remove the goat from the milk stand and return her to the pen.

Making Cheese from Goat Milk

Most varieties of cheese that can be made from cow's milk can also be successfully made using goats' milk. Goats' milk cheese can easily be made at home. In order to make the cheese, however, at least one gallon of goat milk should be available. Make sure that all of your equipment is washed and sterilized (using heat is fine) before using it.

Cottage Cheese

1. Collect surplus milk that is free of strong odors. Cool it to around 40°F and keep it at that temperature until it is used.
2. Skim off any cream. Use the skim milk for cheese and the cream for cheese dressing.
3. If you wish to pasteurize your milk (which will allow it hold better as a cheese) collect all the milk to be processed into a flat bottomed, straight-sided pan and heat to 145°F on low heat. Hold it at this temperature for about thirty minutes and then cool to around 80°F. Use a dairy thermometer to measure the milk's temperature. Then, inoculate the cheese milk with a desirable lactic acid fermenting bacterial culture (you can use commercial buttermilk for the initial source). Add about 7 ounces to 1 gallon of cheese milk, stir well, and let it sit undisturbed for about ten to sixteen hours, until a firm curd is formed.
4. When the curd is firm enough, cut the curd into uniform cubes no larger than ½ inch using a knife or spatula.
5. Allow the curd to sit undisturbed for a couple of minutes and then warm it slowly, stirring carefully, at a temperature no greater than 135°F. The curd should eventually become firm and free from whey.
6. When the curd is firm, remove from the heat and stop stirring. Siphon off the excess whey from the top of the pot. The curd should settle to the bottom of the container. If the curd is floating, bacteria that produces gas has been released and a new batch must be made.
7. Replace the whey with cold water, washing the curd and then draining the water. Wash again with ice-cold water from the refrigerator to chill the curd. This will keep the flavor fresh.
8. Using a draining board, drain the excess water from the curd. Now your curd is complete.
9. In order to make the curd into a cottage cheese consistency, separate the curd as much as possible and mix with a milk or cream mixture containing salt to taste.

Domiati Cheese

This type of cheese is made throughout the Mediterranean region. It is eaten fresh or aged two to three months before consumption.

1. Cool a gallon of fresh, quality milk to around 105°F, adding 8 ounces of salt to the milk. Stir the salt until it is completely dissolved.
2. Pasteurize the milk as described in step 3 of the cottage cheese recipe.
3. Domiati cheese is coagulated by adding a protease enzyme (rennet). This enzyme may be purchased at a local drug store, health food store, or a cheese maker in your area. Dissolve the concentrate in water, add it to the cheese milk, and stir for a few minutes. Use 1 milliliter of diluted rennet liquid in forty milliliters of water for every 2½ gallons of cheese milk.
4. Set the milk at around 105°F. When the enzyme is completely dispersed in the cheese milk, allow the mix to sit undisturbed until it forms a firm curd.
5. When the desired firmness is reached, cut the curd into very small cubes. Allow for some whey separation. After ten to twenty minutes, remove and reserve about a third of the volume of salted whey.

ANGORA GOATS

Angora goats may be the most efficient fiber producers in the world. The hair of these goats is made into mohair, a long, lustrous hair that is woven into fine garments. Angora goats are native to Turkey and were imported to the United States in the mid-1800s. Now, the United States is one of the two biggest producers of mohair on Earth.

Angora goats are typically relaxed and docile. They are delicate creatures, easily strained by their year-round fleeces. Angora goats need extra attention and are more high-maintenance than other breeds of goat. While these goats can adapt to many temperate climates, they do particularly well in the arid environment of the southwestern states.

Angora goats can be sheared twice yearly, before breeding and before birthing. The hair of the goat will grow about ¾ inch per month and it should be sheared once it reaches 4 to 6 inches in length. During the shearing process, the goat is usually lying down on a clean floor with its legs tied. When the fleece is gathered (it should be sheared in one full piece), it should be bundled into a burlap bag and should be free of contaminants. Mark your name on the bag and make sure there is only one bag per fleece. For more thorough rules and regulations about selling mohair through the government's direct-payment program, contact the USDA Agricultural Stabilization and Conservation Service online or in one of their many offices.

Shearing can be accomplished with the use of a special goat comb, which leaves ¼ inch of stubble on the goat. It is important to keep the fleeces clean and to avoid injuring the animal. The shearing seasons are in the spring and fall. After a goat has been sheared, it will be more sensitive to changes in the weather for up to six weeks. Make sure you have proper warming huts for these goats in the winter and adequate shelter from rain and inclement weather.

6. Put the curd and remaining whey into cloth-lined molds (the best are rectangular stainless steel containers with perforated sides and bottom) with a cover. The molds should be between 7 and 10 inches in height. Fill the molds with the curd, fold the cloth over the top, allow the whey to drain, and discard the whey.

7. Once the curd is firm enough, apply added weight for ten to eighteen hours until it is as moist as you want.

8. Once the pressing is complete and the cheese is formed into a block, remove the molds, and cut the blocks into 4-inch-thick pieces. Place the pieces in plastic containers with airtight seals. Fill the containers with reserved salted whey from step 5, covering the cheese by about an inch.

9. Place these containers at a temperature between 60 and 65°F to cure for one to four months.

Feta Cheese

This type of cheese is very popular to make from goats' milk. The same process is used as the Domiati cheese except that salt is not added to the milk before coagulation. Feta cheese is aged in a brine solution after the cubes have been salted in a brine solution for at least twenty-four hours.

Sheep

Sheep were possibly the first domesticated animals, and are now found all over the world on farms and smaller plots of land. Almost all the breeds of sheep that are found in the United States have been brought here from Great Britain. Raising sheep is relatively easy, as they only need pasture to eat, shelter from bad weather, and protection from predators. Sheep's wool can be used to make yarn or other articles of clothing and their milk can be made into various types of cheeses and yogurt, though this is not normally done in the United States.

Sheep are naturally shy creatures and are extremely docile. If they are treated well, they will learn to be affectionate with their owner. If a sheep is comfortable with its owner, it will be much easier to manage and to corral into its pen if it's allowed to graze freely. Start with only one or two sheep; they are not difficult to manage but do require a lot of attention.

Breeds of Sheep

There are many different breeds of sheep—some are used exclusively for their meat and others for their wool. Six quality wool-producing breeds are as follows:

1. Cotswold Sheep—This breed is very docile and hardy and thrives well in pastures. It produces around 14 pounds of fleece per year, making it a very profitable breed for anyone wanting to sell wool.
2. Leicester sheep—This is a hardy, docile breed of sheep that is a very good grazer. This breed has 6-inch-long, coarse wool that is desirable for knitting. It is a very popular breed in the United States.
3. Merino sheep—Introduced to the United States in the early twentieth century, this small- to medium-sized sheep has lots of rolls and folds of fine white wool and produces a fleece anywhere between 10 and 20 pounds. It is considered a fine-wool specialist, and though its fleece appears dark in color, the wool is actually white or buff. It is a wonderful foraging sheep, is hardy, and has a gentle disposition, but is not a very good milk producer.
4. Oxford Down sheep—A more recent breed, these dark-faced sheep have hardy constitutions and good fleece.
5. Shropshire sheep—This breed has longer, more open, and coarser fleece than other breeds. It is quite popular in the United States, especially in areas that are more moist and damp, as they seem to better in these climates than other breeds of sheep.
6. Southdown sheep—One of the oldest breeds of sheep, they are popular for their good quality wool and are deemed the standard of excellence for many sheep owners. Docile, hardy, and good grazing on pastures, their coarse and light-colored wool is used to make flannel.

Housing Sheep

Sheep do not require much shelter—only a small shed that is open on one side (preferably to the south so it can stay warmer in the winter months) and is roughly 6 to 8 feet high. The shelter should be ventilated well to reduce any unpleasant smells and to keep the sheep cool in the summer. Feeding racks or mangers should be placed inside of the shed to hold the feed for the sheep. If you live in a colder region of the country, building a sturdier, warmer shed for the sheep to live in during the winter is recommended.

Straw should be used for the sheep's bedding and should be changed daily to make sure the sheep do not become ill from an unclean shelter. Especially for the winter

months, a dry pen should be erected for the sheep to exercise in. The fences should be strong enough to keep out predators that may enter your yard and to keep the sheep from escaping.

What Do Sheep Eat?

Sheep generally eat grass and are wonderful grazers. They utilize rough and scanty pasturage better than other grazing animals and, due to this, they can actually be quite beneficial in cleaning up a yard that is overgrown with un-desirable herbage. Allowing sheep to graze in your yard or in a small pasture field will provide them with sufficient food in the summer months. Sheep also eat a variety of weeds, briars, and shrubs. Fresh water should always be available for the sheep every time of year.

During the winter months especially, when grass is scarce, sheep should be fed on hay (alfalfa, legume, or clover hay) and small quantities of grain. Corn is also a good winter food for the sheep (it can also be mixed with wheat bran), and straw, salt, and roots can also be occasionally added to their diet. Good food

during the winter season will help the sheep grow a healthier and thicker wool coat.

Shearing Sheep

Sheep are generally sheared in the spring or early summer before the weather gets too warm. To do your own shearing, invest in a quality hand shearer and a scale on which to weigh the fleece. An experienced shearer should be able to take the entire wool off in one piece.

You may want to wash the wool a few days to a week before shearing the sheep. To do so, corral the sheep into a pen on a warm spring day (make sure there isn't a cold breeze blowing and that there is a lot of sunshine so the sheep does not become chilled). Douse the sheep in warm water, scrub the wool, and rinse. Repeat this a few times until most of the dirt and debris is out of the wool. Diffuse some natural oil throughout the wool to make it softer and ready for shearing.

The sheep should be completely dry before shearing and you should choose a warm—but not overly hot—day. If you are a beginner at shearing sheep, try to find an experienced sheep owner to show you how to properly hold and shear a sheep. This way, you won't cause undue harm to

the sheep's skin and will get the best fleece possible. When you are hand-shearing a sheep, remember to keep the skin pulled taut on the part where you are shearing to decrease the potential of cutting the skin.

Once the wool is sheared, tag it and roll it up by itself, and then bind it with twine. Be sure not to fold it or bind it too tightly. Separate and remove any dirty or soiled parts of the fleece before binding, as these parts will not be able to be carded and used.

Carding and Spinning Wool

To make the sheared wool into yarn you will need only a few tools: a spinning wheel or drop spindle and wool-cards. Wool-cards are rectangular pieces of thin board that have many wire teeth attached to them (they look like coarse brushes that are sometimes used for dogs' hair). To begin, you must clean the wool fleece of any debris, feltings, or other imperfections before carding it; otherwise your yarn will not spin correctly. Also wash it to remove any additional sand or dirt embedded in the wool and then allow it to dry completely. Then, all you need is to gather your supplies and follow these simple instructions:

should be removed to keep the ewe healthy and her udder free from infection.

To milk a ewe, secure her to a sturdy pole or hook with a short lead. Wash the utters gently to remove any contaminants. Place the milk bucket below the utters and squeeze the teats downward? rhythmically? until the milk begins to flow into the bucket. Allowing the ewe to eat from a feed bucket while being milked will help to keep her relaxed. Strain and refrigerate the milk immediately.

Diseases

The main diseases to which sheep are susceptible are foot rot and scabs. These are contagious and both require proper treatment. Sheep may also acquire stomach worms if they eat hay that has gotten too damp or has been lying on the floor of their shelter. As always, it is best to establish relationship with a veterinarian who is familiar with caring for sheep and have your flock regularly checked for any parasites or diseases that may arise.

Carding Wool

1. Grease the wool with rape oil or olive oil, just enough to work into the fibers.
2. Take one wool-card in your left hand, rest it on your knee, gather a tuft of wool from the fleece, and place it onto the wool-card so it is caught between the wired teeth of the card.
3. Take the second wool-card in your right hand and bring it gently across the other card several times, making a brushing movement toward your body.
4. When the fibers are all brushed in the same direction and the wool is soft and fluffy to the touch, remove the wool by rolling it into a small fleecy ball (roughly a foot or more in length and only 2 inches in width) and put it in a bag until it is used for spinning.

Note: Carded wool can also be used for felting, in which case no spinning is needed. To felt a small blanket, place large amounts of carded wool on either side of a burlap sack. Using felting needles, weave the wool into the burlap until it is tightly held by the jute or hemp fabrics of the burlap.

Spinning Wool

1. Take one long roll of carded wool and wind the fibers around the spindle.
2. Move the wheel gently and hold the spindle to allow the wool to "draw," or start to pull together into a single thread.

3. Keep moving the wheel and allow the yarn to wind around the spindle or a separate spool, if you have a more complex spinning wheel.
4. Keep adding rolls of carded wool to the spindle until you have the desired amount of yarn.

Note: If you are unable to obtain a spinning wheel of any kind, you can spin your carded wool by hand, although this will not produce the same tightness in your yarn as regular spinning. All you need to do is take the carded wool, hold it with one hand, and pull and twist the fibers into one, continuous piece. Winding the end of the yarn around a stick, spindle, or spool and securing it in place at the end will help keep your fibers tight and your yarn twisted.

If you want your yarn to be different colors, try dying it with natural berry juices or with special wool dyes found in arts and crafts stores.

Milking Sheep

Sheep's milk is not typically used in the United States for drinking, making cheese, or other familiar dairy products. Sheep do not typically produce milk year-round, as cows do, so milk will only be produced if you bred your sheep and had a lamb produced. If you do have a sheep that has given birth and the lamb has been sold or taken away, it is important to know how to milk her so her udders do not become caked. Some ewes will still have an abundance of milk even after their lambs have been weaned and this excess milk

Llamas

Llamas often make excellent pets and are a great source of wooly fiber (their wool can be spun into yarn). Llamas are being kept more and more by people in the United States as companion animals, sources of fiber, pack and light plow animals, therapy animals for the elderly, "guards" for other backyard animals, and good educational tools for children. Llamas have an even temperament and are very intelligent. Their intelligence and gentle nature make them easy to train, and their hardiness allows them to thrive well in both cold and warmer climates (although they can have heat stress in extremely hot and humid parts of the country).

Before you decide to purchase a llama or two for your yard, check your state requirements regarding livestock. In some places your property must also be zoned for livestock.

Llamas come in many different colors and sizes. The average adult llama is between 5½ and 6 feet tall and weighs between 250 and 450 pounds. Llamas, being herd animals, like the company of other llamas, so it is advisable that you raise a pair to keep each other company. If you only want to care for one llama, then it would be best to also have a sheep, goat, or other animal that can be penned with the llama for camaraderie. Although llamas can be led well on a harness and lead, never tie one up as it could potential break its own neck trying to break free.

Llamas tend to make their own communal dung heap in a particular part of their pen. This is quite convenient for cleanup and allows you to collect the manure, compost it, and use it as a fertilizer for your garden.

Feeding Llamas

Llamas can subsist fairly well on grass, hay (an adult male will eat about one bale per week), shrubs, and trees, much like sheep and goats. If they are not receiving enough nutrients, they may be fed a mixture of rolled corn, oats, and barley, especially during the winter season when grazing is not necessarily available. Make sure not to overfeed your llamas, though, or they will become overweight and constipated. You can occasionally give cornstalks to your llamas as an added source of fiber, and you may add mineral supplements to the feed mixture or hay if you want. Salt blocks are also acceptable to have in your llama pen, and a constant supply of fresh water is necessary. Nursing female llamas should receive a grain mixture until the cria (baby) is weaned.

Be sure to keep feed and hay off the ground. This will help ward off parasites that establish themselves in the feed and are then ingested by the llamas.

Housing Your Llamas

Llamas may be sheltered in a small stable or even a converted garage. There should be enough room to store feed and hay, and the shelter should be able to be closed off during wet, windy, and cold weather. Llamas prefer light, open spaces in which to live, so make sure your shed or shelter has large doors and/or big windows. The feeders for the hay and grain mixture should be raised above the ground. Adding a place where a llama can be safely restrained for toenail clippings and vet checkups will help facilitate these processes but is not absolutely necessary.

The llamas should be able to enter and exit the shelter easily and it is a good idea to build a fence or pen around the shelter so they do not wander off. A fence about four feet tall should be enough to keep your llamas safe and enclosed. If you happen to have both a male and female llama, it is necessary to have separate enclosures for them to stave off unwanted pregnancies.

Toenail Trimming

Llamas need their toenails to be trimmed so they do not twist and fold under the toe, making it difficult for the llama to move around. Laying gravel in the area where your llamas frequently walk will help to keep the toenails naturally trimmed, but if you need to cut them, be careful not to cut too deeply or you may cause the tip of the toe to bleed and this could lead to an infection in the toe. Use shears designed for this purpose to cut the nails. Use one hand to hold the llama's "ankle" just above where the foot bends. Hold the clippers in your other hand, cutting away from the foot toward the tip of the nail. The nail's are easiest to clip in the early morning or after a rain, since the wetness of the ground will soften them.

Shearing

It is important to groom and shear your llama, especially during hot weather. Brushing the llama's coat to remove dirt

and keep it from matting will not only make your llamas look clean and healthy but it will improve the quality of their coats. If you want to save the fibers for spinning into yarn, it is best to brush, comb, and use a hair dryer to remove any dust and debris from the llama's coat before you begin shearing.

Shearing is not necessarily difficult, but if you are a first-time llama owner, you should ask another llama farmer to teach you how to properly shear your llama. In order to shear your llama, you can purchase battery-operated shears to remove the fibers for sale or use. Different llamas will respond in different ways to shearing. Try holding the llama with a halter and lead in a smaller area to begin the shearing process. Do not completely remove the llama from any other llamas you have, though, as their presence will help calm the llama you are shearing. It is best to have another person with you to aid in the shearing (to hold the llama, give it treats, and offer any other help). When shearing a llama, don't shear all the way down to the skin. Allowing a thin coating of hair to cover the llama's body will help protect it from the sun and from being scratched when it rolls in the dirt.

Start by shearing a flat top the length of the llama's back. Next, taking the shears in one hand, move them in a downward position to remove the coat. Shear a strip the length of the neck from the chin to the front legs about 3 inches wide to help cool the llama. Shearing can take a long time, so it may be necessary for both you and the llama to take a break. Take the llama for a quiet walk and allow it to go to the bathroom so it will not become antsy during the rest of the shearing process.

Collect the sheared fibers in a container and make sure you are working on a clean floor so you can collect any excess fibers and use them for spinning. Do not store the fiber in a plastic bag, as moisture can easily accumulate, ruining the fiber and making it unusable for spinning.

Caring for the Cria

Baby llamas, called cria, require some additional care in their first few days of life. It is important for the cria to receive the colostrum milk from their mothers, but you may need to aid in this process. Approach the mother llama and pull gently on each teat to remove the waxy plugs covering the milk holes. Sometimes, you may need to guide the cria into posi-

tion under its mother in order for it to start nursing.

Weigh the cria often (at least for the first month) to see that it's gaining weight and growing strong and healthy. A bathroom scale, hanging scale, or larger grain scale can be used for this.

If the cria seems to need extra nourishment, goat or cow milk can be substituted during times when the mother llama cannot produce enough milk for the cria. Feed this additional milk to the cria in small doses, several times a day, from a milking bottle.

Diseases

Llamas are prone to getting worms and should be checked often to make sure they do not have any of these parasites. There is special worming paste that can be mixed in with their food to prevent worms from infecting them. You should also establish a relationship with a good veterinarian who knows about caring for llamas and can determine if there are any other vaccinations necessary in order to keep your llamas healthy. Other diseases and pests that can affect llamas are tuberculosis, tetanus, ticks, mites, and lice.

Using Llama Fibers

Llama fiber is unique from other animal fibers, such as sheep's wool. It does not contain any lanolin (an oil found in sheep's wool); thus, it is hypoallergenic and not as greasy. How often you can shear your llama will depend on the variety of llama, its health, and environmental conditions. Typically, though, every year llamas grow a fleece that is 4 to 6 inches long and that weighs between 3 and 7 pounds. Llama fiber can be used like any other animal fiber or wool, making it the perfect substitute for all of your fabric and spinning needs.

Llama fiber is made up of two parts: the undercoat (which provides warmth for the llama) and the guard hair (which protects the llama from rain and snow). The undercoat is the most desirable part to use due to its soft, downy texture, while the coarser guard hair is usually discarded.

Gathering llama hair is easy. To harvest the fiber, you must shear the llama. However, the steps involved in shearing when you are gathering the fiber are slightly different than when you are simply shearing to keep the llama cooler in the summer months. To shear a llama for fiber collection:

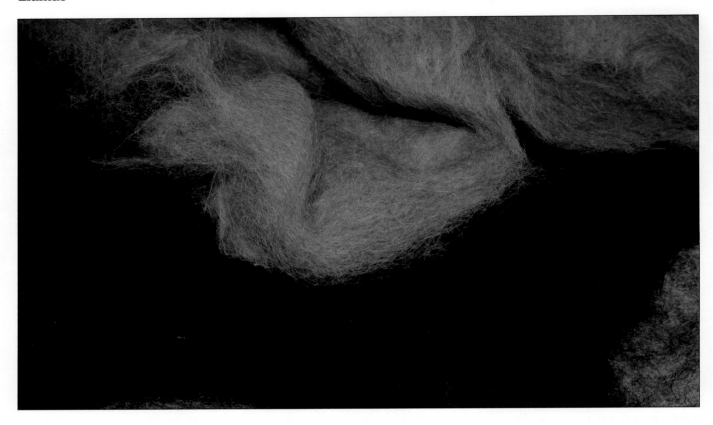

1. Clean the llama by blowing and brushing until the coat is free from dirt and debris.
2. Wash the llama. Be sure to rinse out all of the soap from the hair and let the llama air-dry.
3. You can use scissors or commercial clippers to shear the llama. Start at the top of the back, behind the head and neck and work backwards. If using clippers, sheer with long sweeping motions, not short jerky ones. If using scissors, always point them downward. Leave about an inch of wool on the llama for protection against the sun and insect bites. You can sheer just the area around the back and belly (in front of the hind legs and behind the front legs) if your main purpose is to offer the llama relief from the heat. Or you can sheer the entire llama—from just below the head, down to the tail—to get the most wool. Once the shearing is complete, skirt the fleece by removing any little pieces or belly hair from the shorn fleece.

The fiber can be hand-processed or sent to a mill (though sending the fibers to a mill is much more expensive and is not necessary if you have only one or two llamas). Processing the fiber by hand is definitely more cost-effective but you will initially need to invest in some equipment (such as a spinning wheel, drop spindle, or felting needle).

To process the fiber by hand:

1. Pick out any remaining debris and unwanted (coarse) fibers.
2. Card the fiber. This helps to separate the fiber and will make spinning much easier. To card the fiber, put a bit of fiber on one end of the cards (standard wool cards do the trick nicely) and gently brush it until it separates. This will produce a rolag (log) of fiber.
3. Once the fiber is carded, you can use it in a few different ways:
   a. Wet felting: To wet felt, lay the fiber out in a design between two pieces of material and soak it in hot, soapy water. Then, agitate the fiber by rubbing or rolling it. This will cause it to stick together. Rinse the fiber in cold water. When it dries, you will have produced a strong piece of felt that can be used in many crafting projects.
   b. Needle felting: For this type of manipulation, you will need a felting needle (available at your local arts and crafts or fabric store). Lay out a piece of any material you want over a pillow or Styrofoam piece. Place the fiber on top of the material in any design of your choosing. Push the needle through the fiber and the bottom material and then gently draw it back out. Continue this process until the fiber stays on the material of its own accord. This is a great way to make table runners or hanging cloths using your llama fiber.
   c. Spinning: Spinning is a great way to turn your llama fiber into yarn. Spinning can be accomplished by using either a spinning wheel or drop spindle, and a piece of fiber that is either in a batt, rolag, or roving. A spinning wheel, while larger and more expensive, will easily help you to turn the fiber into yarn. A drop spindle is convenient because it is smaller and easier to transport, and if you have time and patience, it will do just as good a job as the spinning wheel. To make yarn, twist two or more pieces of spun wool together.
   d. Other uses: Carded wool can also be used to weave, knit, or crochet.

If you become very comfortable using llama fiber to make clothing or other craft items, you may want to try to sell these crafts (or your llama fiber directly) to consumers. Fiber crafts may be particularly successful if sold at local craft markets or even at farmers' markets alongside your garden produce.

Cows

Raising dairy cows is difficult work. It takes time, energy, resources, and dedication. There are many monthly expenses for feeds, medicines, vaccinations, and labor. However, when managed properly, a small dairy farmer can reap huge benefits, like extra cash and the pleasure of having fresh milk available daily.

Breeds

There are thousands of different breeds of cows, but what follows are the three most popular breeds of dairy cows.

The Holstein cow has roots tracing back to European migrant tribes almost two thousand years ago. Today, the breed is widely popular in the United States for their exceptional milk production. They are large animals, typically marked with spots of jet black and pure white.

The Ayrshire breed takes its name from the county of Ayr in Scotland. Throughout the early nineteenth century, Scottish breeders carefully crossbred strains of cattle to develop a cow well suited to the climate of Ayr and with a large flat udder best suited for the production of Scottish butter and cheese. The uneven terrain and the erratic climate of their native land explain their ability to adapt to all types of surroundings and conditions.

Ayrshire cows are not only strong and resilient, but their trim, well rounded outline, and red and predominantly white color has made them easily recognized as one of the most beautiful of the dairy cattle breeds.

The Jersey breed is one of the oldest breeds, originating from Jersey of the Channel Islands. Jersey cows are known for their ring of fine hair around the nostrils and their milk rich in butterfat. Averaging to a total body weight of around 900 pounds, the Jersey cow produces the most pounds of milk per pound of body weight of all other breeds.

Housing

There are many factors to consider when choosing housing for your cattle, including budget, preference, breed, and circumstance.

Free stall barns provide a clean, dry, comfortable resting area and easy access to food and water. If designed properly, the cows are not restrained and are free to enter, lie down, rise, and leave the barn whenever they desire. They are usually built with concrete walkways and raised stalls with steel dividing bars. The floor of the stalls may be covered with various materials, ideally a sanitary inorganic material such as sand.

A flat barn is another popular alternative, which requires tie-chains or stanchions to keep the cows in their stalls. However, it creates a need for cows to be routinely released into an open area for exercise. It is also very important that the stalls are designed to fit the physical characteristics of the cows. For example, the characteristically shorter Jersey cows should not be housed in a stall designed for much larger Holsteins.

A compost-bedded pack barn, generally known as a compost dairy barn, allows cows to move freely, promising increased cow comfort. Though it requires exhaustive pack and ventilation management, it can notably reduce manure storage costs.

Grooming

Cows with sore feet and legs can often lead to losses from milk production, diminished breeding efficiency, and lameness. Hoof trimming is essential to help prevent these outcomes, though it is often very labor intensive, allowing it to be easily neglected. Hoof trimming should be supervised or taught by a veterinarian until you get the hang of it.

A simple electric clipper will keep your cows well-groomed and clean. Mechanical cow brushes are another option. These

⌃ To milk a cow, sit on a stool and wash the udders with warm water. Place a pail under the teats and begin squeezing the top of the teat with the thumb and forefinger. Tighten the other fingers, gently squeezing downward. Release and repeat rhythmically until the milk slows and the udder is soft to the touch.

brushes can be installed in a free-stall dairy barn, allowing cows to groom themselves using a rotating brush that activates when rubbed against.

Feeding and Watering

In the summer months, cows can receive most of their nutrition from grazing, assuming there is plenty of pasture. You may need to rotate areas of pasture so that the grass has an opportunity to grow back before the cows are let loose in that area again. Grazing pastures should include higher protein grasses, such as alfalfa, clover, or lespedeza. During the winter, cows should be fed hay. Plan to offer the cows 2 to 3 pounds of high-quality hay per 100 pounds of body weight per day. This should provide adequate nutrition for the cows to produce 10 quarts of milk per day, during peak production months. To increase production, supplement feed with ground corn, oats, barley, and wheat bran. Proper mixes are available from feed stores. Allowing cows access to a salt block will also help to increase milk production.

Water availability and quantity is crucial to health and productivity. Water intake varies,

however it is important that cows are given the opportunity to consume a large amount of clean water several times a day. Generally, cows consume 30 to 50 percent of their daily water intake within an hour of milking. Water quality can also be an issue. Some of the most common water quality problems affecting livestock are high concentrations of minerals and bacterial contamination. Send out 1 to 2 quarts of water from the source to be tested by a laboratory recommended by your veterinarian.

If you intend to run an organic dairy, cows must receive feed that was grown without the use of pesticides, commercial fertilizers, or genetically-modified ingredients along with other restrictions.

Breeding

You may want to keep one healthy bull for breeding. Check the bull for STDs, scrotum circumference, and sperm count before breeding season begins. The best cows for breeding have large pelvises and are in general good health. An alternative is to use the artificial insemination (A. I.) method. There are many advantages to A. I., including the prevention of spreading infectious genital diseases, the

early detection of infertile bulls, elimination of the danger of handling unruly bulls, and the availability of bulls of high genetic material. The disadvantage is that implementing a thorough breeding program is difficult and requires a large investment of time and resources. In order to successfully execute an A. I. program, you may need a veterinarian's assistance in determining when your cows are in heat. Cows only remain fertile for twelve hours after the onset of heat, and outside factors such as temperature, sore feet, or tie-stall or stanchion housing can drastically hinder heat detection.

Calf Rearing

The baby calf will be born approximately 280 days after insemination. Keep an eye on the cow once labor begins, but try not to disturb the mother. If labor is unusually long (more than a few hours), call a veterinarian to help. It is also crucial that the newborns begin to suckle soon after birth to receive ample colostrum, the mother's first milk, after giving birth. Colostrum is high in fat and protein with antibodies that help strengthen the immune system, though it is not suitable for human consumption. When you choose to wean the calf will depend on whether the calf is being raised for dairy or meat, forage availability, and the condition of both the mother and calf. It's important to research the breed you are

raising and consider the situation of your own farm and the needs of the particular animals when deciding on when to wean and whether to separate the mother from the calf. contracting any germs from other animals.

You can teach the calf to drink from a bucket by gently pulling its head toward the pail. A calf should consume about 1 quart of milk for every 20 pounds of body weight. A calf starter can be used to help ensure proper ruminal development. You can find many types of starters on the market, each meeting the nutritional requirements for

calves. Begin milking young cows as soon as they are separated from the mother. This will get them used to the process while they are small enough to be more manageable.

Common Diseases

Pinkeye and foot rot are two of the most prevalent conditions affecting all breeds of cattle of all ages year-round. Though both diseases are non-fatal, they should be taken seriously and treated by a qualified veterinarian.

Wooden tongue occurs worldwide, generally appearing in areas where there is a copper deficiency or the cattle graze on land with rough grass or weeds. It affects the tongue, causing it to become hard and swollen so that eating is painful for the animal. Surgical intervention is often required.

Brucellosis or bangs is the most common cause of abortion in cattle. The milk produced by an infected cow can also contain the bacteria, posing a threat to the health of humans.

Pigs

Pigs can be farm-raised on a commercial scale for profit, in smaller herds to provide fresh, homegrown meat for your family or to be shown and judged at county fairs or livestock shows. Characterized by their stout bodies, short legs, snouts, hooves, and thick, bristle-coated skin, pigs are omnivorous, garbage-disposing mammals that, on a small farm, can be difficult to turn a profit on but yield great opportunities for fair showmanship and quality food on your dinner table.

Breeds

Pigs of different breeds have different functionalities—some are known for their terminal sire (the ability to produce offspring intended for slaughter rather than for further breeding) and have a greater potential to pass along desirable traits, such as durability, leanness, and quality of meat, while others are known for their reproductive and maternal qualities. The breed you choose to raise will depend on whether you are raising your pigs for show, for profit, or to put food on your family's table.

Eight Major U.S. Pig Breeds

1. Yorkshire—Originally from England, this Large White breed of hog has a long frame, comparable to the Landrace. They are known for their quality meat and mothering ability and are likely the most widely distributed breed of pig in

PIG TERMINOLOGY

pig, hog, or swine	Refers to the species as a whole or any member thereof.
shoat or piglet (or "pig" when species is referred to as "hog")	Any unweaned or immature young pigs.
sucker	A pig between birth and weaning.
runt	An unusually small and weak piglet. Often one per litter.
boar or hog	A male pig of breeding age.
barrow	A male pig castrated before reaching puberty.
stag	A male pig castrated later in life.
gilt	A young female not yet mated (farrowed) or has birthed fewer than two litters.
sow	An active breeding female pig.

the world. Farmers will also find that the Yorkshire breed generally adapts well to confinement.

2. Landrace—This white-haired hog is a descendent from Denmark and is known for producing large litters, supplying milk, and exhibiting good maternal qualities. The breed is long-bodied and short-legged with a nearly flat arch to its back. Its long, floppy ears are droopy and can cover its eyes.

3. Chester—Like the Landrace, this popular white hog is known for its mothering abilities and large litter size. Originating from cross breeding in Pennsylvania, Chester hogs are medium-sized and solid white in color.

4. Berkshire—Originally from England, the black and white Berkshire hog has

fatback are cut off in one piece, parallel with the back just below the tenderloin muscle on the rear part of the middling. Remove the fat on the top of the loin, but do not cut into the loin meat. The lean meat is excellent for canning or it may be used for chops or roasts and the fatback for lard. The remainder should then be trimmed for middling or bacon. Remove the ribs cutting as close to them as possible. If it is a very large side, it may be cut into two pieces. Trim all sides and edges as smoothly as possible.

Ham

Cut off the foot 1 inch below the hock joint. All rough and hanging pieces of meat should be trimmed from the ham. It should then be trimmed smoothly, exposing as little lean meat as possible, because the curing hardens it. All lean trimmings should be saved for sausage and fat trimmings for lard. The other half of the carcass should be cut up in similar manner.

Loins

Separate the loin from the belly by sawing through the ribs, starting at the point of greatest curvature of the fourth rib. Skin the fat from the loin, leaving the lean muscle barely covered with the fat. The loins can be boned out and used for sausage if a large amount of that is desired, or if the weather will not permit holding them as fresh meat they can be given a middle cure as boneless

loins. The loin is best adapted for the pork (loin) roast or for pork chops. The latter are cut in such a way as to have the rib end in each alternate piece or chop.

Meat trimmings and fat trimmings

After the carcass has been cut up and the pieces are trimmed and shaped properly for the curing process, there are many pieces of lean meat, fat meat, and fat which can be used for making sausage and lard. The fat should be separated from the lean and used for lard. The meat should be cut into convenient-sized pieces to pass through the grinder.

Rendering lard

The leaf fat makes lard of the best quality. The back strip of the side also makes good lard, as do the trimmings of the ham, shoulder, and neck. Intestinal or gut fat makes an inferior grade and is best rendered by itself. This should be thoroughly washed and left in cold water for several hours before rendering, thus partially eliminating the offensive odor. Leaf fat, back strips, and fat trimmings may be rendered together. If the gut is included, the lard takes on a very offensive odor.

First, remove all skin and lean meat from the fat trimmings. To do this cut the fat into strips about 1 ½ inches wide, then place the strip on the table, skin down, and cut the fat from the skin. When a piece of skin large enough to grasp is freed from the fat, take it

in the left hand and with the knife held in the right hand inserted between the fat and skin, pull the skin. If the knife is slanted downward slightly, this will easily remove the fat from the skin. The strips of fat should then be cut into pieces 1 or 1 ½ inches square, making them about equal in size so that they will try out evenly.

Pour into the kettle about a quart of water, then fill it nearly full with fat cuttings. The fat will then heat and bring out the grease without burning. Render the lard over a moderate fire. At the beginning the temperature should be about 160°F, and it should be increased to 240°F. When the cracklings begin to brown, reduce the temperature to 200°F or a little more, but not to exceed 212°F, in order to prevent scorching. Frequent stirring is necessary to prevent burning. When the cracklings are thoroughly browned, and light enough to float, the kettle should be removed from the fire. Press the lard from the cracklings. When the lard is removed from the fire allow it to cool a little. Strain it through a muslin cloth into the containers. To aid cooling, stir it, which also tends to whiten it and make it smooth.

Lard which is to be kept for a considerable time should be placed in air-tight containers and stored in the cellar or other convenient place away from the light, in order to avoid rancidity. Fruit jars make excellent containers for lard, because they can be completely sealed. Glazed earthenware containers, such

⧩ **Bacon comes from the pork belly.**

≪ A slab of pork lard.

as crocks and jars, may be also be used. All containers should be sterilized before filling, and if covers are placed on the crocks or jars, they also should be sterilized before use. Lard stored in air-tight containers away from the light has been found to keep in perfect condition for a number of years.

When removing lard from a container for use, take it off evenly from the surface exposed. Do not dig down into the lard and take out a scoopful, as that leaves a thin coating around the sides of the container, which will become rancid very quickly through the action of the air.

Curing Pork

The first essential in curing pork is to make sure that the carcass is thoroughly cooled, but meat should never be allowed to freeze either before or during the period of curing.

The proper time to begin curing is when the meat is cool and still fresh, or about twenty-four to thirty-six hours after killing. See page 145 for pork curing suggestions.

⩗ An old-fashioned kettle for rendering lard.

SELECTING QUALITY MEAT

As a general rule, the best meat is that which is moderately fat. Lean meat tends to be tough and tasteless. Very fat meat may be good, but is not economical. The butcher should be asked to cut off the superfluous suet before weighing it.

1. *Beef*. The flesh should feel tender, have a fine grain, and a clear red color. The fat should be moderate in quantity, and lie in streaks through the lean. Its color should be white or very light yellow. Ox beef is the best, heifer very good if well fed, cow and bull, decidedly inferior.

2. *Mutton*. The flesh, like that of beef, should be a good red color, perhaps a shade darker. It should be fine-grained, and well mixed with fat, which ought to be pure white and firm. The mutton of the black-faced breed of sheep is the best, and may be known by the shortness of the shank; the best age is about five years, though it is seldom to be had so old. Whether mutton is superior to either ram or ewe, and may be distinguished by having a prominent lump of fat on the broadest part of the inside of the leg. The flesh of the ram has a very dark color and is of a coarse texture; that of the ewe is pale, and the fat yellow and spongy.

3. *Veal*. Its color should be white, with a tinge of pink; it ought to be rather fat, and feel firm to the touch. The flesh should have a fine delicate texture. The leg-bone should be small; the kidney small and well covered with fat. The proper age is about two or three months; when killed too young it is soft, flabby, and dark colored. The bull-calf makes the best veal, though the cow-calf is preferred for many dishes on account of the udder.

4. *Lamb*. This should be light-colored and fat, and have a delicate appearance. The kidneys should be small and imbedded in fat, the quarters short and thick, and the knuckle stiff. When fresh, the vein in the fore quarter will have a bluish tint. If the vein look green or yellow, it is a certain sign of staleness, which may also be detected by smelling the kidneys.

5. *Pork*. Both the flesh and the fat must be white, firm, smooth, and dry. When young and fresh, the lean ought to break when pinched with the fingers, and the skin, which should be thing, yield to the nails. The breed having short legs, thick neck, and small head, is the best. Six months is the right age for killing, when the leg should not weigh more than 6 or 7 lbs.

Part 6

The Workshop

"Every nail driven should be as another rivet in the machine of the universe."

—Henry David Thoreau

Geometrical Tools	342
Holding Tools	352
Workshop Furniture	359
Houses, Runs, and Coops for Poultry	371
Gates and Fences	381
Sheds, Tool Houses, and Workshops	385
Kitchen Furniture	391
Bedroom Furniture	399

Geometrical Tools

Woodworking is one of the most basic and most useful skills a person dedicated to self-sufficiency can master. Once you understand the basics of what tools to use and how to use them, there's almost no limit to what you can construct. Here you will find a wide selection of the hand tools, materials, and processes used in woodworking, as well as diagrams and instructions for dozens of specific projects. Much of this section is adapted from the writings of Paul N. Hasluck, whose wealth of knowledge on the subject has served as the foundation for generations of woodworkers.

Classifying Tools

Tools can be categorized by their functions, as follows:

1. Geometrical tools for laying off and testing work: they include rules, straight-edges, gauges, etc.
2. Tools for holding and supporting work: such tools are benches, vices, stools, etc.
3. Paring or shaving tools, like chisels, spokeshaves, planes, etc.
4. Saws.
5. Percussion or impelling tools, such as hammers, mallets, screwdrivers, and (combined with cutting) hatchets and axes.
6. Boring tools, including gimlets and brace-bites, etc.
7. Abrading and scraping tools, such as rasps, scrapers, glasspaper, and implements such as whetstones, for sharpening edged tools.

The most useful tools in these categories are discussed in the following pages.

Tools for Marking and Scribing

The simplest of these is the pencil. Sharpen a flat oval section to a chisel edge; if sharpened to a point, the pencil wears away quickly and will only mark a fine, solid line for a few minutes. The greater surface area of lead in the chisel edge makes it last for longer before it requires resharpening. Steel scribing and marking tools are illustrated by Figures. 1 to 3. The chisel end marking awl (Figure 1) and the striking knife (Figure 2) are used for all purposes of scribing and marking smooth work, where an indented line works better than a black line, the scratch providing a good starting point for edge tools. A striking knife (Figure 3) can be made by grinding down an old table knife.

Figure 1—Chisel-end Marking Awl

Figure 2—Striking Knife and Marking Awl

Straight-edge

A straight-edge 15 feet long, 6 inches wide, and 1¼ inches thick is large enough for all practical purposes of the joiner, mason, bricklayer, engineer, etc. If you're making your own, the best material is pine, as it is the least affected (permanently) by change of temperature or weather. The pine board must be cut from a straight-grown tree, since a board from a crooked trunk will not keep parallel and straight for any length of time, owing to the grain crossing and recrossing its (thickness) edge. Straight-edges are made from all parts of boards cut from whole logs, but you cannot assume they will keep perfectly straight and true for any length of time.

To test the truth of a straight-edge this size, get a clean board 1 foot longer and about 7 inches or 8 inches wide. Lay the straight strip at about the center of the board, and with a sharp pencil draw a line on the board along the trued edge of the strip, keeping the side close to the board, and making the line as fine as possible. Now turn the strip of the line, and if the trued

Figure 3—Homemade Striking Knife

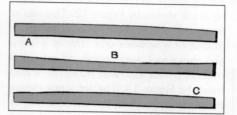

Figure 4—Whitworth Method of Testing Straight-edges

edge is perfectly straight the line also will appear so. If the line is wavy, the edge must be planed until only one line is made when marked and tested from each side; mark a fresh line for each test, otherwise you can become confused and inaccurate. With one edge now perfectly true, you can proceed with the other edge. Set a sharp gauge to the required width, and mark the second edge lightly on each side of the rule, working the gauge from the true edge; then the wood is planed off to the gauge marks, and the second edge tested as to its being true with the first one, using the pencil line, as before. An even more precise test than the gauge line for parallelism can be done by using a pair of calipers. The points of the calipers are drawn along the edges, and if they are perfectly parallel, there will be no easy or hard places, the presence of which might possibly not be detected by the pencil line. If the edges will stand both these tests, the strip is perfectly straight and parallel.

Whitworth Method of Testing Straight-edges

Sir J. Whitworth's famous method of trueing engineers' straight-edges should interest any woodworker. Three straight-edges are prepared individually, and each is brought to a moderate state of accuracy; two of them, A and B (Figure 4), are compared with each other by placing them

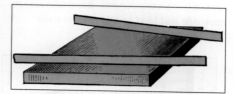

Figure 5—Testing Surface with Straight-edges

edge to edge, and any irregularities found are removed. The process is then repeated until A and B fit each other perfectly. The third straight-edge, C, now is compared with both A and B, and when it fits both perfectly, then there is no doubt whatsoever that the three are straight. Why this is the case becomes obvious when you remember that though A may be rounded instead of being straight, and B may be hollow sufficiently to make them fit each other perfectly, it is impossible for C to fit both the rounded and the hollow straight-edge.

Testing Surfaces with Straight-edges

How surfaces are tested for winding with straight-edges is shown by Figure 5, from which it is obvious that if the work has warped ever so slightly a true straight-edge must disclose the fact, as it could not then lie flat on its edge across the work. If the board is in winding, each straight-edge will magnify the error. If the winding is wavy, the edges will touch at certain points, and in other places light will be seen between them and the work. Taking a sight from one straight-edge to the other is another test you can perform.

Rules

A 2-feet four-fold boxwood rule (Figure 6) is the best for the all-round purposes of the joiner; and for those who can use the slide rule, the tool shown in Figure 7 would be handy. A simple 2-feet two-fold rule (Figure 8) is useful, but the rule with double arch joints shown by Figure 9 will be your best bet. The average worker will find a simple rule preferable to an elaborate one. Figure 10 shows a combined rule and spirit level, the rule joint also being set out to serve as a protractor. This tool may prove useful in special circumstances, but its use as a spirit level is not recommended. It is preferable to have rule and level as two distinct tools.

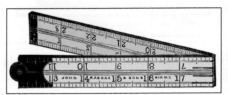

Figure 6—Two-feet Four-fold Rule

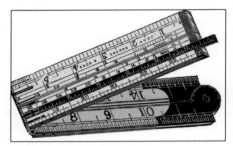

Figure 7—Rule with Brass Slide

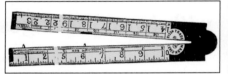

Figure 8—Two-feet Two-fold Rule

Dividing a Board with a Rule

A simple way to divide a board of any width into any number of parts is illustrated by Figure 11. Suppose a board of 9 inches is to be cut into six equal parts; place the 1-foot rule so that its ends touch the opposite edges of board, as shown in Figure 11; draw a line right across, and upon this line mark off from the rule every 2 inches, as 2, 4, 6, etc. Remove the rule, and draw lines parallel with the edge of the board, intersecting with the marks upon the oblique line, thus obtaining six parts, each really 1½ inches wide. The principle of this is simple: 2 inches is the one-sixth part of 1 foot, and whatever be the slant of the rule across the board (and the narrower the board, the greater the slant) each 2-inch mark must denote a one-sixth less than 24 inches length or width is to be divided into eight parts; then as 3 inches is

one-eighth of 2 feet, use a 2-foot rule in the same way as before, and mark off at every 3 inches.

Squares and Bevels

Woodworkers constantly use squares for setting out and testing work. The simplest is the try square. A combination try and miter square is shown by Figure 12. This has an iron stock hollowed out to lower its weight to that of a wooden one. This is a useful and cheap tool, very unlikely to get out of truth. A patent adjustable try square is illustrated by Figure 13. The set screw clamps the blade in the stock just where it may be most convenient for awkward work such as puttingbutts, locks, and other fittings on doors and windows. The graduated blade is very useful. The sliding bevel is handy for setting off angles in duplicate, since by using the set screw, the blade can be made to assume any angle with the stock. Figure 14 shows a bevel with a simple ebony stock, and Figure 15 shows one with an ebony stock framed in brass, this protection keeping the edges true for an almost unlimited period. The joiner's steel square is a mere right angle of steel, sometimes nickel plated, graduated in inches, ¼ inch, and ¹/₁₆ inch.

Testing and Correcting Try square

A carpenter's try square that is thought to be untrue may be tested in the following way. Get a piece of board with an edge that has been proven to be straight, apply the square as shown at A (Figure 16), and draw

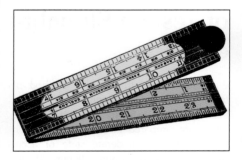

Figure 9—Rule with Double Arch Joints

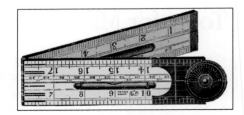

Figure 10—Rule with Spirit Level

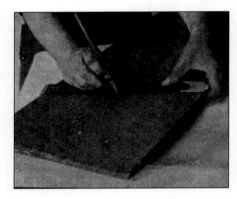

Figure 11—Dividing with Rule

a line; then turn the square as at B, and if it is true the blade should fit the line; if it is less than a right angle it will be as seen in C and D (Figure 16), and if more than a right angle the defect will be as indicated at E and F (Figure 16). If the blade has moved or has been knocked out of truth through a fall, it should be knocked back into its proper position, and, when true, the rivets should be tightened by careful hammering. If the blade is too fast in the stock to be knocked back, it should be filed true.

Crenellated Squares

A crenellated square has a tongue in which there is a series of crenellations or notches at the graduations. It is especially

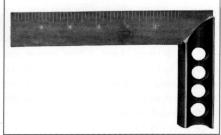

Figure 12—Iron Frame Try Square

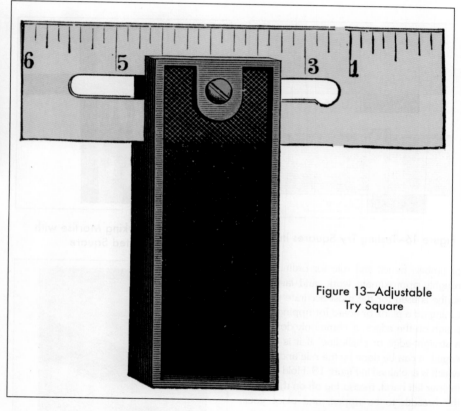

Figure 13—Adjustable
Try Square

useful in marking off mortises, though it is available for all other ordinary applications. Three sides of a piece of timber can be set out without moving the work. To use this square, say in marking out a mortise or tenon, take it in the left hand and lay its tongue upon the surface of the work, as in Figure 17. The lower end of the main arm is lowered for 2 inches or so from the surface to get a better purchase, and then an awl, held in the right hand, is placed in a notch at the correct distance from the edge to mark the left-hand edge of the mortise or left hand face of the tenon as the case may be. Then push the square forward, pressing it down gently on the work to make one mark. Next, replace the square and place the awl in another notch at the thickness of the tenon or width of the mortise to make a second mark. Horizontals are drawn by means of the smooth edge of the tongue, as shown in Figure 17.

Marking Work for Sawing

The chalk line, pencil and rule, and scribe are all used for marking lines to guide the saw. The chalk line is used for long pieces

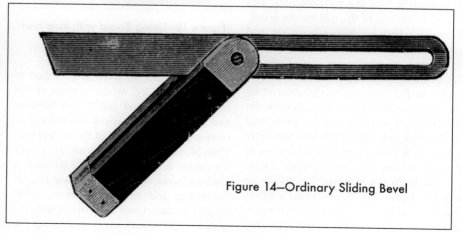

Figure 14—Ordinary Sliding Bevel

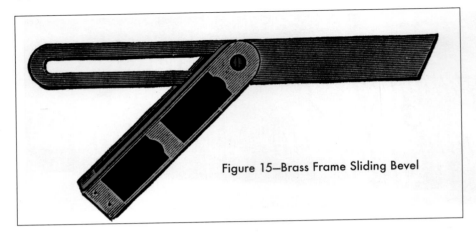

Figure 15—Brass Frame Sliding Bevel

and tenons. The saw can then be made to cut close outside the scribed line, allowing for sufficient margin of material to be removed with the plane; or the saw can pass along the scribed line, as in cutting dovetails and tenons, with no after-finish required. In either case, use the scribed line over the pencil-marked one, because the cutting can be done much more accurately in the first case than the latter. Also, when the end of a piece of timber has to be squared with the plane, there is, besides having greater accuracy, much less risk of spalting or breaking out of the grain occurring with scribed lines than with pencil-marked lines.

Marking and Cutting Gauges

The carpenter draws a line at a short distance from, and parallel to, the edge of a board by using a rule and pencil, the method previously described and seen in Figure 18. The use of the pencil or marking

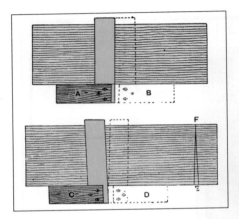

Figure 16—Testing Try Squares in Truth

Figure 17—Marking Mortise with Crenellated Square

of timber; pencil and rule for ordinary and roughly approximate work, and lastly, the scribe is used for the most accurate sawing. Lining off a plank or board for ripping, when rough on the edges, is commonly done with a straight-edge or chalk line. If it is square-edged, it can be done by the rule and pencil, which is explained in Figure 18. Hold the rule in your left hand, measuring off on the board the breadth to be ripped, and place your forefinger against the edge to act as a fence. The pencil is held in your right hand and placed at the end of the rule on the board. Then move both hands simultaneously, and you can trace the required line backward or forward, as may be the case. Lines for cross-cutting, when square across or at right angles to the edge, are easily obtained by the square, as long as you keep its blade flat on the board or plank and the stock hard to the edge (see Figure 19). Use the miter square for lines at an angle of 45 degrees to the edge (Figure 20), and for other angles, set and apply the bevel-stock in the same way (see Figure 21). The bevel-stock differs from the square only in that it has a moveable blade and is capable of being adjusted at any desired angle with the stock by using

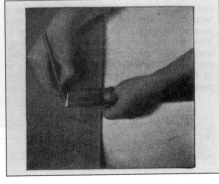

Figure 18—Lining Board with Rule and Pencil

a screw. To use the chalk line method of marking (Figure 22), whiten a piece of fine cord with chalk and pull it taut between two points whose positions are marked to correspond with the end of the cutting line. Then lift the chalk line vertically at or near the center, and as it's suddenly released, it chalks a perfectly straight and fine line on the timber, and furnishes a correct guide for you to saw. Lines are marked with the timber scribe in such cases as squaring the ends of planed stuff or in marking dovetails

Figure 19—Squaring Line on Board

Figure 20—Marking Miter Line on Board

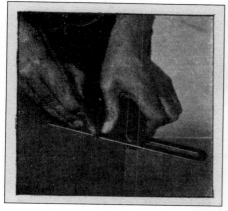

Figure 21—Using Sliding Bevel

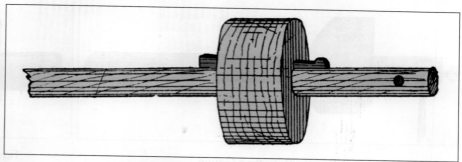

Figure 23—Pencil Gauge with Round Stem

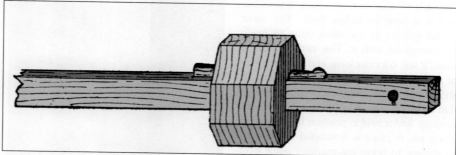

Figure 24—Pencil Gauge with Square Stem

gauge is an advantage over this method. Figures 23 and 24 show that there are two ways to make the pencil gauge. It can be made of any hard wood, such as beech. The stem can be round (Figure 23) or square (Figure 24), and the head may be round or octagonal. Make sure the head can slide up and down the stem easily, but without side-play. The gauge may be made to use up odd pieces of lead pencil, and these should be sharpened (with a chisel) to a wedge-shaped point. Figures 25 and 26 show a pencil gauge made from a broken rule fitted into a block so as to run easily, and secured at any distance (as indicated by the rule's edge) using a thumbscrew. A represents a block of birch wood, 1½ inches by 1 inch by 1 inch, mortised so it receives the rule. B is a 5-inch length of an ordinary rule, with a slot C just large enough to fit screw D, which is fixed to block A. The thickness of the wood between the washer and the rule should be only 1⅛ inches, to allow a little pliability. You can make a cutting or scratch gauge similarly by inserting a pin at E, exactly over the first ¹⁄₁₆ inch, as that

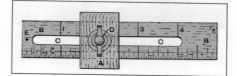

Figure 25—Rule Pencil and Cutting Gauge

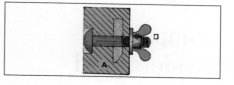

Figure 26—Section through Rule Gauge

Figure 27—Improved Pencil Gauge

Figure 28—Ordinary Marking Gauge

Figure 22—Using Chalk Line

distance is always allowed for. Store-bought marking and cutting gauges are illustrated by Figures 27 to 32. A beechwood pencil gauge is shown by Figure 27, a marking gauge with a steel point by Figure 28, an improved cutting gauge for scribing deep lines in Figure 29, and mortise gauges for scribing mortise holes and tenons by Figures 30 to 32. The mortise gauges are of ebony and brass, the one illustrated by Figure 32 having a stem of brass. The ordinary marking gauge is shown in Figure 33.

Panel Gauges

A panel gauge (Figure 34) is used to mark a line parallel to the true edge of a panel, or any piece of wood that is too wide for the ordinary gauge to take inches. The stock is of maple, beech, or similar wood. It is 1 inch thick, and has a ³⁄₈-inch by ³⁄₈-inch rebate at the bottom. A mortise is made for the stem to pass through, and another is made at the side for the wedge. The wedge should

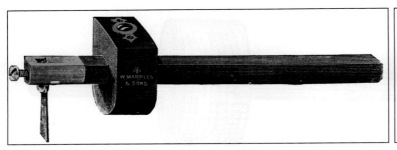

Figure 29—Cutting Gauge

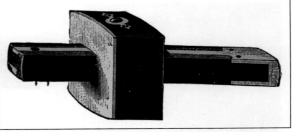

Figure 30—Square Mortise Gauge

be made of box-wood or ebony if possible, and is a bare ¼ inches thick. The taper of the mortise in the stock must be made to correspond with it. The stem should be about 2 feet 6 inches long, and may be made of a piece of straight-grained mahogany. It should not fit the mortise too tightly, just so that it can be moved with the hands without tapping, and is held in position by the wedge when set. A piece is dovetailed in the end, as shown, to bring the marking point level with the bottom of the rebate. The stem may be made square if possible, or if the rounded mortise presents a difficulty. Make sure the stock is well finished and nicely polished.

Compasses, Dividers, and Calipers

Joiners and cabinet workers have a multitude of uses for the above tools, which are of the simplest construction. The ordinary form

Figure 31—Oval Mortise Gauge

Figure 32—Brass Stem Mortise Gauge

Figure 33—Using Marking Gauge

Figure 34—Panel Gauge

of wing compasses is such that the wing (the curved side projection) forms one with the left leg, while the right leg has a slot that lets it slides up and down the wing, and the set screw is tightened when the legs need to be fixed at a certain distance apart. For very accurate work, compasses with the sensitive adjustment at side, can be very useful. They are used for stepping off a number of equal distances, for transferring measurements, and for scribing. Calipers are used for measuring diameters of round pins, circular recesses, etc.; for the former purpose use outside calipers and inside calipers for the latter purpose.

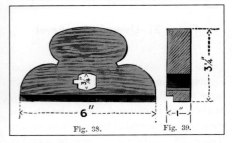

Fig. 38.   Fig. 39.

Figures 35 and 36—Elevation and Section of Panel Gauge Stock

Appliances for Mitering

The technical term *miter* is applied usually to the angle between any two pieces of wood or molding where they join or intersect, for example the angles in a picture frame. In this instance, the joint would be a true miter—that is to say, it would be 45° or half the right angle (90°) formed by the two inner edges of the frame. Although the term miter is generally understood to apply to a right angle, any angle, acute or obtuse, can also be called a miter.

Miter Blocks

There are various appliances employed in cutting miters, but the simplest is the miter block. Lay the project on the rebate, shown by c (Figure 41), and use the saw kerfs A and B as a guide for the tenon saw. The best form of miter block is made from a piece of dry beech, about 16 inches long, 6 inches wide, and 3 inches thick. Then cut a rebate c to about the size shown in Figure 41. Make sure that the angle is perfectly true. Lines A and B are set out to an angle of 45°, and then square down the rebate

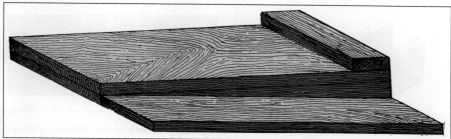

Figure 40—Shooting Board Giving Oblique Planing

Figures 37 and 38—Frame of Improved Shooting Board

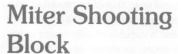

Figure 39—End Elevation of Improved Shooting Board

and back of the block. Cut the lines down with a tenon saw, and remember that the accuracy of the saw depends on the value of the finished miters. Figure 42 is a section of the miter block commonly used by joiners. This is simply two pieces of wood planed up true and screwed or nailed together. This plan answers very well, as when it becomes worn and out of truth another can be made inexpensively. The block shown by Figure 43 has a ledge on the bottom as shown; owning to the inward slant the project is more easily held.

Miter Box

Figure 44 shows a miter box which serves the same purpose as the block. This is made with three pieces of deal about 1 inch thick, nailed together at the bottom as shown. Miter boxes for heavy work require a strengthening piece on top to hold the sides together (see Figure 45). Sometimes three pieces may be necessary (see Figure 46). Both these illustrations show pieces of molding in position for miter-cutting.

Miter Shooting Block

Figures 47 and 48 show a miter shooting block for shooting or planing the edges of objects sawn in the miter block or box. In Figure 47 the bottom piece is of dry red deal, 2 feet 6 inches long, and rebated. For the top piece, select a hard material, such as mahogany or beech, or whichever you prefer. Make sure it is planed perfectly true, and cut at the ends to a "true miter" (45°); Then firmly screw it to the bottom piece. It is better to fix the ledger pieces across the bottom, to keep the board from warping. Figure 62 shows that the bottom piece is made of two separate boards.

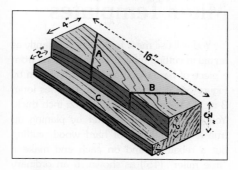

Figure 41—Miter Sawing Block

Figure 42—Section of Miter Block

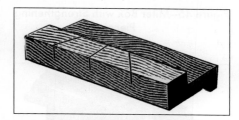

Figure 43—Inclined Miter Sawing Block

Combination Shooting Board

Notwithstanding the large number of patented mitering machines in the market, skilled jointers, when any particularly good piece of work is in hand, still prefer to use the ordinary homemade wooden shoot. The machines, while new and in good condition, are without a doubt faster, but if they are carelessly used, they are apt to get out of order, and then their work is far from satisfactory; while the wood shoot will stand a deal of rough usage, and is also easily repaired, Figure 49 illustrates several improvements on the old form of shoot. A miter-shoot, square-shoot, and joint-shoot are combined in one board, which will prove very handy when you only want to use these appliances occasionally. The shoot consists of a top board of seasoned yellow deal 3 feet by 9 inches by 1½ inches, slot-screwed to an under-frame of teak, made up of the plane bed B, 3 feet by 2¾ inches by 1½ inches, into which are framed three

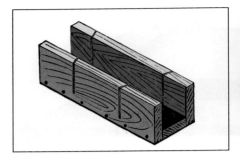

Figure 44—Miter Box

Figure 45—Miter Box with Strengthening Piece

Figure 46—Miter Box with Strengthening Pieces

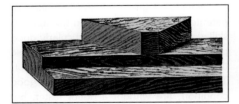

Figure 47—Miter Shooting Block with Solid Base

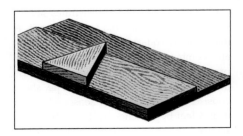

Figure 48—Miter Shooting Block with made-up Base

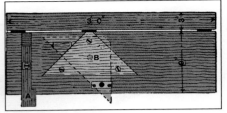

Figure 49—Combination Shooting Board

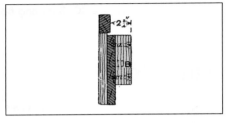

Figure 50—End Elevation of Combination Board

cross rails 2¼ inches by 1 inch flush on the under-side. The center of the top board is the miter-block, a piece of dry oak 2 inches thick cut with two of its sides exactly square with each other, and at an angle of 45° with the third. This block, instead of being fixed in the general way, should be mounted on a pivot in its center, as shown at B (Figures 49 and 50), and is capable of adjustment either as a miter-shoot, as shown in the full lines, or as a square-shoot, as indicated by the dotted lines in Figure 49; it is firmly secured in either position by means of three ½-inch by 3-inch square-head screw bolts. Be sure that the grain of the block runs parallel with the plane bed so shrinkage will not alter its shape. The board should be arranged for jointing by removing the miter-block and working the boards against the adjustable stop A. This stop works in an undercut groove, and is secured in any requited position by the screw bolt; the projection at the end prevents the boards from slipping while being planed.

Miter Templates

You will use miter templates constantly as an aid in cutting miters. They are made from a piece of hard wood, in the form shown by Figure 54; it is usually about 4 inches long, 3 inches wide either way, and ½ inch thick. You can make a true miter by planing up true a square block of hard wood, cutting out a rebate B, and on each end make a "true miter" (45°) as shown. If an ordinary cupboard framing is examined at the junction

of the rail with the jambs, it will be seen that each of the molded edges has been mitered as shown in Figure 55. To obtain this miter, you can use the template. Figure 56 and Figure 57 show the template applied to the edge being held by the left hand, while the right guides the chisel A.

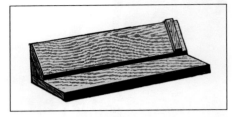

Figure 51—Donkey's ear Shooting Block

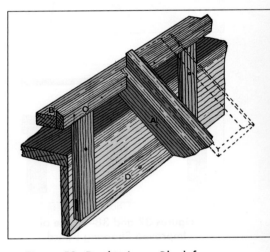

Figure 52—Donkey's-ear Block for Shooting Wide Surfaces

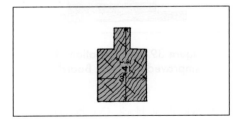

Figure 53—Rest of Donkey's-ear Block

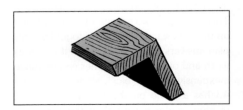

Figure 54—Miter Template

Figure 55—Molding with Mitered Joint

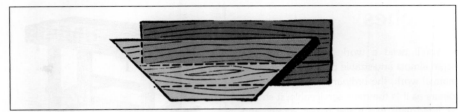

Figure 56—Application of Miter Template

Spirit Level

The spirit level is used to determine the plane of the horizon—that is the plane forming a right angle to the vertical plane. A frame holds a closed glass tube nearly filled with anhydrous ether or with a mixture of ether and alcohol. Make sure your spirit level comes with a graduated scale engraved on the glass tube or on a brass or steel rule attached to the frame beside it, so as to mark the position of the bubble, the tube being so shaped that when the level is lying on a flat and horizontal surface the bubble occupies the center of the tube. Many levels have provision for altering the length of the bubble. Figure 58 is a view of an ordinary spirit level, and its construction is made quite clear by the sectional view, Figure 59. To use the spirit level, apply it to the work twice, reversing it at the second application, and find the mean of the two indications when you are finished. Spirit levels are made in many shapes and sizes, but the method of construction is always the same. A serviceable tool has a narrow shape, about 10 inches long and its greatest breadth being $1^1/_{16}$ inches, and diminishing to ½ inch at the ends.

The frame is of any hard, tough wood, like box, ebony, lignum-vitae, birch, beech, walnut, or oak. At the back of the tube should be silvering, which shows up the bubble and enables side lights to be dispensed with. The tube is set in plaster-of-paris, and has a brass cover. Store-bought spirit levels are constructed generally in ebony or rosewood, with better qualities having a metal protection for the edges and faces. This will preserve the truth of the instruments for a longer time. A serviceable American level has a mount entirely of steel, which is hexagonal in section, and has rounded ends. Another handy form is the one mounted wholly in brass; this has a revolving protector over the bulb opening, and there is provision for adjustment should the level after a time wear out of truth. A very convenient form of level is the one with a graduated screw slide that allows the fall per foot to be shown at a glance.

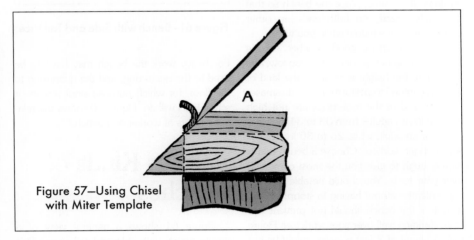

Figure 57—Using Chisel with Miter Template

Figure 58—Spirit Level

Figure 59—Section of Spirit Level

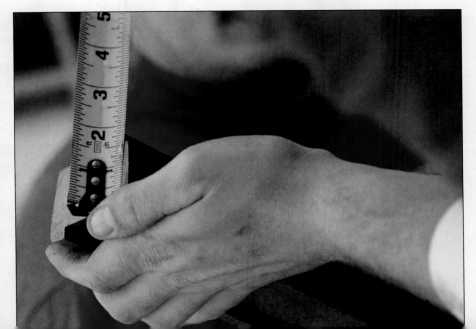

Holding Tools

Benches

You'll need a work bench before you begin almost any sizable project. For general manual work, the ordinary bench is used by joiners as it is the most serviceable; it should be fitted with two wood bench screws and wood vice cheeks, one at each left-hand corner of the bench, to accommodate two workers. If possible, place the bench so that light falls directly on both ends—in other words, face the window while you work.

There are many good benches on the market, but do not get one that is too low, and make sure the height is right for the kind of work you intend to perform on it. The smaller benches sold at the tool-shops are not high enough for an adult—from 33 to 34 inches is great for an adult, while 26 to 30 inches will suit younger workers. Choose a bench that is high enough to give you the most command over your tools. You should be able to work conveniently without having to stoop; but the height of the bench should not prevent you from standing well over your work (see Figure 60). A height of 2 feet 6 inches might be just right for occasional use, but too low to work at for any length of time. A simple method of raising it slightly from the floor is to put a piece of quartering under each pair of legs.

Figure 61—Bench with Side and Tail Vices

For heavy work the bench may have to be fixed to the quartering, and the quartering to the floor, for which purpose stout screws or screw bolts will do. Figure 60 shows the relative heights of worker and bench.

Various Kinds of Benches

Figure 61 shows a general view of a simple bench with side and tail vices. This form is extremely useful for cabinet making and similar work, where it is helpful to hold pieces of material that may have to be planed, molded, chamfered, mortised,

Figure 60—Workman and Bench

grooved, etc., without using a bench knife or similar method of fixing. The material could be held between stops, one being inserted in one of the holes in the top of the bench and another in the hole made in the cheek of the tail vice. The following dimensions are only suggestions and the bench may be made longer or shorter, narrower or wider, to meet your needs: Top, 5 feet by 1 foot, 9 inches, and 2 inches thick. Height, 2 feet, 7 inches. Distance between legs, 3 feet, 2 inches lengthwise, and 1 foot, 3 inches

Figure 62—Double Bench with Vice at each end

Figure 65—Cabinet-worker's Bench

Figure 63—Folding Bench in use

Figure 64—Folding Bench not in use

sideways. Legs, 3 inches by 3 inches. Your bench should be constructed of hard wood, such as beech or birch, and in any case it will be best to have hard wood for all the parts forming the top, side cheeks, and cheeks of vices, as these are the main parts of the bench; consider choosing red deal for the framing of the legs, rails, etc. A simple bench is illustrated by Figure 62; this is a suitable model for general carpentry and joinery. The framework is made of a thoroughly seasoned dry spruce fir or red pine, and the top of birch or yellow pine.

The folding bench illustrated in Figure 63 will be found useful in cases where a portable bench is required for occasional use only. When not using the bench, the screw, screw cheek, and runner can be taken out, the legs folded on to the wall, and the top and side folded down as seen in Figure 64. You will find a more elaborate bench for cabinet work in Figure 65; it consists of two principle parts, the underneath framework and support, and the top. The former has two standards joined by two bars. On the feet of the standards rests a board which serves to hold heavy tools and other articles. There is a rack for small tools and underneath this a band, tacked at short intervals, for other tools. The front rail has holes on its top face 1 inches by ¾ inches for holding bench stops, while in the front face of the rail are round holes for holding other pins, illustrated by T, 1½ inches square at one end, but made round at the other end to fit tightly into the holes. The pin T and the block V (Figure 65), screwed on the end of the moveable jaw of the vice, serve to hold wood during the process of edge planing. Holes in the back rail receive pins W which are convenient for cramping up joints. A kitchen table bench is shown in Figure 66. The end of the table employed is not the one containing the usual drawer. Two blocks of wood A B, 3 inches square, are attached to the table top by two cramps embedded in the ends of one of the pieces. Through mortise holes C C are inserted slats glued and wedged to block A, but running loosely in holes in B. S is a screw, and the two parts of the bench serve the purpose also of vice cheeks; though if desired the two blocks can be screwed together solid.

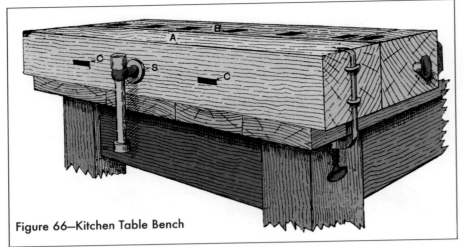

Figure 66—Kitchen Table Bench

Figure 70—Bench Screw Vice

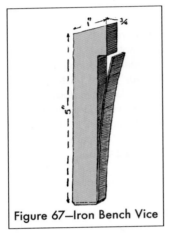

Figure 67—Iron Bench Vice

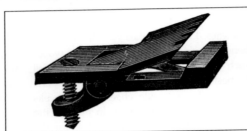

Figure 68—Adjustable Bench Stop

Figure 69—Hinge used as Bench Stop

Common Bench Screw Vice

A common form of joiner's bench screw is shown in general view by Figure 70 and Figure 71 is a sectional view. D is the side or cheek of the bench to which the wooden nut A is screwed. The box B, which accurately fits the runner shown inside it, is fixed to the top rail connecting the legs, and to the top and side of the bench. Be careful to keep the runner at right angles to the vice cheeks. To fasten the vice outer cheek and screw together so that upon turning the cheek, the screw will follow, cut a groove shown by Figure 71. Then from the under edge of the cheek, make a mortise and drive in a hardwood key, F, so that it fits fairly tightly into the mortise and its end enters E. The screw cheek is usually about 1 foot, 9 inches long,

Bench Stops

Generally, store-bought benches are provided with holes for the reception of stops that allow for work to be held for planing, etc. These stops are made of iron and shaped as in Figure 67, and have springs at their sides by means of which they are held tightly and at any required height in the bench holes. You will find an adjustable stop for screwing to the bench in Figure 68. For a temporary stop some workers drive a few nails into the bench end, leaving the heads projecting enough to hold the wood. A better substitute can be made out of an ordinary butt hinge, one end of which should be filed into teeth so as to hold the wood better. Be sure to leave this end loose, and screw the other end down tightly to the bench end as shown by Figure 69. A long, light screw through the middle hole in the loose side will afford sufficient adjustment for thin or thick work. When you are finished, the hinge can be taken up and put away.

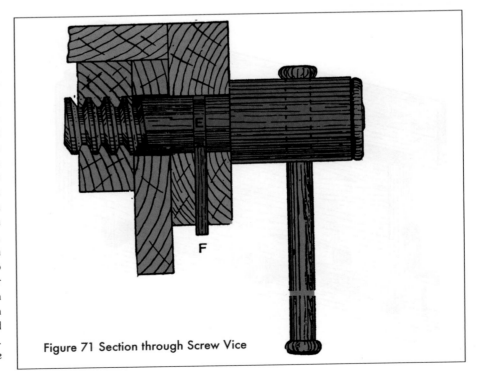

Figure 71 Section through Screw Vice

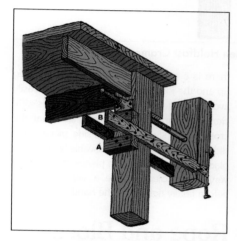

Figures 72 and 73—Kitchen Table Screw Vice

Figure 77—End Elevation of Sawing Stool

Figure 79—Braced Sawing Stool

Figure 78—Three-legged Sawing Stool

Figure 80—Bolted Sawing Stool

Figure 74—Side General View of Kitchen Table Vice

Figure 75—Common Sawing Stool

Figure 76—Front Elevation of Sawing Stool

9 inches wide, and 2 inches to 3 inches thick. The runner is about 3 inches by 3 inches and 2 feet long. You can also buy the wooden screws and nuts ready made.

Figure 81—Sawing Horse

Screw Vice for Kitchen Table

Figures 72 to 74 show a simple device for fixing a screw vice to a kitchen table, the vice being detachable for removal as required. The device illustrated does not cause the least degree of damage to the table. A hole is made in the table leg for the screw to pass through, the nut or box of which is fixed to the back of the leg as shown. Two hardwood runners, 2 inches by ¾ inch by 1 foot, 8 inches, should be made and dovetailed into the screw cheek, which is 2¼ inches thick, 1 foot, 3 inches long, and has its breadth regulated by the size of the leg. The distance between runners should be the same as the thickness of the leg. The runners are kept in position by two blocks A and B, which are screwed to the back of the leg. An adjustable pin C made from a piece of ½-inch round iron, will be required, and must be sufficiently long to pass through both runners. Screw a block, D (Figure 73), to the leg, the face of the block being flush with the front edge of the top.

Sawing Stools or Trestles

Figure 75 shows a standard sawing stool, Figure 76 is a side elevation, and Figure 77 an end elevation. Suggestive sizes are figured on the drawings. The thickness of the material can, of course, be increased or decreased according to requirements. The simplest sawing stool, but the least reliable, is the one with three legs shown by Figure 78, but this is of little service and almost useless. Better and more usual forms are shown by Figures 79 and 80, these being about 20 inches high, firmly and stiffly made. In Figure 79 all the parts are mortised and tenoned together, and strutted to give strength, but in Figure 80 the legs are simply shouldered and bolted into the sides of the top. The cross stretchers are

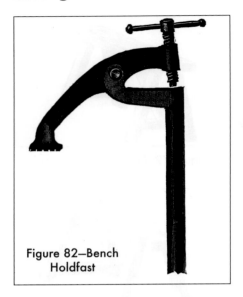

Figure 82—Bench
Holdfast

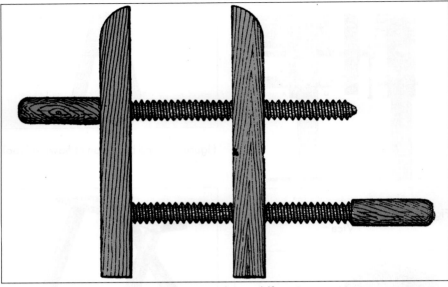

Figure 83—Wooden Holdfast Cramp

Figure 84—G-cramp

slightly shouldered back and screwed or
bolted to the legs.

Cramps

Cramps are used to hold work on the bench,
to hold together work in course construction,
to facilitate the making of articles in which
tight and accurate joints are essential, and to
hold together glued joints until the glue is dry
and hard. A holdfast for temporarily securing
work to the bench is shown in Figure 82. This
ranges in length from 12 inches to 16 inches.
The old-fashioned holdfast cramp is illustrated
by Figure 83; this is made entirely of wood,
and the cheeks of the cramp range in length
from 6 inches to 16 inches. Iron cramps
are shown by Figures 84 and 85, where
Figure 84 is the ordinary G-cramp. Figure
85 shows one of Hammer's G-cramps with

instantaneous adjustment, this being an
improved appliance of some merit. The
screw is merely pushed until it is tight on a
work held in the cramp, and a slight turn of
the winged head then tightens up the screw
sufficiently. The sliding pattern G-cramp is
illustrated by Figure 86, this possessing an
advantage similar to, but not as great as,
that of Hammer's cramp. Sash cramps and
jointers' cramps (non-patent) resemble Figure
87. There are several makes and many differences in detail, but Figure 87 illustrates the
type. There are a number of patent cramps for
sashes and general joinery, but Crampton's
appliance (Figure 88) is typically sufficient.
You can set the right-hand jaw at any position on the rack. After you insert the work,
push the right-hand saw against it tightly and
the lever handle adjusts instantly. When you
joint up a thin project with an ordinary cramp,

there is a great risk of the material buckling
up and the joint being broken. You can easily
avoid this risk by using a cramp which is sufficiently explanatory when it is said that the
cross pieces slide upon the side pieces, one
sliding bar being made immovable by iron pins
placed in holes in the side pieces. In cramping
very thin projects, place a weight upon it
before finally tightening the hand screw.

Rope and Block Cramp

To cramp up boards with ropes and
blocks, place the wood blocks A about 4
inches long, and 1½ inches square, on the
edges of the boards B, and place a rope
around them twice and make a knot. Then

Figure 85—Hammer's G-cramp

Figure 86—Sliding G-cramp

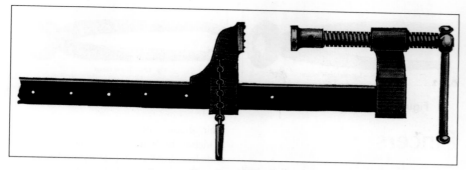

Figure 87—Sash Cramp

Figure 88—Crampton's Patent Cramp

Figure 89—Dog, Round Section

Figure 90—Dog, Square Section

are shown respectively by Figures 89 and 90. The boards being already close together, the dog is inserted across at a right angle to the line of joint, one point being in one board and one in the other. The further you hammer in the dog, the closer the boards are cramped together. Floor boards can be tightened up without using a floor dog with the method illustrated in Figure 91. The board next to the wall should be well secured to the joists, and then three or four boards can be laid down and tightened up by means of wedges, as shown. The following is the method of procedure: Place a piece of quartering about 2 inches by 3 inches next to the floor board, as at C. Cut a wedge, and place it at B, then nail down a piece of batten to the joists, as at A (both this and the wedge can be cut out of odd pieces of floor board). The wedge B should be driven with a large hammer or axe until the joints of the board are quite close.

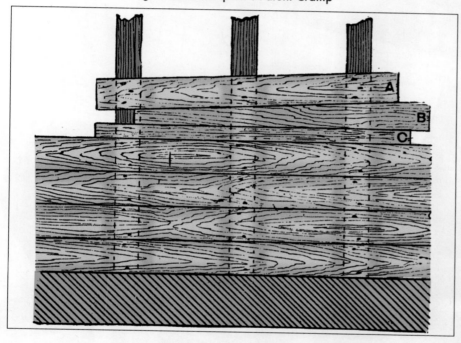

Figure 91—Wedge Cramp for Floor Boards

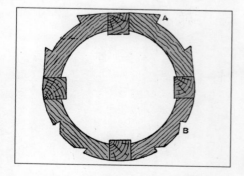

Figure 92—Circular Seat with Cut Cramping Pieces

place a small piece of wood between the two strands of rope and twist it around. This twisting draws the rope tighter on the blocks, thereby cramping the boards together. Three of these sets would be sufficient to cramp a number of long boards.

Cramping Floor Boards

Floor boards are commonly cramped or tightened up using "dogs," of which two forms

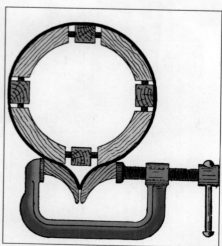

Figure 93—Circular Seat with Flexible Cramp

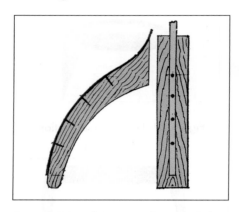

Figures 94 and 95—Wood Horn of Flexible Cramp

Fig. 128.—Lancashire Pincers.

Figure 96—Lancashire Pincers

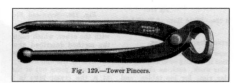

Fig. 129.—Tower Pincers.

Figure 97—Tower Pincers

Pincers

Pincers are used for extracting and beheading nails, and for other purposes where a form of hand vice is wanted for momentary use. There is little variety in their shape, and they range in size from 5 inches to 9 inches Usually one handle ends in a small cone (see Figure 96) or ball (See Figure 97), and the other in a claw for levering out nails, etc. Figure 96 shows Lancashire pincers, and Figure 97 Tower Pincers.

Workshop Furniture

Woodworker's Tool Chest

A woodworker's first job, when he has gained enough skill, should be the creation of a good tool chest. This will offer a chance to practice using a range of tools, and the chest itself will keep those tools under lock and key when not in use. An ideal tool box should have a specific place for everything, as the chest illustrated by Figure 98 does. The letter references in Figures 98 and 100 are:

A. bottom plinth
B. top plinth
C. rim round lid
D. compartment for bead-planes, plough, etc.
E. compartment for various tools, planes, etc.
F. compartment for saws
K. bottom till
J. second till
H. top till
L. sliding-board to cover compartment E
M. cleats to hold division between E and F
N. cleats to hold division between D and E
P. runners for sliding-board L
R. runners for tills
S. runners for tills K

The length of the chest must be sufficient to accommodate a rip-saw, so the chest is 33 inches long internally; and if it is made 20 inches wide by 21 inches deep, you will find it convenient for all purposes. For the outside case, use white deal no less than 1 inch thick. In gluing up the front, back, and ends to obtain the necessary width, tongue and dowel the joints, the former being the better method. In dovetailing the framework of the chest, make the pins small, and have them no more than 1½ inches apart; make sure that the joints in the front and back do not come immediately opposite those in the ends, or at some future time the chest may break in two. Figure 99 is a transverse section through the chest, Figures 100 to 103 show the details of construction. The plinths A and B run all round the chest, and are 6 inches and 2½ inches wide, respectively, and 1 inch thick, with the top edge of A and the bottom of B finished with a plain bevel; the tip edge of a ¼-inch bead being worked on it also. The plinths may be mitered at the corners, but it is better to dovetail them, and so obtain extra strength and good appearance. The plinth B is kept down about ¾ inch from the top of the chest to form a rebate for the lid to shut upon. The bottom of the chest is formed with boards 1 inch thick, tongued and grooved, and nailed on crossways—that is, the grain runs from front to back of the chest. The lid also is of 1 inch deal, with the joints tongued and grooved, and the ends clamped. It overhangs the chest all round in about ¹⁄₁₆ inch, and is hung with a pair of strong brass butts, and the self-acting spring lock is put on; then the rim C can be dovetailed together at the corners, and nailed to front and ends. This should result in a good fit where the rim of the lid meets the plinth B. Now you have finished the skeleton of the chest.

Tool Chest Partitions

For the inside fittings of the tool chest use good yellow deal or pine, which you can finish by staining, although sometimes a more fancy wood is used. From Figure 99 it is seen that the chest is divided in its width into three parts: D is for bead-planes, plough, etc.; this is 7 inches wide, and is covered by the sliding tills; E holds miscellaneous tools, best planes, or anything not in everyday use; and F (3½ inches wide inside) is the saw till. These are divided by the two partitions shown, that between D and E are 9 inches high, and between E and F are 14 inches. The three tills H, J, and K slide to and fro to give access to compartments beneath, and when in place at the back of the chest form a covering for compartment D; and a sliding-board beneath the tills, when pulled out as shown by dotted lines in Figure 99, covers

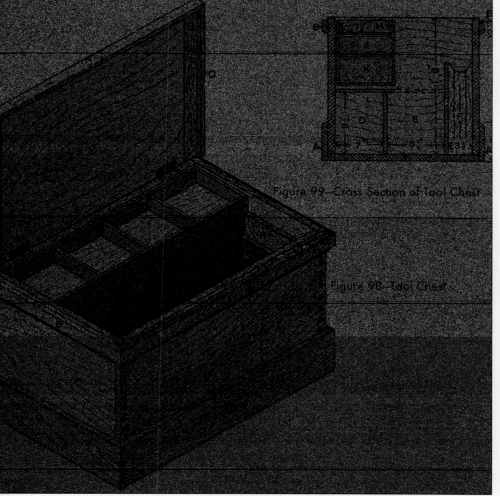

Figure 98—Tool Chest

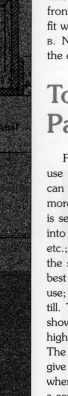

Figure 99—Cross Section of Tool Chest

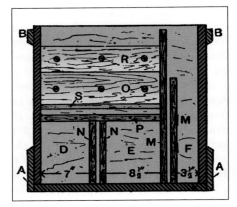

Figure 100—End of Tool Chest with Cleats and Runners

compartment E. The bench-planes, which are in everyday use, can be packed away on the sliding-board between the tills and the highest partition. Figure 100 shows one end of the chest with the cleats about 1 inch wide by ½ inch thick; between these the partitions fit. The cleats holding the partition between E and F are fixed first, ½ inch apart, and are as shown at M M (Figure 100), the one nearer the back of the chest being continued nearly to the top, the other, nearer the front, stopping at the same height as the partition, namely 14 inches. The cleats N must be 8½ inches high from the bottom of the chest, and ¾ inch apart. The back partitions having been placed in position, fix the ledges P, with their top edges 9½ inches from the bottom of the chest; then they run from the back to the long upright cleat M, and on them works the sliding-board L, 9 inches by ¾ inch; this is clamped at the ends for the sake of strength and to make it slide more easily. It must be

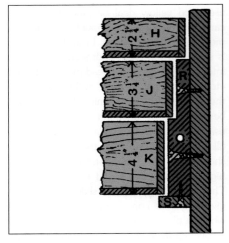

Figure 101—Section of Tool Chest Till Runners

a good fit endways to avoid jamming against the ends of the chest. The runners for the tills (Figure 101) are made long enough to reach from the back of the chest to the long upright cleat M (Figure 100), and should be made of hard wood. The principle piece R, which forms the runners for the two top tills, is 7½ inches wide by 1 inch thick, rebated to half its thickness at O for a depth of 3¼ inches. A piece of hard wood S, 1½ inches by ½ inch is screwed on to the thick edge of R, and forms the runner for the bottom till. These runners can be fixed in position one on each end of the chest, leaving about ⅛ inch clearance between the bottoms and the top of sliding board L. The partition between compartments E and F can be made and fitted between the cleats M M; along its upper side is a strip of 1½ inches by ½ inch deal, cut to fit between the cleats on each end of the chest, fixed level with the top edge on the side nearest the front of the chest and packed off about 1/16 inch. The slot thus formed can be used as a rack for squares, the stocks resting on top of the partition, and the blades hanging down out of the way inside the saw till.

Racks for Saws and Chisels

The saw racks in the chest, as shown in Figures 102 and 103, are 14 inches long, 3½ inches wide, and 1 inch thick, shaped at the top ends. Each has three slots, or rather, saw kerfs, and in one rack (Figure 102) the middle kerf runs from the top to within 3 inches of the bottom, the others stopping the same distance from the bottom, and about 1½ inches from the top. In the other racks (Figure 103) the middle slot is stopped at both top and bottom, the others being open at the top end. These two racks are fixed at about 8 inches from each end by screwing through the horn at the top of each to the front of the chest. The partition being then put into its place, screws can be put through it into each saw-rack, which will hold all in place. When placing the saws in the racks, the points are inserted in the closed slots of racks, and the handle ends dropped into the open slots, two saws pointing one way and one the opposite way. To take chisels, a piece of hard wood 2 feet long, 1 inch square, with a series of notches 1 inch apart cut into it wide enough to take the tools, can be screwed to the front

Figures 102 and 103—Saw Racks.

of the chest just above the top of the partition; this leaves an equal space at each end to allow the hand to be inserted to remove saws form the rack. The handles of the larger chisels will be just inside the front of the chest, convenient for withdrawal when wanted for use, and the blades with hang out of the way in the saw till.

Tool Chest Tills

The three sliding tills for the inside of the chest are all that remains. They will all be 9 inches wide outside, but varied in depth, as shown by Figure 101, on which dimensions are marked. They should be of ¾ inch stuff, with ½ inch bottoms and divisions, the rims dovetailed together, and the fronts and back rebated to receive the bottoms, the grain of which should run across the width of the tills; and at each end the bottom should be of hard wood. The divisions should be trenched into the sides, forming in K, J, and H two, three, and four compartments respectively. One of the bottom divisions should be fitted up for the brace and bits, with racks for the bits fitted round the brace. Other divisions can be fitted with racks for small chisels, gouges, gimlets, bradawls, and various other tools, the aim throughout being to have a place for all, so that nothing can roll about and get damaged. Turn-buttons to take the tenon and dovetail saws can be screwed on to the

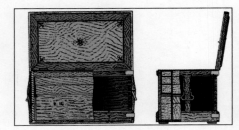

Figure 104 and 105—Elevation and Sections of Tool Chest

under-side of the lid, so that when it is closed they will be in position between the top till and the front of the chest. The purpose of the cleat M, running up higher than its fellow, is to stop the tills from coming into collision with the stocks of squares when in their rack. The sliding-board L can be grasped underneath with the fingers when it is desired to draw it forward, and it should have a couple of thumb-holes cut in its top as a means of pushing it back. Each till should have a pair of flush-rings inserted in the front, so that either can be pulled forward and its contents exposed without the necessity of touching the others. A strong iron handle on each end of chest will now make it complete.

Tool Chest Lid

For the lid a piece of pine 2 feet, 11½ inches by 1 foot, 9 inches by ⅞ inch must be made. In some cases the ends are clamped, but the lid will stand better if properly cross battened. When the lid has been planed, the inside should be roughed with the toothing plane and two or three battens screwed across the grain on the other side to prevent warping. The inside can then be veneered with a center panel and a banding about 2½ inches wide as shown by Figure 104. Use Spanish mahogany veneer for the center panel, and a light or dark fancy wood veneer for the line, which may be about ⅜ inch wide, and for the center and corner inlays. Some workers veneer the center, and have a margin about ¼ inch thick, as shown on the underside of Figure 106, a planted molding being used for covering the edge of the veneer. When the glue is thoroughly dry the battens may be removed from the back, and the mahogany plinth A (Figure 106) screwed to the front and the ends; the parts seen when the lid is open should be polished. Pieces of 1¼ inch by ⅛ inch hoop iron B (Figure 106) can now be screwed round the top of the lid to protect

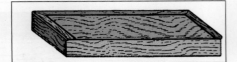

Figure 107—Top Tray of Tool Chest

the edges, and the space between filled in with ⅜ inch deal boards C, screwed across the grain of the lid; the ends can be rounded down to the hoop iron to strengthen it and also to prevent warping. The lid can now be hung with three 2¾ inches by ¾ inch brass butt hinges as shown in Figure 104. A strong lock can then be let in the front and a sash lid D (Figure 106) screwed to the plinth at the front. Figure 105 shows that the top plinth at the back is kept above that of the sides and front, to support the lid when open. In addition to the lock, one or two holes should be bored through the lid at both ends and countersunk in the hoop iron, so that the lid can be screwed down for traveling, etc. The outside corners of the chest should be protected with angle plates on the plinth as shown on the right-hand side of Figures 104 and 105, and these may be made by bending pieces of 1½ inches No. 16 b.w.g. iron 6 inches long to aright angle, punching the holes and countersinking for No. 10 screws.

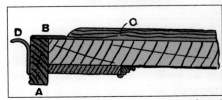

Figure 106—Section through Front of Tool Chest Lid

Inside Fittings of Tool Chest

For the interior fittings of the chest a small nest of drawers at the back is sometimes used, but some prefer trays, as shown, as the drawers are liable to stick if a tool gets misplaced. Also they take a lot of material and labor, and the drawers are most difficult to secure than trays when the chest is packed. The trays should be of ⅜ inch mahogany, dovetailed together like a drawer, the top trays being fitted with lids, as shown in Figure 107; the total depth over all is 2¼ inches. Figure 108 shows one of the lower trays, and these do not have lids. The cross divisions in the trays may be made to meet requirements, but the following plan of dividing is a good one: Top tray at the back, space 1 foot, 4 inches long at the center with divisions about 8 inches at each end; second tray, the same as the top; and the bottom tray, one division in the center. For the narrow trays, the top one may be divided the same as the top one at the back; the second tray, with a partition 7½ inches from one end; and the bottom tray, with a division 10½ inches from the opposite end to the tray

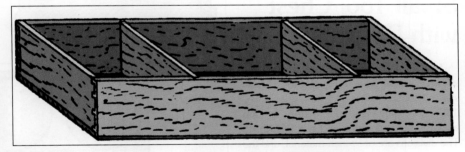

Figure 108—Second Tray of Tool Chest

above. The space in the chest below the trays may be divided longitudinally into three compartments. The boards to form the divisions are fixed to an upright piece of wood ⅝ inch thick, B, secured to the ends of the chest. This part of the text is just deep enough to take small planes placed on end. A saw rack may with advantage be fitted in the space under the front trays, and the center space is covered with a board A, which slides back under the back trays. The inside of the chest should be French-polished, and the outside should have three or four coats of good paint.

Figure 109—Tool Chest with Drawers

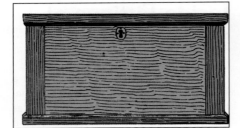

Figure 110—Front View of Tool Chest

Packing Tool Chest for Transit

In packing the chest for traveling, the bottom divisions should be filled first, and the heaviest tools placed in the center portion. The slide can then be pulled over and fixed with a small screw at one end. The trays can then be filled, and secured either by means of strips fixed across the ends, or by filling the space between them with soft material that will not damage the polish. The lids of the top trays can then be fastened by placing across them two strips at the ends that will just fill up the space between the tops of the trays and the lid, when the latter can be locked and screwed.

Small Tool Chest with Drawers

For the small tool chest with drawers, shown in several views by Figure 109, a handy size is 1 foot, 9 inches by 1 foot, 2 inches by 1 foot deep. The sides, ends, bottom, and top are of red deal finishing about ¾ inch thick. The divisions are of ½ inch material, and the drawer fronts of ⅝ inch material. The sides, backs, and bottoms of the drawers are of ⅜ inch material, but of course these dimensions may be varied to meet your requirements. Figure 109 shows that the front is hinged on the bottom, so as to drop down and to allow of ready access to the drawers. To keep the front from twisting and warping, it must be clamped as shown, and when the front is closed up it is secured to the lid by a lock (see also Figure 110). In addition, a hook A (Figures 109 and 111) and eye B (Figure 111) may be used. The bottom is finished off with a plinth, which is rebated as illustrated in the section (Figure 112) will have extra strength. The lid should be stiffened by a 1¼ inches by ½ inch rim. The well c and the space D

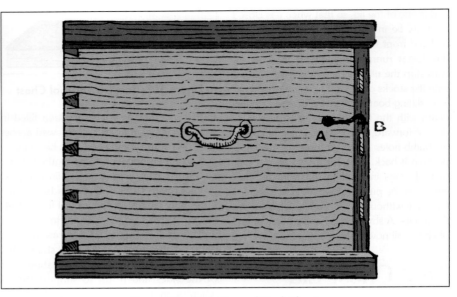

Figure 111—End View of Tool Chest

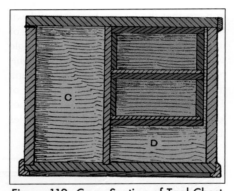

Figure 112—Cross Section of Tool Chest

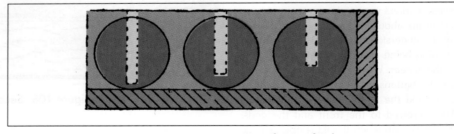

Figure 113—Cross Section of Part of Lid

under the drawers will be found very useful for large tools.

Utilizing Tool Chest Lids

The insides of tool chest lids are adapted readily to hold hand saws and tenon saws, the ends of which are held in wooden clips. The handle can be fastened by means of a

button B, this method being just as suitable for hand saws as for the tenon saw shown. When the button is moved into position, it will allow for the saw to be taken out.

Simple Tool Chest

The simple chest shown in longitudinal section by Figure 118 must be long enough to take the rip saw, and it is as light as possible consistent with strength. Figure 119 is a cross section. The yellow pine is ⅞ inch thick for the body, 1 inch for the lid, and ⅝ inch for the outside plinth and facings;

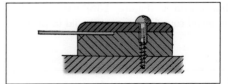

Figure 114—Try-Square Holder

the bottom is of ¾-inch red pine. The plinth has an ovolo molding on it, but you will commonly see an ogee molding. The top facing is in two parts, one being screwed to the edge of the lid and rounded on the top;

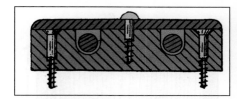

Figure 115—Holder for Pincers

the other one, upon which are run a bead and a chamfer, is merely nailed to the box. The body of the box is dovetailed and glued, but the plinth and top facing are mitered and nailed. The bottom has ploughed and cross-tongued joints waterproofed by painting with white lead. Clamp the top to prevent warping, and screw the facings on after the lid had been fitted to the size of the box. Battens of well painted red or yellow pine should be screwed to the under-side of the bottom, to keep the box clear of wetness. Figure 119 shows the inside arrangement. At the back, a space for smoothing planes, rebate planes, casements, etc., is formed by nailing fillets to the ends of the box, to which the pieces of pine A is screwed. Narrow fillets are nailed to the ends of the box outside of the piece A, and the piece B is screwed to them. A space is thus provided for the tenon saws and hatchet. The method o fixing the hand saws to the inside of the lid is shown by dotted lines in Figure 118, and is on the same principle as that already described. A piece of wood, the thickness of the saw

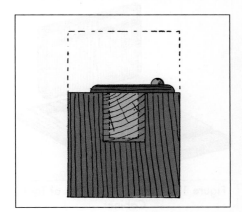

Figure 116—Elevation of Mallet Holder

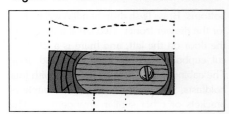

Figure 117—Plan of Mallet Holder

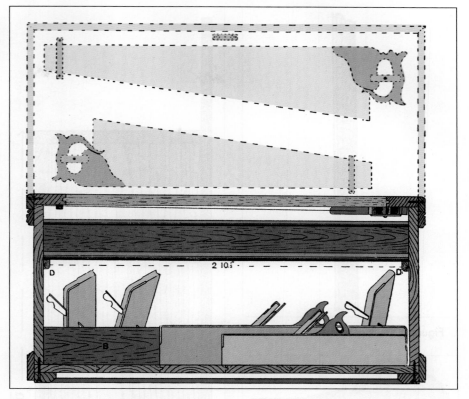

Figure 118—Section of Simple Tool Chest

handle, is fitted in the hole, and screwed to the lid. A piece of sheet brass to form a long button then is screwed to the block of wood; this button, when turned round as in Figure 118, prevents the handle of the saw from leaving the lid. The hardwood clip to hold the point of the saw has no recess for the back as there shown. The method of packing the chisels and gouges is seen in Figure 119. A small fillet, with a strip of leather glued to the top edge, is nailed to the bottom, and a thin piece of pine, projecting about 1 inch above the leather, is nailed to the fillet. This receives the points of the chisels and gouges. Another piece, with various sizes of holes cut out, is screwed about 3 inches or 4 inches up, to keep the top part steady. Figure 120 shows the piece with the holes checked out and a thin piece screwed to the front; this is much easier than mortising the holes. The tray C (Figures 118 and 119) is a box the whole length of the inside, lap dovetailed at the back. It is divided into various compartments (two small ones and a large one will be found very handy) by thin pieces of pine, either raggled into the front and back or merely butted and nailed. The bottom, which is ⅝ inches thick, is screwed up. It is very common to have a hardwood flap on the tray, as shown, but this can be dispensed with at will. A back stile is

screwed to the back of the tray, and the flap is hinged to it. Fillets D are screwed to the ends of the box, on which the tray slides to and fro.

Tool Cabinet

You can see a useful tool cabinet with paneled doors and three drawers illustrated in Figure 121. It would have a neat appearance if it were made in oak or other hard wood and polished; or even if made of deal, if stained and varnished. The leading dimensions figured in the vertical section (Figure 122), and in the horizontal section (Figure 123) are only suggestive. The sides, top,

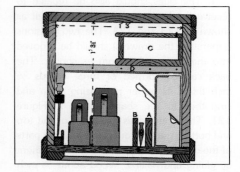

Figure 119—Cross Section of Simple Tool Chest

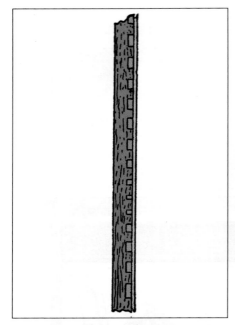

Figure 120—Tool Chest Chisel Rack

Figure 121—Tool Cabinet

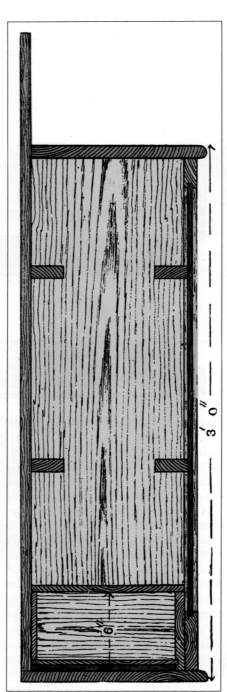

Figure 122—Vertical Section of Tool Cabinet

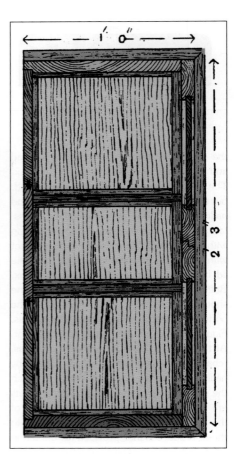

Figure 123—Horizontal Section of Tool Cabinet

Figure 124—Housing for Sides of Tool Cabinet

and bottom are of ¾-inch material, grooved and tongued together, as indicated in Figure 124. The sides should also be rebated to receive the back, and grooved for shelf as shown in 123 and 124. The two divisions (separating the drawers) should be grooved into the shelf and bottom as shown. The back can be formed of three boards ⅝ inch thick, its upper part being sawn and smoothed to the shape shown in Figure 121. The front edges and ends of the top and bottom may be rounded. Fit all the parts of the case together, and finally secure them by gluing and nailing. Wood about ⅞ inch thick will be required for the stiles and rails of the doors; the panels may be about ⅜

inch thick. The doors should be mortised and tenoned together and ploughed to receive panels; they should be finished by being glued, wedged, planed off, fitted to the case, and rebated together (see Figures 121 and 123), after which they can be hung with 3-inch butts. The drawers should be properly dovetailed; ¾-inch wood will do for the fronts, and ½-inch for the sides, backs, and

bottoms. Brass flush drop handles will be best for the drawer fronts. Two small bolts secure the door on the left, and there is a 2½-inch cut cupboard lock on the right-hand door. The cabinet could be fixed to a wall with four holdfasts, or it might rest upon a couple of brackets or other similar arrangement. The inside can be fitted with racks, according to requirements.

Strong Stool or Work Bench

A stool or bench is useful for various purposes, both in the household and in the workshop. It can, without much trouble, be taken to pieces, so that it may be conveniently stowed away when not in use. When using the stool for domestic use, the sizes of the timber should be about 2 inches by 2 inches for all parts of the frame. The top consists simply of a slab about 1 inch thick, having four holes bored in it to fit over the dowels or pins A. Suitable measurements for the finished article are: Length of top, 2 feet, 9 inches ; width of top, 1 foot, 3 inches; and height from ground,

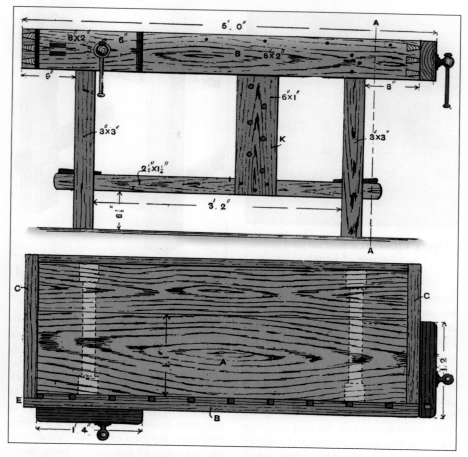

Figures 128 and 129—Side Elevation and Plan of Bench with Side and Tail Vices

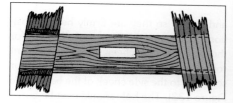

Figure 125—Bottom Rail of Work Bench

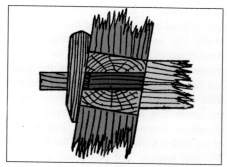

Figure 126—Joint between Rail and Bottom Stretcher

Figure 127—End of Rail in Leg

about 1 foot, 10 inches. For a work bench, however, you can increase these measurements, being sure to increase the framework proportionately, and the top, instead of fitting over pins or dowels, should be secured with nuts and bolts. The spread of the legs at the bottom should be such that they would occupy the four corners of a rectangle, equal and similar to that of the top; this prevents all tilting, and secures stability for the bench, a point that is often overlooked. Before marking out the framework, make a mold; take a piece of wood, about ⅜ inch thick, and square off one end, as at B, and make the other cut to the desired bevel or splay of the legs, as shown at A; a narrow strip is fastened to the edge, so as to form a fence. With this tool, you should not experience any difficulty in marking out in a proper manner the lines for the necessary joints. Figures 125 to 126 show those in the lower part. These figures are sufficiently explanatory in themselves, and need no further comment. Make all the joints by gluing and wedging, and cut the

movable key–wedges from hardwood. A stool or bench made on this principle from good dry wood will stand any amount of rough usage, and should last as long as the timber from which it is made, there being no nails or other source of weakness to lessen its durability.

Bench with Side and Tail Vices

The general view of a bench with side and tail vices is given by Figure 61 and the construction of the bench is dealt with below. Figure 128 is a side elevation, Figure 129 a plan, and Figure 130 a section on A A (Figure 128). Having sawn out the pieces, next plan them true. Then set out the legs and rails, the latter for mortising, and the former for tenons. The mortises go right through, producing a much firmer result than when the tenons are only stubbed in half-way. The

Figure 130—Part elevation and cross Section of Bench

Figure 131—Joints of Rails and Legs of Bench

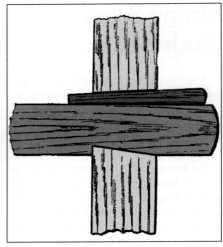

Figure 132—Section showing Rail wedged in Bench Leg

haunched mortise and tenons between the top rails and legs, with the tenons of the cross rails through the legs, are shown in Figure 131 and 132; the tenon of the rail is firmly held in position by a wedge, which must be released, and the tenon of the rail lifted up, before it can be withdrawn. The side rails have a bare-faced tenon—that is, have a shoulder on the inside only. When these joints fit suitably, the legs and cross rails should be glued together and cramped up, and the tenons fixed by wedges, which should be glued before insertion. The top should be planed to breadth and thickness, and then the ends cut off and planed square and to length. The front of the back and end cheeks C (Figure 129) should next be carefully set out and worked. At the front end of the side cheek B, the thickness for dovetailing is not the full 2 inches, but is less by ¾ inch than the breadth of the pin hole, as shown at E (Figure 129). After the side cheeks have been dovetailed and fitted together, groove the front cheek on the back for receiving the stop (see Figures 129 and 133). The inside edge of the top should be rebated as shown at F (Figure 130) to receive the well board. This should fit just tight between the end cheeks, the front and side back cheeks being firmly secured to the top plank and well board. Four-inch screws may be used for the front and side cheeks, and 2½-inch screws for the back, the heads being sunk a little below the surface. Glue the side cheeks to the main board of the top. Then mortise and tenor the cheeks and ends of the runners together as in Figure 134, the top of the runner being kept at the same distance from the top of the cheek as the

thickness of the top plank; two tenons may be more troublesome to make, but the result will be stronger than when only one tenon is used. Make sure these joints are firmly glued and wedged together, with the runner at right angles to the cheek. In Figures 130 and 133 the construction of the guide boxes for the runners is clearly illustrated, the pieces G, a trifle deeper than the thickness of the runner, being firmly fastened to the top plank with 3½-inch screws. The bottom is formed of ¾ inch boarding screwed to the guides. The box for the tail runner extends from the top rail to the inner surface of the end cheek. You will find wrought-iron bench screws about 18 inches by ⅞ inch, having split collars, to be the most satisfactory, and in fixing them into their places, push in the cheek

and runner so they are firmly held in position; then mark the center of the hole for the screw in the cheek, leaving sufficient room for the flange of the box (or nut) for screwing to the side cheek of the bench (see Figure 133). The hole should next be bored through the cheeks of the screw and bench with a bit slightly larger than the diameter of the screw. Then the collars and boxes can be fixed in position, and the framework of the legs and top fitted together. Notch the top rail of the back legs for the runner, shown by H (Figure 131), and if the work has been done accurately the top will just slide on the upper part of the legs. When the parts are adjusted, the front cheek should be secured to the legs, and the top of the bench to the top rails of the legs with 3½-inch screws. The peg board K (Figure

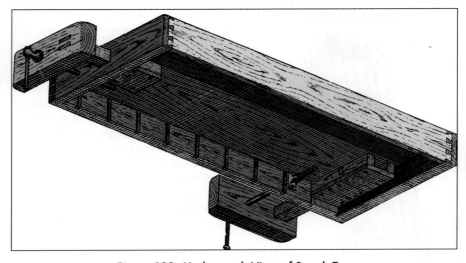

Figure 133—Underneath View of Bench Top

Figure 134—Bench Vice Cheek and End of Runner

128) should be screwed to the front of the bottom rail and to the back of the front cheek. The following are the net sizes of the pieces required (a little in excess of these dimensions should be allowed for waste in working): Top board, 2 inches by 13¼ inches by 4 feet, 8 inches; well board, ¾ inches by 8 inches by 4 feet, 8 inches; peg board, ⅞ inch by 6 inches by 1 foot, 9 inches ; runners, 2 inches by 2½ inches by 2 feet, 3 inches; runner guides, 1½ inches by 2 inches by 3 feet, 2 inches; guide box bottoms, ¾ inch by 5½ inches by 1 foot, 7 inches; screw cheeks, 2½ inches by 6 inches by 2 feet, 7 inches; front and end cheek, 2 inches by 6 inches by 9 feet; back cheek, 1 inch by 6 inches by 5 feet; legs, 3 inches by 3 inches by 9 feet, 8 inches; top rail (front end), 2 inches by 4 inches by 1 foot, 9 inches; bottom rails (ends), 2 inches by 2 inches by 3 feet, 6 inches ; and bottom rails (front and back), 1¼ inches by 2½ inches by 8 feet, 2 inches.

Portable Folding Bench

The portable folding bench shown ready for use by Figure 63, and with the flap down by Figure 64 is illustrated in side elevation by Figure 135, and in end elevation by Figure 136. An end elevation of the bench when folded is given by Figure 137. Sizes that will meet all ordinary requirements are indicated in the illustrations, which show the construction so clearly that only the leading points need description. The legs and rails are jointed together by plain halving and dovetail-halving. The top is at least 1½ inches thick, and is formed of two boards jointed; to keep it true it should be clamped. The top should be hinged to the rail marked A, and the side of the bench hinged to the top as represented by B (Figures 136 and 137), 3-inch butt hinges being used for this purpose. The wall-piece C should be firmly screwed to the rail of the top A. The legs should be hinged at the top of this piece, and also at the bottom to the strop marked D, which should be sufficiently thick to project from the wall to the thickness of the wall-piece C. The piece C can be attached to the skirting board with a few screws. The wall-piece C, if against a lath-and-plaster partition, can be firmly and easily fixed to two or three of the studs of the partition with half a dozen screws; if it is against a brick wall, drill a few holes into the wall and drive in hardwood plugs; or, better

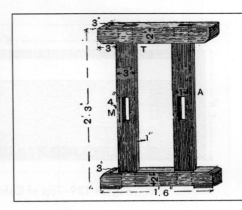

Figure 138—End Framework of Cabinetworker's Bench

still, probe the wall with a long fine bradawl until the joints are found (if this is done carefully, little damage will be suffered by the paper), and then with a steel chisel cut some holes about ¾ inch square and about 3 inches or 4 inches deep. These holes may then be fitted with hardwood plugs, into which screws are inserted through the wall-piece. The fitting-up of the screw, cheek, and runner (the last named being of hardwood) is not difficult. The leg to which the screw is attached is larger than the others. The side and top of the bench when folded up can be kept in position by a hook and eye as shown. The bench may be made additionally firm by inserting a few screws through the side into the legs, and through the top into the rails. When it is required to remove the bench, all that is necessary is to withdraw these screws.

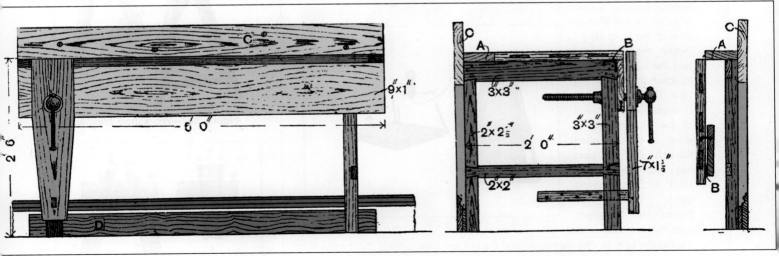

Figures 135 and 136—Elevations of Folding Bench
Figure 137—Folding Bench with Flap Down

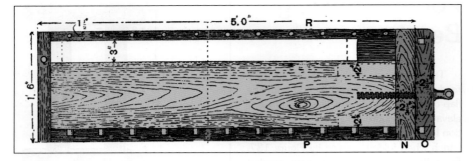

Figure 139—Top of Cabinet-Worker's Bench

THE JUNIOR HOMESTEADER

Make a Birdhouse!

Teach kids basic woodworking skills with simple, fun projects. This is a small birdhouse that will hang from a tree branch. It's perfect for wrens.

Materials

- Large tin can
- Wooden board about 7 inches square
- Carpet or upholstery tacks
- Earthen flowerpot
- Small cork to plug up the flowerpot hole
- Eye screw
- Short stick
- Wire
- Small nails

Directions

1. Mark the doorway on the side of the can and cut the opening with a can opener.
2. Fasten the can to the square baseboard (A) by driving large carpet tacks through the bottom of the can into the board.
3. Invert the flowerpot to make the roof. Plug up the drain hole to make the house waterproof (use a cork or other means of stopping up the hole) (B).
4. Screw the eye screw into the top of the plug to attach the suspending wire. Drill a small hole through the lower end of the plug so that a short nail can be pushed through after the plug has been inserted to keep it from coming out.

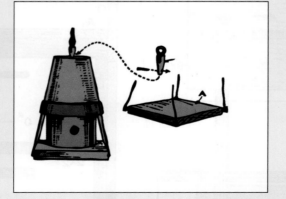

5. Fasten the flowerpot over the can with wire, passing the loop of wire entirely around the pot and then running short wires from this wire down to small nails driven into the four corners of the base (A).
6. Now the bird temple can be painted and hung on a tree.

Sawing Stools and Workshop Trestles

The joint most generally used for connecting the legs of the sawing stool to the top beam is shown in Figure 140. These joints may be fastened with nails, but a stronger method is to glue and screw them together. A serviceable trestle for workshop and general use is shown in Figure 141. Quartering of light or heavy scantling will be appropriate, based on the purpose you are using the trestle, and the method of framing it together will change as well. It is mortised and tenoned as follows. The four legs A are mortised into the top B, as shown in Figures 142 and 143. The mortise in B is cut longer than the width of the tenon of A, to allow for driving in wedges c. The complete

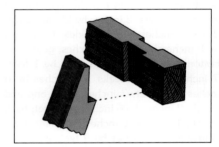

Figure 140—Joint for Sawing Stool

Figure 141—Workshop Trestles

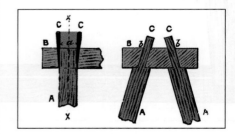

Figures 142 and 143—Legs Mortised and Tenoned into Top of Trestle

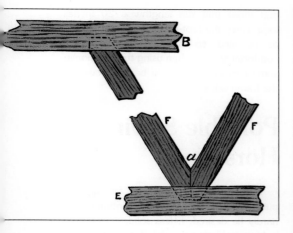

Figure 144—Struts Stump-tenoned into Stretcher

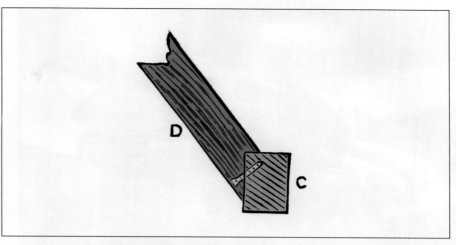

Figure 146—Strut Shouldered upon Stretcher

joint is dovetailed, wider at the top than the bottom, and the legs cannot fall out. Drive in the wedges as shown, against the ends of the tenon and the end grain of the top B, and not against the flanks b of the tenons. If they were driven against the flanks b, they would split the top B. The short stretchers D (Figure 141) are mortised into the legs a, and wedges are driven in against the end grain, as in the previous instance. The long stretcher E (Figure 141) between the short ones is mortised in the same fashion. A trestle made thus with close joints will stand much rough usage. If the legs were short and the scantling of large section, as with a sawing stool, strutting would not really be necessary. But it is better to strutt high trestles made of slight scantling, say not exceeding 2 inches by

2 inches, or 2 inches by 2½ inches cross section. In Figure 141 the struts at F F are tenoned into b and e, but the tenons do not pass through, and are not wedged. It is quite enough to stump-tenon the ends of F F (Figure 144) at top and bottom. The two struts about at a, and further steady the framing. A simpler method of framing trestles of this type together is shown in Figure 145. The only members that are mortised are the legs A, into the top B. The cross stretchers c are simply let for about ½ inch into the legs, and screwed or bolted. The struts D are stump-tenoned into the top, but at the other end they are merely shouldered back to fit over the stretchers c (see also Figure 146), and screwed or bolted. For a somewhat heavy trestle this simpler method is quite good enough; but for a lighter trestle, like previously described, the method of framing together with long bottom stretchers and mortised joints throughout makes a firmer job.

Sawing Horse

For sawing wood, and more especially for sawing firewood, a horse is a great help. There are more ways than one of making it; but this is best for ordinary use. This is mainly built of scantling 3 inches square, and is neat, strong, firm, and serviceable. The four pieces forming the legs are about 2½ feet long, and they are so arranged that the upper limbs of the cross have only half the length of the lower ones; They are halved at the intersection, and strongly nailed together, the nails being driven from the outer side and sent well home; for, should they project, they

would be likely to catch and blunt the teeth of the saw. All the parts have to be so arranged as to leave nothing which can interfere with the free play of the saw, especially no iron. For this reason the central piece a, which chiefly serves to tie the two pairs of legs together, is kept below the intersections, so that it may fit up closely between the legs. Its two ends are, for a length of 3 inches, cut as shown in section by Figure 148. This piece is of the same scantling as the legs to which it is nailed. For ordinary work, 18 inches will be a good length for it. The upper cross rails B B give support to this piece, and are, as is shown, cut away to receive its lower angle at each end. These are of 1-inch wood, 2½ inches wide, and about 1 foot long, and are nailed to the inner sides of the legs. A part only of one of these rails is seen in Figure 147, but you can see its extent by the dotted lines. Now nail on the foot-rails c c, 1 inch by 2½ inches to the legs 2 inches from their bottoms. The cross foot-rails are 2 feet long,

Figure 145—Workshop Trestle

Figure 147—Sawing Horse

Figure 148—End of Longitudinal Rail of Sawing Horse

and those that run lengthwise 20 inches long. Keep the foot-rails as near the ground as indicated, so that your foot can rest comfortably on it when using the saw horse; some part of your weight being thrown on the frame will steady it.

Portable Sawing Horses

A better portable horse for general purposes than the one just mentioned can hardly be desired. If less solid, it would be lacking in firmness, and if too strong, it would be liable to be shaken to pieces by the constant jarring to which it is subjected in use. But sometimes a lighter and more portable horse is desired, and to meet this requirement a shut-up horse may be made as follows: After halving the two pairs of legs as above, cut mortises through them at the intersections, say 1 inches wide by 1½ inches beyond them. In each tenon will be a hole into which a pin, removable at pleasure, can be driven to fasten the frame together for use. In convenience and stability this horse is of course greatly inferior to that described before.

Fixel Sawing Horse

The most simple as well as the firmest of all sawing horses is, however, the primitive fixed one. The legs of this should be about 1 foot or 15 inches longer than those shown in Figure 147, and, pointed at the ends like stakes, they are driven into the ground before being nailed together at their intersections. To connect the pairs, all that is needed is a tie nailed to the legs at the point D (Figure 147), that is, just below the upper limb and on the side upon which the sawyer will stand. Such a horse has the merit of complete immobility; but, of course, it will not serve every purpose.

Houses, Runs, and Coops for Poultry

Hen Coop with Chicken Run

The coop shown by Figure 149 is intended for use in rearing early chickens. The front has a hinged flat which may rest on the top of the run as shown to shelter it partially during the daytime, or it may be

Figure 149—Hen Coop with Run

lifted higher and secured with a hook and staple. At night you can remove the run and let the flap down to keep the brood warm and ward off cats and rats. It the latter are troublesome, the holes over the flap may be covered with wire netting The construction is more clearly shown in the longitudinal section (Figure 150), and in the view of the front of coop with the run removed (Figure 151). To make the coop, first prepare the boards to fold the sides; put them together and nail to ledges of 1½ inches by ¾ inch boards, as shown at the top and bottom of Figure 150. Sound ¾ inch deal should be used throughout, and the joints of the boards should be tongued and grooved for the sheeting. Nail on the boards to form the back, putting a strip up the corners if needed, and get out a rail A (Figure 150), 2 inches by ¾ inch, notching it for the front

rails to fit in at the bottom, and secure it at the sides. Fit another rail across the top as shown, then put on the roof. Next fit up the front (Figure 151), mortising the middle rail A through the roof to allow of its being lifted to release the hen. Make the hinged flap for the front by cutting three or four boards to length and cross-battening them with a couple of ledges; then prepare a rail 2 feet long by 1½ inches by ¾ inch, and secure the flap to this with a pair of butt or tee hinges. This rail should be secured to the front of the coop with screws, so that it can be removed easily with the flap with not required. You can make the run by cutting two 9-inch or 10-inch boards 3 feet or more long to form the sides, and a piece 1 foot, 10½ inches by 9 inches wide for the front. Upright pieces may be nailed on to strengthen the corners B (Figure 150), and a cross rail C must be used at the side against the front of the coop to carry the wire netting. A hinged flap at the front end of the run will be found useful when supplying soft food and water for the chickens, and the top of the run may be covered with ¾-inch or 1-inch mesh wire netting, secured to the sides and ends with small staples. A few center-bit holes may be bored in the coop through the top of the sides to ensure thorough ventilation.

Portable Fowl-House and Run

The house and run illustrated by Figure 152 occupies but a small space while forming the cheapest possible pen in which fowls may be kept in health and comfort. The house and run are made separately, so that they are easily removable to fresh ground when

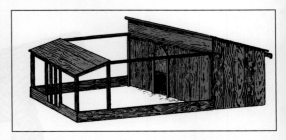

Figure 152—Portable House and Run

the current space becomes foul, and the run is covered with wire netting, which is omitted in Figure 152 to prevent confusion. For the house illustrated, the run is 8 feet or 9 feet long by 5 feet wide, with a height of 2 feet, 9 inches. At the end farther from the house the run is roofed over for a couple of feet, to provide shelter in case of rain at times when it is not desirable for the birds to crowd into the house. Frame the run together with 1½ inches of square deal scantling, with 7 inches by ¾ inch boarding round the bottom, and the joints between the uprights and rails may be either half-lapped or mortised and tenoned. In the center of the back end, four or more bars may be framed as shown, to give the birds access to a trough of soft food or clean water, and enable them to take the one or the other without upsetting or fouling it. You can have the two center bars be moveable for feeding purposes. For a pen of seven or eight birds the dimensions of the house may be 5 feet long, 3 feet, 6 inches wide, 3 feet, 8 inches high in front, and 2 feet, 10 inches at the back. It may be built throughout with ¾ inch matchboarding, strengthened with framing pieces of 2-inch by 1-inch deal. Figure 153 is an elevation of the front of the house, the run being removed, and shows the doorway by which the fowls have access and is fitted with a sliding shutter, which may be closed at night and be held open in the daytime by means of

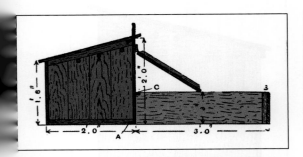

Figure 150—Longitudinal Section of Coop and Run

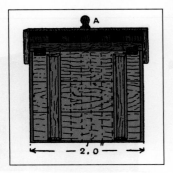

Figure 151—Front of Coop

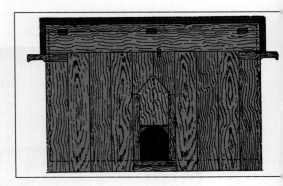

Figure 153—Front of Portable House

Figure 154—Fowl-house Ventilator

a cord and pulley, or a hole and iron pinches In the top portion a hinged flap is provided for the ventilation, and this may be opened and closed by means of a common iron stay as used for greenhouse ventilation.

Interior Arrangements of Fowl-house

Figure 155 is a longitudinal section of the house, showing the arrangement of the perches, floor, and nesting boxes. The perches are made by sawing a 2½-inch or 3-inch round pole through the center, or instead a couple of 2½-inch by 1½-inch rails may have the corners rounded off on one side. For carrying the perches a couple of notched fillets A (Figure 156) are nailed to the front of the nesting boxes at one end, and a similar pair may be nailed inside the opposite end of the house; these should be arranged at a height to keep the perches about 1 foot above the ground. The floor is made by nailing 1-inch boards to three or four 1½-inch square ledges, and it should be well coated with a mixture of hot tar and quicklime sprinkled with as much sand as will lie on it. The floor is simply a platform quite separate from the house, and will fall away when the house is lifted up. Care must be taken that there is not sufficient space between the sides of the house and the floor for the birds to get their feet caught and jammed. At one end, as shown on the left of Figure 155, a shelf is fixed across and secured to the sides by means of fillets at about 16 inches or 18 inches above the floor, according to the breed of the birds, and the space under the shelf is divided into three compartments, as shown in Figure 156, to serve as nesting boxes. A strip about 2½ inches or 3 inches wide is nailed across the front at the bottom, as shown in Figures 155 and 156, to keep the nesting material in position, and pieces 2 inches wide are nailed up the sides on the outside, pieces 5 inches wide being put on the divisions between the boxes to separate the nests from the rest of the house. For removing the eggs, fit the end of the house behind the nesting boxes with a ledged door, the outside of which is shown in Figure 157. This is hung at the top by a couple of cross-garnet hinges, and secured with a turn-button at the bottom, or if necessary by a more secure fastening. For a larger house it may be advisable to have a couple of doors over the boxes. For the purpose of lifting the house, handles are screwed to the top rails, back and front. The roof should be made watertight by nailing strips over the joints or covering with felt, and the house can be painted or tarred outside and well limewashed inside. The woodwork of the run should also have three coats of paint or a coat of tar. Four ventilation holes are shown at the top of Figure 157; these should be bored with a 1-inch center-bit at each side.

Figures 155 and 156—Sections of Portable House

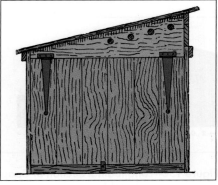

Figure 157—End of Portable House

Sitting Box for Broody Hens

For rearing chickens under a hen, it is necessary to provide a quiet nest where the mother bird will not be disturbed by the other fowls; and when a special shed would be too expensive, the sitting-box illustrated by Figure 158 will be useful. It may be made of ¾-inch deal boarding, or if some sound packing cases can be obtained, they may be used without much alteration, an inch or so in the dimensions either way not being of much importance. To make the box, first prepare the sides A (Figure 158), cutting the boards 1 foot, 4 inches long and making up the width to 1 foot, 4 inches, the strips B (Figure 159) being nailed at the top and bottom to hold the sides together. These strips or ledges may be 1½ inches wide by ¾ inch thick. On the front edges, notch out a piece at c (Figure 159) 1½ inches by ¾ inch, and 1½ inches from the top, and a piece at the bottom D, 2½ inches by ¾ inch, and on the inside nail a ledge F (Figure 160), 1½ inches by ¾-inch mesh wire netting a little larger than the inside of the box, and secure it to the ledges round the bottom of the box with wire staples, so that it will sag in the middle and nearly touch the ground. Make a door for the front by nailing ledges G and H (Figure 160) across a couple of boards fitted into the opening, the top ledge G being kept a little longer than the width of the door to form a stop. The door may be hung with a pair of butt hinges, or with a couple of pieces of leather fixed to the bottom rail, and secured at the top with a turn-button. To complete the box ¾-inch holes should be bored through the sides and back, just below the top ledge, and either a leather strap, as shown, or a common iron handle secured on the top in the center for carrying

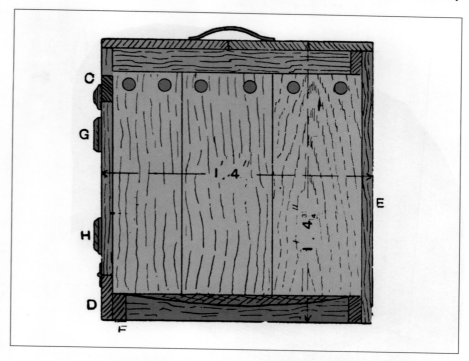

Figure 159—Part Elevation and Section of Box

the box. Figure 158 shows the box with the lid let down to allow the fowl to walk out of the nest; Figure 159 is a front view of the box with a portion of the right-hand side in section; and Figure 160 is a section of the box from the front to the back, showing the wire netting at the bottom.

Lean-to Poultry House and Run

Figure 161 is a general view of a poultry house and run, Figure 162 being a sectional plan. Front and end elevations are given by Figures 163 and 164 respectively. The run is continued under the roosting shed,

which is an advantage where space is a consideration. The run can be made to any extra length desired, in which case it may be necessary to provide one or more intermediate uprights to the front and cross pieces to the roof to carry the wire netting. Timber about 3 inches by 3 inches will be most suitable for the general framework, and ¾-inch matchboarding for the roof. As most of the leading dimensions are given on the accompanying illustrations, the sizes and number of the pieces required can be readily seen. The quantity of wood required will be like so:

1. Framework, about 80 feet of 3 inches by 3 inches
2. Matchboarding for sides and roof, 130 feet of 6 inches by ¾ inch
3. Floor boarding, 40 feet of 6 inches by ⅞ inch grooved and tongued
4. Door to run, 10 feet of 3 inches by 1½ inches, and 2 feet of 4½ inches by 1½ inches

Roosting Shed of Lean-to House

After the pieces for the framework of the roosting shed have been cut, plan them up and set out the four posts. Mortises are made for the tenons of the bottom rails and the

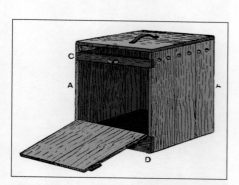

Figure 158—Sitting Box for Broody Hen

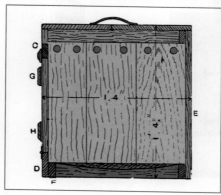

Figure 160—Cross Section of Sitting Box

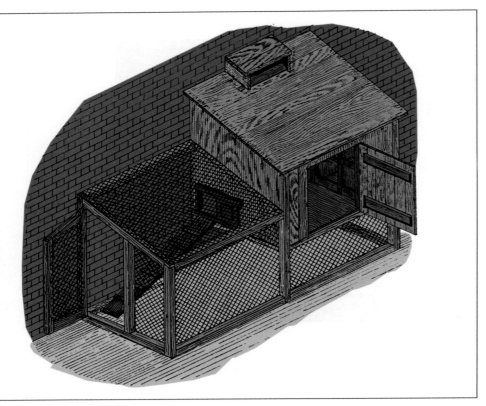

Figure 161—Poultry House and Run

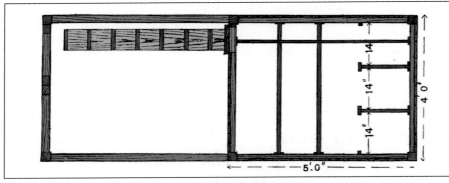

Figure 162—Sectional Plan of Poultry House and Run

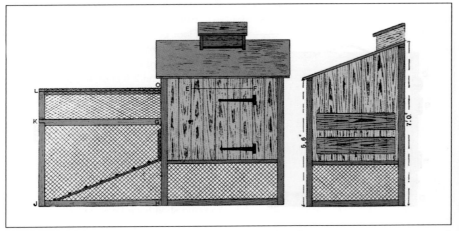

Figure 163 and 164—Elevations of Poultry House and Run

floor, at the same time making mortises for the rails of the run in two of the vertical posts. At this point, the three rails at the bottom, and the four making the floor support, should be set out, and the tenons and shoulders should be cut in the usual way. Cut the mortises in the posts carrying the bottom rails about 2 inches from the bottom, which lets a part of the posts go in the ground, and makes the structure more rigid. The two top rails A B are planed on the top edge to the slope of the roof. A simple but suitable form of joint for the top of the posts and the top rail of the front, and top rail of the back. Next, fit the frame and nail the joints together; a little paint applied to the parts making the joints before fastening will add to their durability and appearance. For the floor, tongued and grooved floor boards (or even thick matchlining) is fitting. The floor P R should be nailed to the front and sides, some strips or fillets, about 1¼ inches by 1 inch, being nailed to the inside of the posts as shown, care being taken to keep their outer edges flush with those on the floorboard. The matchboarding to the front and sides can then be nailed in position, keeping the frame true while fixing. The door is best made afterwards by nailing the top of the boards and cutting along the line E F (Figure 163). Start the cut at L before the boards are placed in position. The two ledges for the back are of the same thickness, and 3½ inches wide, and are secured to the back of the boards by nailing through the front side and clinging the nails on the back of the ledges (See Figure 161). You can hinge the door with two 15-inch cross-garnets, shown in Figure 163. To keep out the draught, and also to form stops for the door and support the pieces of board over the top of the door,

three strips of wood, about 2 inches wide and ¾ inch thick, should be nailed round the inside of the opening, and should project about ½ inch, so that the door shuts against them. Before constructing the roof it will be best to fit up the nests; ¾-inch boarding will do for this purpose. The best means of fixing the shelves and divisions will be to nail fillets to the sides and floor, to which the boarding can be nailed. Matchboarding will be best for the roof, since rafters will not be required, one bearer, going from side to side c, being enough to support the roof. The boards in this case run from front to back. The entrance holes for the fowls can now be cut out, and two rebated slides and a sliding door made. A hinged flap is fixed on the outside of the slide, covering the holes of the nests, so that eggs can be removed without opening the door of the roost shed. Cut out a hole in the roof for ventilation, and fix a ventilator as illustrated. The flap D is hinged to the roof, and can be opened or closed to regulate ventilation. You should cover the roof with felt to make it watertight in case of rain.

THE JUNIOR HOMESTEADER

Build a treehouse!

Building a treehouse requires a lot of work and some good planning. Be sure to build the house in trees that are large and sturdy. If you do not have such old trees in your yard, you can always modify these plans to build treetop havens on wooden platforms raised above ground.

Low Two-tree Treehouse

This treehouse can be constructed out of ordinary boards and timber. It does not sit up quite as high in the trees but is still elevated above the ground to give a good view of the yard and surrounding area.

Directions

1. Select a location between two trees that are roughly 6 to 8 feet apart. The trees should have fairly straight trunks and should be at least 15 inches in diameter. Make sure they are healthy and sturdy—not decaying in any way.
2. Using an axe, clear off the brush and small branches up to 20 feet on the tree trunks (or to the height of where the treehouse will be located).
3. Take four or five pieces of spruce (from a lumberyard or home center) that are 2 inches thick, 8 inches wide, and 16 feet long. Saw off and nail two of the pieces to the trunks of the trees 8 feet above the ground. First cut away some of the bark and wood to make a flat surface on the trunk. You will need 16-inch steel-wire nails to anchor the boards to the trees.
4. Cut two timbers 6 feet long and the other two the length of distance between the tree trunks. In the 6-foot pieces, cut notches on the underside. The ends of the bracket timbers will fit into these notches.
5. Cut the ends of the timbers to form a square frame so that they dovetail. Spike in 6-foot timbers to the tree trunks so that they will rest on the first two timbers that were

nailed to the trees (see image on top right).
6. Place the remaining two timbers in position so that the ends fit into those fastened to the trees. Nail them well.
7. Support the first timbers that are spiked to the tree trunks with 15-inch blocks nailed below them. The cross timbers and last ones form the frame. Place the frame into the dovetailed joints at the ends.
8. Cut two more timbers and lay them across the supporting timbers, nailed to the trees, so they will fit inside the front and back timbers, and secure them with long nails. The floor frame is now complete.
9. Construct a frame 7 feet high at the front and 6 feet high in the back out of 2 x 3-inch spruce. Spike the side timbers, forming the top, to the insides of the tree trunks (see bottom right image). Mount the bottoms of the uprights on the corners of the floor frame and use four long nails to hold them into place.
10. Now, cut two timbers and arrange them in an upright position at the front 30 inches apart. The door will be here. Halfway between the floor and the top of the frame-work, construct timber all around except between the door timbers. This will add strength and will allow the sheathing boards to be nailed. It will also make one more anchoring beam between the tree trunks.
11. Then nail the side rails into the tree trunks in a corresponding way to the top (roof) strips.
12. Make the floor from lumber 4, 6, or 10 inches wide. The boards should be planed on both sides.
13. Construct the roof of the same boards. You can lay tarred paper over them and fasten it to the edges with nails. This will help waterproof the roof at least for one year. To make the roof last longer, you can shingle it.

14. Windows can also be made in the side and back walls. These should be about 24 inches square. The door can be constructed out of boards held together with battens. A lock can also be furnished to keep out unwanted visitors.

The treehouse will need a ladder in order to access it. This can either be purchased from a yard sale or can be made out of hickory poles and cross-sticks 20 inches wide. To keep the ladder from slipping while ascending and descending, affix loops to the top of the ladder; these will fit over large, sturdy nails driven into the doorsill, and the ladder will be relatively stable.

A flexible ladder can also be made out of ropes and hung much the same way as the wooden ladder. This type of ladder, though not as sturdy, can be drawn up when people are in the treehouse so no one else can enter.

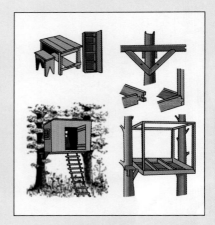

≈ **Refer to these illustrations when constructing the low two-tree treehouse.**

Inside the treehouse, small chairs and other seats can be constructed and used for relaxing. Narrow shelving can be made and fastened over the windows with brackets. Small things can then be housed on these shelves. A small table may also be housed in the treetop shelter.

High Treehouse in One Tree

If you have a tree large and strong enough (oaks are very good for this), a treehouse can be successfully built in its branches. For this plan, the treehouse will be 25 feet above the ground, and below it is a landing from which a rope ladder can be dropped to the ground. A more solid, wooden ladder connects the landing with the deck of the treehouse and it can be situated through a hole in the deck of the house.

Since every tree is different, it is difficult to give exact dimensions of the frame of this treehouse and how many floorboards should be used. But the construction of a single-tree treehouse is, in many ways, similar to that of the low two-tree treehouse. The trunk of the tree will have to project up through the treehouse and the out-spreading branches will need to support the lower parts of the floor frame. This treehouse can have either a peaked or a flat roof,

depending on the structure of the treehouse and the space allowed for a roof within the treetop.

Brace the floorboards well to the main trunk of the tree with long and short brackets or props. These will help make the house secure in the tree. Drive large spikes into the tree where the lower ends attach to the trunk. Nailing cleats or blocks under these will help to support and strengthen the structure.

˅ You can even build a deck for your treehouse, as long as you're building on a very sturdy tree.

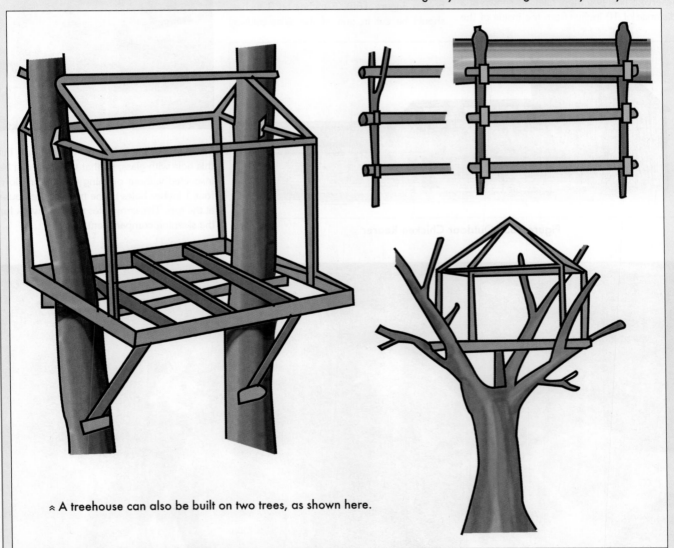

˄ A treehouse can also be built on two trees, as shown here.

˄ Plans for building a high treehouse in one tree.

Outdoor Chicken Rearer

Figure 165 shows an outdoor rearer for accommodating fifty chickens. The house or sleeping compartment A is 2 feet square, made of ¾-inch floor boards, and is 24 inches high at the front and 18 inches at the back, thus giving a good slope to the roof. Two fillets of wood E (Figure 166), 1 inch by ¾ inch, are nailed at a di stance of 2 inches from the ground, and on this fillet the bottom is fixed. The two sides are shaped at the bottom as shown to allow for air to get to the lamp. In the center of the bottom a 6-inch circular hole is cut to accommodate the lamp reservoir. Nail two more fillets as shown at a distance of 10 inches from the inside of the

bottom. On these rest the inner lid, made of ¾-inch material. it fits easily inside the case, and has handles for convenience in lifting out, and a 3-inch hole should be cut in the middle to allow the waste heat to escape. The four corners should be blocked up with pieces of wood 5 inches wide, to keep the chicks from overcrowding there, and the rearer bottom will have a hexagonal shape. The outside lid overlaps all round by about 1½ inches, and should be hinged at the side. On the front of the house fillets measuring 1½ inches by 1 inch are fixed; they should be 9 inches from the bottom at the ends, and should have a rise of 9 inches at the middle. These support one end of the glass run. Cut a small hole in the front of the house for the doorway and attach a ladder here. Another hole D (Figure 165), 5 inches by 3 inches, should be cut in one of the sides midway

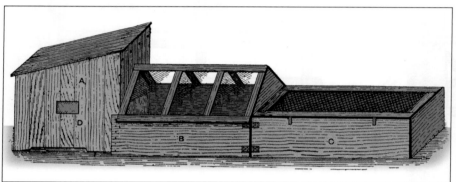

Figure 165—Outdoor Chicken Rearer

between the inner lid and the bottom. Fill this hole with glass, so that the lamp can be inspected without opening the rearer. Bore four 1-inches holes in the front of the house at the top. This completes the woodwork for the sleeping compartment.

Runs for Chicken Rearer

The runs are made to fold up for convenience in packing when the rearer is not in use. For the sides of the runs, you will need four pieces, two pieces for the glass run B (Figure 165) 2 feet 8 inches long by 9 inches wide by ⅝ inch thick, and for the wire run C two pieces 3 feet, 6 inches by 9 inches by ⅝ inch ; one piece for the end of wire run 2 feet by 9 inches by ⅝ inch; and one middle partition made of ¾-inch floor board, which should be cut to the same eave as the

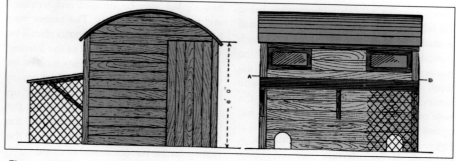

Figures 167 and 168—Side and End Elevations of Fowl House with Semicircular Roof

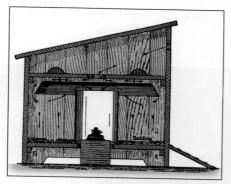

Figure 166—Section of Sleeping Compartment of Rearer

fillets on the front of the house. In making the whole up, hinge together the sides of the glass and wire runs, leaving a ¾-inch space between the two. This space is to support the middle partition, which just slips in. Two pieces of material 9 inches by 1½ inches by 1½ inches are nailed at the ends of the wire runs, and to these pieces the end is fixed using screw-eyes. The sides of the glass run overlap the house by 2 inches, and are again fastened with screw-eyes to the house. One side of the glass lights is fastened to the house using brass plates which are screwed to the frame, one at the top and one at the lower end of the frame. A frame of wood should be made to fit flush on the sides of the wire run, and four pieces of iron 2 inches

long by ½ inch wide should be screwed to the sides of this frame to keep it from slipping off. To the under-side of this frame a piece of wire netting, 1 inch mesh, is fixed.

Fowl House with Semicircular Roof

Figure 167 shows a side elevation, Figure 168 an end elevation, and Figure 169 a plan and part section of a simple fowl house with a semicircular roof. Figure 169, showing a section on A B (Figure 168), makes clear that the house is divided into two compartments, with a passage along the back. The compartments are for the poultry, and could house two separate types. Each is complete with nests, roost, and trap, as shown at N, K, T (Figure 169). Any suitable boxes will do for nests, and some may be nailed to the sides at convenient heights, instead of being all on the floor. The passage is for attendance, and the fowls should not have access to it. Plain ledged doors are put in the partition between the passage and compartments; there is also a plain ledged door on the passage, the front of which is shown in Figure 167. The back of one of the inside doors is shown in Figure 170. These inside doors are not boarded

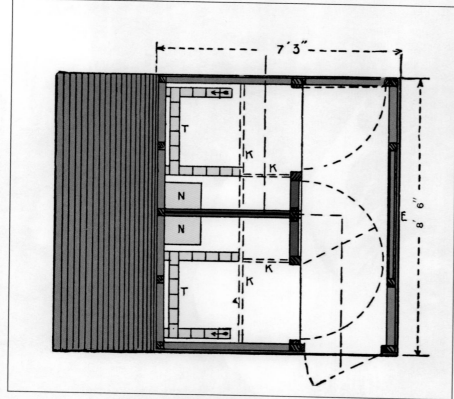

Figure 169—Sectional Plan of Fowl House

Figure 170—Inside Door of Fowl House

379

Houses, Runs, and Coops for Poultry

close, a space of 1½ inches or so being left between for light and ventilation. Install a small glazed window in the front of each compartment, and another in the back of the passage. A covered run stretches along the front, lined up on the side and ends with wire netting. A semicircular-headed hole at the base of each compartment gives ingress and egress to the run. The roof is, as shown in Figures 167 and 171, semicircular in outline; its boards are ⅝ inch thick, and are covered with felt-cloth tarred. The run is also boarded and similarly covered. The roof ribs or spars will do at 2 feet centers, one of course to be at each end; 1½ inches thick will be enough. The framing is put together with butt joints nailed diagonally. The boarding is rough and laid horizontally, overlapping as shown in Figure 172. The window-frames are square

arrised, but may be dressed and mortised and tenoned at the joints. The timber may be red or white pine. Corner posts should be 3 inches by 3 inches, the rest of the framing 3 inches by 2 inches. The posts may be let into the ground, but it is best to cut them on the bottom sill, which should be leveled up a few inches. The floor may be simply the ground leveled; but to keep out vermin the floor should be asphalt or concrete, and the netting on the run continued all round.

Figure 172—Boarding on Fowl House Framing

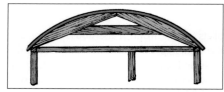

Figure 171—Semicircular Roof

Gates and Fences

Field Gate

Well-seasoned oak is best material for a field gate, but larchwood is cheaper, and is suitable when not too ripe and when felled at the right season. Figures 173 and 174 illustrate one form of braced field gate, 9 feet long and 4 feet 6 inches, the slamming stile 3½ inches by 3 inches, the top and bottom rails 3½ inches by 2¼ inches, and the intermediate rails and braces 3 inches by ⅛ inch. In making this gate, first cut the different pieces a little larger than the finished sizes, and plane them true to breadth and thickness. Set out the stiles, and place them with face sides together and face edges outwards. Then mark off the positions for the mortises and square down the lines both face edges at once, as shown in Figure 175. As all the mortises are to go through, take each stile separately and continue the lines across the face side and down the opposite edge, as in Figure 176, which shows the stile set out, Figure 177 showing it mortised. All the rails should be placed one on top of the other and

marked off for the shoulders and the center muntin. As the mortises for this will only be stubbed, lines will only be required on the edge. Each end of the top and bottom rail should have the shoulder lines set out and cut. The three intermediate rails will require the shoulder line to be marked across one face only, as these will be barefaced tenons. The stiles may now be gauged for the mortises, those for the top and bottom rails should be in the center of the stiles, as in Figures 176 and 177, however in the intermediate rails, one side of each mortise will be in the center of the thickness of the stile. Then gauge the rails for the barefaced

tenons. The next process will be marking the mortise and cutting the tenons and shoulders. The framing prepared so far should be fitted together, and the whole held together by nailing on a couple braces. Gates are,

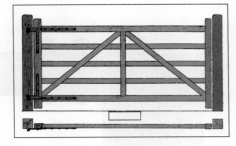

Figures 173 and 174—Elevation and plan of Field Gate

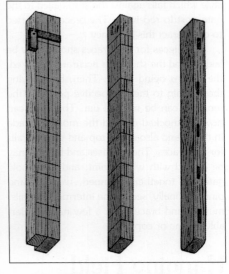

Figures 175 and 176—Setting out Field Gate Stiles
Figure 177—Field Gate Stile Mortised

or should be, braced diagonally; and you will find it easier to put in a single brace, extending it from the lower left-hand corner to the upper right-hand corner. The weight of the gate has a tendency to pull it down toward the ground at the falling style—at the style which falls against the stop nailed to the gate post to receive it. The brace is inserted to counteract this tendency.

The pieces for the braces should now be laid on and the shoulders accurately marked, the tenons being made. Then, by applying these again to their respective positions the mortises can be marked out. The stiles can now be knocked off and the mortises made in them, and also in the top and bottom rails for the braces. The mortises and tenons may be coated with white paint, and the whole gate put together, cramped, wedged, and pinned. Finally, secure the intermediate rails, muntin, and braces with a few nails, preferably of zinc or copper.

Hanging Field Gates

The majority of field gates are hung with ordinary hooks and rides, and by a little knowledge of certain natural laws these hinges can be so arranged as to give an automatic self-closing function to a gate from any part of the half-circle to which it may be opened. Gates should be perfectly plumb and in the same vertical plane. When hung, the gate's line of repose should be close to the striking-post. If hung properly this is the only place within the half-circle at which it will remain stationary. In the method of hinging adopted for the gate shown by Figure 178, the top hinge should be 3 feet long, made of 2¼-inch by ½-inch iron with a gradual thickening over the harr to the eye, shown in Figure 180. This hinge should be about 3½ inches over the harr, so that the hook may be fixed soundly in the post. The

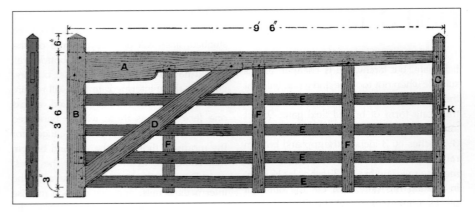

Figure 179—Sussex Field Gate

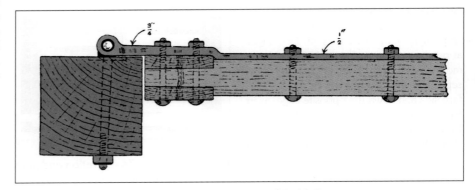

Figure 180—Top Hinge of Field Gate

bottom hinge is a short one (see Figure 181), and is fixed to the harr about 8 inches from the bottom end with two 5-inch by ½-inch bolts. The hooks for the hinges should have screwed and bolted ends. A good kind of hook for this type of work is represented by Figure 182. A different method of hinging is shown in Figure 182, the first showing the top hinge and the second the bottom. The top hinge is a double strap, and therefore the strain is not put on the bolts, or the beams of the gate. The neck of the hinge at G should be at short as possible, and the hook should reach through the post and fasten with nut and screw, as shown. The bottom hinge is short at the sides, and is fixed with two countersunk nails on each side. Its hook should be square and short, as in H; the shank being nearly parallel, it then holds tightly

and remains firm. This method of having a post with a gate hung on hooks that are in the same vertical and horizontal planes with each other is not recommended because the center of gravity is always the same in the vertical plane, no matter what arc of the circle the gate may be moved. The consequence is that it requires the application of

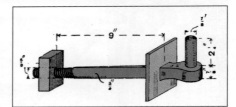

Figure 182—Hinge Hook for Field Gate

the same amount of power to shut it as to open it, when usually, it shuts automatically.

Repairing Field Gates

Figure 183 shows an ordinary field gate repaired in all the places where it is likely to need it. Assume that K is a broken rail that must be spliced. The old bar or rail is sawn

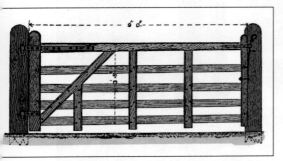

Figure 178—Five-bar Field Gate with Upright Braces

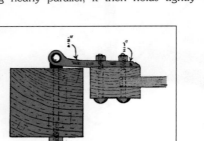

Figure 181—Bottom Hinges

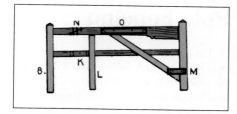

Figure 183—Field Gate Repaired

off about a foot from the downright L, shown by the dotted lines. A short piece of new material is then driven into the mortise head b, and cut off the right length so that it will fit close up to the downright, the two then are nailed together as shown. Figure 184 shows this splicing on plane, the shaded parts being sections of the head and downright. Of course, longer splicings may be needed. It is a bad plan to cut the rail halfway on a downright,

Figure 184—Spliced Gate Rail

and butt the other up to it; this can cause the downright to split down the middle. One of the first places for a gate to rot is at the junction of the brace and harr. This is caused by

Figure 185—Spliced Gate Beam

the wet lodging there, and eventually finding its way into the mortise, and there being no outlet it has to stay there until it rots its way out. The only way to repair this is, as shown at M (Figure 183), by a piece of oak about 4½ inches by 1½ inches nailed firmly to the brace and harr. This should be placed as low down as the bottom ride will allow, unless the wood is rotten, so that the nails will not hold, in which case it must be placed where the solid wood is; but to be efficient it must be placed low down. At N (Figure 183) is shown the beam spliced. This should never be done as shown here,

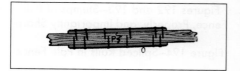

Figure 186—Scarfed Gate Beam

unless it is broken at the small part; if broken farther back, this method of splicing is useless. First cut the splicing on the beam, and then cut the tenon of the new piece you are putting on so it fits in the head; it can then be put in its correct place and marked exactly by the beam, so that it will fit the first time. The longer the splicing is made, the stronger the joint will be; it should never be less than 9 inches long, and must be nailed from both bottom and top, and

Figure 187—Wicket Gate

a couple of large nails passed right through and clinched. This splice can be made stronger by wrapping hoop-iron round it, as shown in Figure 185, and nailing well on each side of the splice. The iron can easily be bent round close by fixing one end first and then pulling it over with one hand and tapping it with a hammer at the same time. In the case of a gate beam being broken in the thicker parts, which very frequently happens, the only way to repair it is to scarf it with a piece of oak about 1½ inch thick nailed on each side. This is shown by O (Figure 183), and in plan at Figure 186. Chamfer off the edges so that they will not injure the cattle, and if nailed on well, the gate will be almost as strong as a new one. P (Figure 186) shows the fracture of the beam, o being the scarfing pieces, which you should make of seasoned oak. The only other repairs likely to be needed are new downrights and heads. The former only require driving in and nailing, and the latter have only to be mortised the same way as the old ones and pinned.

Figure 188—Joints in Wicket Gate

Wicket Gates

The simplest and smallest gate that you can make is a wicket gate. It has the average size of about 3 feet, 6 inches to 2 feet, 6 inches to 3 feet width—enough in fact, to admit access to any ordinary garden, court, paddock, or enclosure of no great size. Figure 187 shows the plainest form of the wicker gate of this type. The back stile and the head should be made of oak, the rails, brace, and pales of deal. To frame this wicket, all the rails should be mortised through the stiles with a bareface tenon, and pinned. Wedging is of no use for this kind of work. The brace (which should never be left out) should be mortised and pinned into the back stile and the under side of the top rail, as shown at Figure 188.

Wood for Fencing

For ordinary fencing, English oak is preferred; no other wood lasts so long in the ground, with the exception of Spanish chestnut, and this, when grown to a large side, is apt to be brittle, and not suitable for fencing. For pole fencing, almost any kind will do for the rails, though chestnut works best; the stumps should be either chestnut or oak, but oak when small is nearly all sapwood, and rots quickly. Larch is sometimes used, but it is very poor quality, and

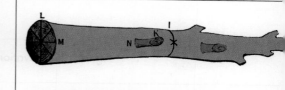

Figure 189—Oak ready for Cross-cutting and Cleaving

hardly worth putting up, especially where the fence has any strain to bear, as in fields where the cattle rub against it. Wrought-iron rose-head nails should also be used in every case, common cut nails are useless. The length of time that a post will last in the ground depends upon the nature of the wood, and the kind of earth it stands in, sand being the worst, and damp clay the best; some people remove the sand and surround the posts with clay, as this is believed to be the best preservative.

Barways

Barways are rough fence with movable rails, and are made simply by cutting gaps in the hedge, and putting up an ordinary fence post at one side, and a running bar-post on the other, with a distance of 9 feet between them. The principle of the bars is this: Apertures are cut in the opposite posts to receive the ends of the bar, which can be made of material 3 inches deep and 2 inches thick. The depth of each aperture must be sufficient to receive the end of the bar that is put into it, to a depth that will allow for ½ inch to 1 inch between the mouth of the aperture on the opposite side

and extreme end of the bar. Supposing the space between the aperture is 3 feet and the depth of each hole is 4 inches, the bar must be of a length not more than 3 feet 3½ inches. The bar is marked and placed in by running the large end through the mortise in the running-post, and drawing it back, so that the other end enters the mortise in fence post, and they are removed in the same way. To prevent cattle from forcing their way behind the posts, short rails are placed in the remaining mortises, the other ends resting in the hedge. Figure 191 shows a complete barway. All the back rails should be pinned, but, of course, the pins must not be allowed to catch the bars.

Pole Fencing

The above description applies to the principal kinds of rough fencing; you will not find the others to be of great difficulty. The most troublesome may be the pole fence, formed by stumps driven in the ground and rails nailed to them. It is used primarily for parting fields and can be put up very cheaply, as it doesn't require post-holes or mortises. The stumps must be cut from the largest poles, and should not be less than 3 inches in diameter, and for a fence 4 feet high they should be cut off 5 feet, 6 inches long; the ends should be pointed, as shown in Figure 192, but not too sharply, and tapering them on four sides so that the pointed part is of a square section. They will drive into the ground relatively easy, and will hold tight, but if pointed as shown in Figure 193 they are driven in with difficulty and do not hold tight for any length of time. The poles which serve as rails are from 12 feet to 19 feet long and about 2 inches in diameter at the larger end, which should be chopped flat on one side, where it fits on the stump, as it can then be fixed with shorter nails. In putting up this kind of fence, drive in a row of stumps at the right distance apart to take the ends of the rails—for instance, if the rails are 12 feet long, the stumps should be 11 feet between; if the lengths vary, as they most likely will, they must be laid out on the ground, and the stumps driven in accordingly, allowing 6 inches at each end for overlapping. When the row of stumps is in, drive in a stump midway between each two, or the fence will be very weak. The rails can then be nailed on, the top one first, and from that gauging with the eye the positions of the others; the stoutest should be at the top, and the larger ends should be kept all one way.

The splicings are nailed as in Figure 194, the thicker end always being under the other.

Figure 190—Elevation of Rough Fencing

Figure 191—Barway with Back Rails

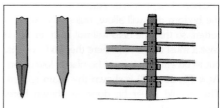

Figures 192 and 193—Stumps of Pole Fence, Properly, and Importantly Sharpened
Figure 194—Spliced Rails in Pole Fence

Sheds, Tool Houses, and Workshops

Open Shed or Hay Barn

The shed or hay barn described below can be made by anyone able to use an axe and saw. Figure 195 is an isometric sketch of the frame, and Figure 196 is an end view. The posts A go into the ground about 4 feet, rise above it about 15 feet, and are about 8 inches square at the bottom and taper to 6 inches at the top. Each post can be sawn or chopped and must have a tenon about 4 inches long, 3 inches wide, and 1½ inches thick, cut on the top end, or the tenon may be left the whole

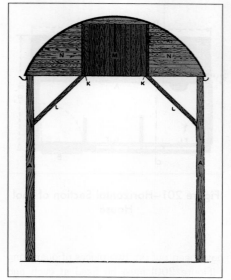

Figure 196—End Elevation of Hay Barn

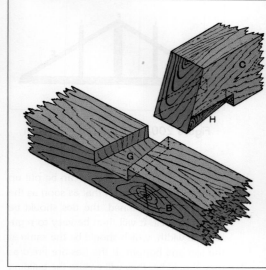

Figure 197—Tie Dovetailed to Plate

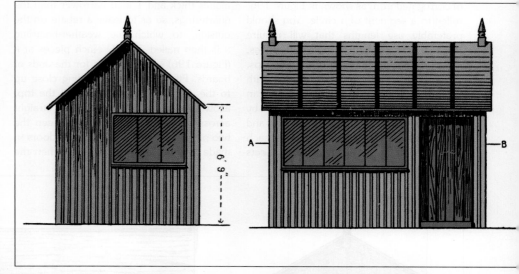

Figures 198 and 199—Elevations of Gable-roof Tool House

Figure 195—Framing for Hay Barn

width of the post. Plates B, preferably of yellow deal, should be about 6 inches by 4 inches, and mortised to fit the tenons, allowing them to reach about 6 inches over the end posts. If the proposed building is of such a great length that the plates have to be spliced, the splicing should be done by making a halved joint, not less than 1 foot in length, the joint, of course, to rest on one of the posts. Ties C should be from 6 inches by 4 inches to 9 inches by 4 inches, according to the length of the building. The former dimensions will work for a structure up to 14 feet wide and the latter will do up to 20 feet wide. The king-post D should be 6 inches by 4 in, no smaller, whether the building is wide or narrow. It should be tenoned into the center of the tie, as illustrated. The braces E, 3 inches square, can

be placed as shown in the two end trusses, or the middle ones can be shorter. These don't need to be mortised or tenoned if they are strongly nailed. The ridge F, 3 inches by 3 inches, should lie in slots cut in the tops of the king-posts. If it has been spliced, the splicing should be at one end of the posts, as in the case of the plates. The ties C should be dovetailed to the plates B, as in Figure 197, which shows the plate cut at G and dovetail H on the tie. These should all be fitted and numbered before you put up the posts. The king post, braces, and ridge should also be fitted in their respective places, but as yet nothing should be permanently fixed.

Erecting the Hay Barn

After digging the necessary holes, the posts should be placed along one side and propped up to prevent them from falling. The plate for that side can then be put on the tenons at the top and pinned, the pin-holes having already been bored. The two end posts should then be adjusted so that they stand upright and parallel, and that the plate is about level from end to end. Afterwards, ram the two posts up tightly, and then raise or lower the intermediate ones

Figure 200—End Truss of Hay Barn

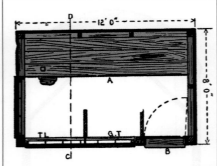

Figure 201—Horizontal Section of Tool House

Figure 202—Portable Span-roof Workshop

as required. The other posts can be put up in the same way, except that as soon as the plate is on and pinned, the ties should be put on as well. It will then be easy to regulate the width, which should be the same at the top and bottom. If the ties are fitted as they should be, they will ensure the building being square at the angles when finished. After the king-posts, braces, and ridge have been properly adjusted, the structure will be ready for the roof, which you should make of corrugated iron, as shown in Figure 196, rolled to a segment of a circle. You should preferably use lengths that will require three to reach over from eaves to eaves, allowing 6 inches for each of the two laps. These sheets must be fixed together with short bolts while on the ground. They can be placed in position to form the roof by nailing or screwing them to the ridge and plates. Let them overhang 6 inches at each end of the building. When the roof sheets

have been put on, two pieces of quartering I (Figure 200) must be fixed at each end of the barn. They can be tenoned into the ties, and the braces or rafters cut away so that the upright pieces fit inches On these two pieces are nailed two other pieces, 1¼ inches thick and 1 inch narrower than the quarterings, so as to leave a rebate on the outside, to which the weather-boarding N is then nailed. The 1¼-inch pieces at K (Figure 196) serve as stops for the ends of boards. Fit the weather-boarding close up to the corrugated-iron sheets—at the top, allowing about two boards to run straight across as shown. The space between the two pieces K is filled in with folding-doors M, made to open outwards. These, when the

barn is full of hay, can be opened to allow a current of air to pass through , and also to ventilate the building. Braces L (Figure 196) can be fixed next, not only at the ends but between each pair of posts. Mortise them into the posts with a stub tenon, and nail them to the ties and plates at the top. This method will be stronger and more convenient than nailing at both ends. If a large barn is built, it should be made high enough to admit of loaded wagons being drawn in between the side posts.

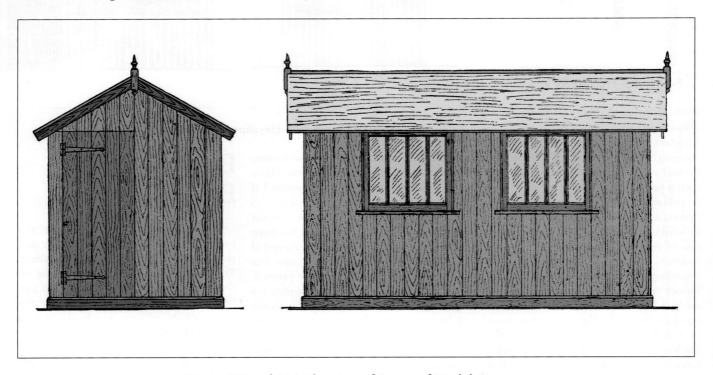

Figures 203 and 204—Elevations of Span-roof Workshop

Portable Span-roof Workshop

Figure 202 is a perspective view of a portable workshop constructed so that the floor forms one piece with the joints and curb, the ends and sides each form a section, and each side of the roof can be of one or more pieces as desired. The ends are attached to the sides by the angle posts meeting and being held together by three ½-inch bolts. For neatness, all the wood should be planed. Figures 203 and 204 are elevational view, and at Figure 205 the skeleton framework is shown. The quantities of material required are as follows:

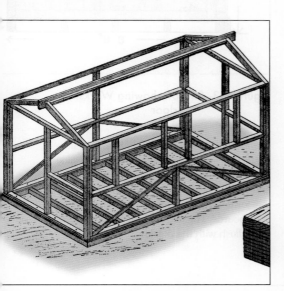

Figure 205—Workshop Framing

1. Curb, 38 feet of 3¼-inch by 3¼-inch
2. Joists, 48 feet of 3-inch by 2-inch
3. Uprights, 77 feet of 2½-inch by 1½-inch
4. Rafters, braces, and bottom rails, 41 feet of 2½-inch by 1½-inch
5. Ridge, 13 feet of 6-inch by 1-inch
6. Bargeboards, 16 feet of 3½-inch by 1-inch
7. Plinth, 38 feet of 5-inch by ⅞-inch
8. Floorboarding, 132 feet of 7-inch by 1-inch
9. Matchboarding, 638 feet of 6-inch by ¾-inch
10. Finials, 3 feet of 2½-inch by 2½ inch
11. Ridge fillets, 26 feet of 2½-inch by 1-inch
12. Roof ledges, 52 feet of 4-inch by 1-inch
13. Door ledges, 7 feet 6 inch of 5-inch by 1-inch

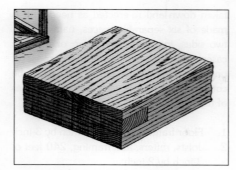

Figure 206—Corner Joint of Curb

14. Window sills, 7 feet 6 inch of 4-inch by 1½-inch
15. Window stiles and top rail, 18 feet of 2-inch by 1½-inch
16. Window bars, 15 feet of 2-inch by 1¾-inch
17. Middle rails, 33 feet of 2½-inch by 2-inch
18. Bottom and end rails, 48 feet 2½-inch by 1½-inch
19. Side plates, 24 feet of 3-inch by 2½-inch
20. Outside bead, 12 feet of ¾-inch by ¾-inch
21. Inside bead, 24 feet of 1-inch by ⅝-inch

Beginning with the floor, the curb forming the outside frame should be halved together, as seen in Figure 206. The joists should be notched into the curb, and when fitted these parts should be secured by nails. Probably the best material for the floor will be 1-inch grooved and tongued floorboards cut to length and nailed down with close joints, the boards extending to the outside of the curb all round (See Figures 206 and 207). The framework of each end should be mortised and tenoned together. The top plate of the front and back is splayed off to the same angle as the roof. The outside of the framing should be covered with ¾-inch machine-prepared matchboarding, and to improve the appearance of the inside the backs of the boards should be smoothed over before being fixed to the framing, the beaded sides being outward. Figure 208 shows the boards of the sides A and ends B projecting over the angle posts. The door may be formed of five boards with three ledges nailed across the inside, and can be hung with a pair of 18-inch cross garnets (See Figure 203). Any suitable fastener may be used. Two pieces, 4 inches by 1½-inches, should be prepared to the section at C (Figure 207) and cut to fit between the window posts, projecting

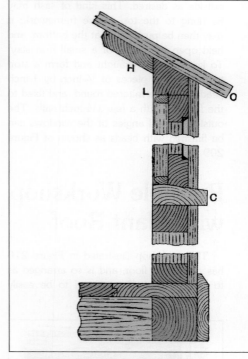

Figure 207—Section of Workshop Window

at each end as shown by Figures 202 and 204. Two sashes with three bars may be used; They are of very simple construction, being mortised and tenoned together, and rebated, either on the inside or on the

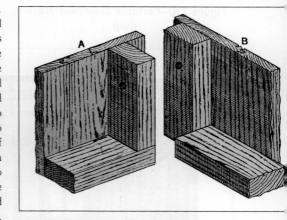

Figure 208—Corner Posts, Sills, and Boarding

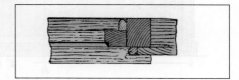

Figure 209—Part Horizontal Section of Workshop Window

outside as desired. This kind of sash may be hung to the top of the framework; it may then be pushed out at the bottom, and held open by means of a small iron stay. To keep out the draught and form a stop to the sashes, pieces of ⅝-inch by 1-inch bead should be mitered round, and fixed to the framing with a few 1½-inch nails. The outside vertical angles of the windows can be finished with beads as shown at Figure 209.

Portable Workshop with Slant Roof

The workshop illustrated in Figure 210 has a boarded floor, and is so arranged as to be a tenant's fixture, and to be easily

Figure 210—Portable Slant-roof Workshop

taken down and re-erected as needed. It is made of six separable pieces, namely roof, two sides, two ends, and floor; or in some cases, the wall might serve the purpose of one side, and only five pieces will need to be constructed. The following are the quantities of timber needed:

1. Floor frame, 38 feet of 3-inch by 3-inch
2. Joists, rafters, and framing, 240 feet of 3-inch by 2-inch
3. Floorboards, 156 feet of 6½-inch by ⅞-inch
4. Side frame, 24 feet of 3-inches by 1½-inch
5. Matchboarding, 466 feet of ¾ inch
6. Sashes and skylight, 34 feet of 1½-inch by 9-inch board

As the drawings show the construction very thoroughly, it is not necessary to give lengthy description here, but only the particulars of construction. Figure 211 is an inside elevation of one side, Figure 212 a section of line A A in the previous figure showing the inside of one end, Figure 213 a section on line B B showing the inside of the end containing the door, and shows at A a half-plan with joist and framing of floor, and at B a half-plan of roof with rafters, part of skylight, and part of boarding. The four outside pieces of the floor should be formed of 3-inch by 3-inch timber, half lapped together at the angles. The joints may be 3 inches by 2 inches, notched into the outside pieces. All the joints should be firmly fixed together with

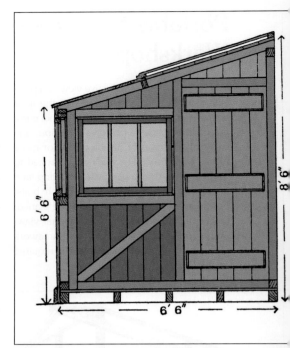

Figure 213—Section showing Workshop door

2½-inch nails. Prepared floor-boards will be preferable for the floor, and if you are willing to pay an extra expense, have them grooved and tongued. The boards should be fixed to the joists and sills with 2-inch or 2½-inch floor-boards. The edges of the boards should finish flush with the outside.

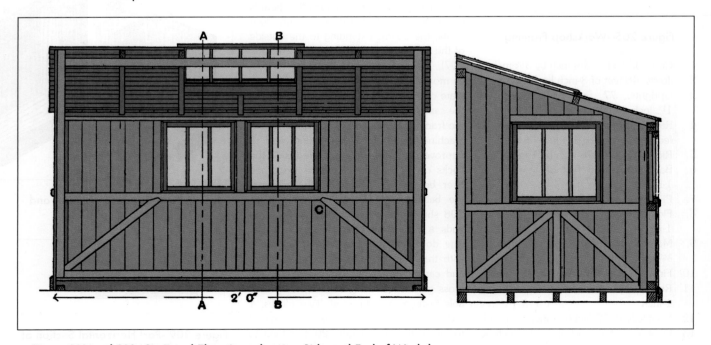

Figure 211 and 212—Sectional Elevations showing Side and End of Workshop

Log Huts or Cabins

In some surroundings, if rough timber is cheap, a log cabin or hut would make a strong tool shed. Log huts are built all over the world in different degrees of finish and comfort, according to their purpose and to the ingenuity of the owner. As a rule, the logs are notched with an axe at a short distance from each end, and built up in alternate pairs to form the walls, and cut short wherever door- or floor-frames occur. The interstices between the logs may be filled up with branches, cemented with clay, and finished with a finer plaster; but in more pretentious buildings the crevices are blocked up neatly with triangular pieces split from logs and nailed to the walls; the inside can then be close-boarded. If a fireplace is required, the best material for it is stone; however earth or bricks work as well. The chimney-flue may be formed outside the hut with a wood shafeet In the backwoods the stripped trunk of a tree is used as a core for the flue, inserted temporarily in the center of the shaft; clay is then rammed in round it, and when the flue is formed the tree trunk is drawn out at the

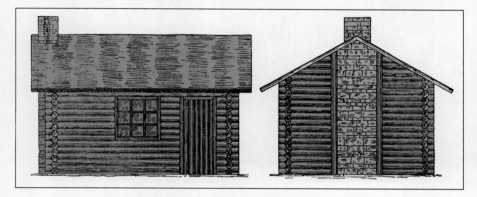

Figure 214—Section of Log Hut Roof

top. The fireplace is made by cutting a square hole in the logs just above the floor, forming a wood shaft at the bottom of the flue, and lining inside with clay to a good shape. Then all the clay is coated with a plaster of gravel and clay, cow-dung, and water; this could be improved upon owing to other and better materials being available. As for the roof, it should be high pitched, and have deep eaves. Figure 214 shows a section, A indicating tie-beams, 2 feet or 3 feet apart; B, wall-plates notched and pinned on to beams; C, rafters, same

Figures 215 and 216—Elevations of Log Hut or Cabin

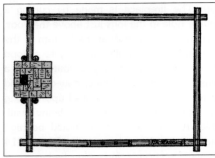

Figure 217—Plan of Log Hut or Cabin

distance apart as the tie-beams, and secured to them; D, collars spiked to rafters; E, ridge-pole tied with withies to the rafters. Nail battens on the rafters, and cover with shingles. Figures 215 and 216 show a common form of log cabin built up in much the same way. The logs are notched together and further secured by boring with an auger through each log into the last log fixed, and driving in a wooden pin or nail. If you require a chimney, it should be built of some local stone or similar material, but stoves with piping are often made to serve the purpose. The roof is often formed by splitting the logs into two and is covered with material that will render the roof waterproof. Figure 217 is a plan of the log hut.

Kitchen Furniture

Kitchen Corner Cupboard

A small corner cupboard as illustrated in Figure 218 will be found very useful, and can be quickly built. The body, outer frame, and door frame can be made from ¾-inch prepared material, to save planning up. You won't need any mortises, grooves, or rebates. Figure 218 gives a general view of the body. To make the body, cut from some 8-inch by ¾-inch material four 2-feet 6-inches lengths. Nail small cleats or battens across them (See c, Figure 219); these will make the two parts that form the back of the cupboard. Figure 220 shows a plan of top and bottom; these should be cut from ¾-inch stuff. Two shelves must be cut from ½-inch or ⅝-inch prepared material to the shape of the top and bottom, but ⁵⁄₁₆ inch less from front to back. This will allow the door frame to come flush with the outer frame when the door is closed. The four pieces referred to being cleated, and the top, bottom, and shelves cut to their proper shape and size, nail the two parts that form the back to the top and bottom; Then place

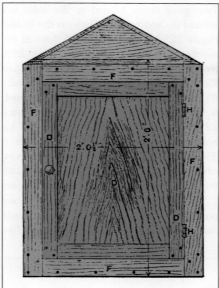

Figure 218—Kitchen Corner Cupboard

the shelves on the cleats c c, nail with small wire nails, and the body is finished, except for planning the front edges of the back a little, so that the outer frame may lie flat and even against it.

Outer Frame of Corner Cupboard

The outer frame is made from ¾-inch material like so: Cut off two 2-feet, 6-inches lengths, and two 2-feet, ½-inch lengths; these should be 2½ inches wide. Halve the ends, and nail these together, and the frame F (Figure 218) is made. A small portion of the frame is cut away with a chisel to receive two butt hinges. This frame is nailed against the front of the body, and the outer edges of the frame are planed to correspond with the angle of the back of the cupboard. The door frame is made from material 2 inches wide,

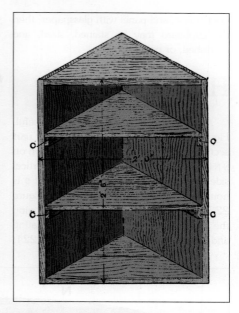

Figure 219—Shelves of Corner Cupboard

and the ends are halved and put together as the outer frame, which should thus fit nicely within the outer frame. On the inside of the door frame a piece of pine of suitable length and width, and about ⁵⁄₁₆ inch thick, nicely planed up, is nailed to the frame, seen in Figure 221; This piece forms the door-panel and at the same time tends to strengthen the door frame, and looks as well from the outside as if the frame had been mortised and tenoned together, and the panel let into a rebate in the frame. Secure two 1½-inch butt hinges, as shown. Hang the door D (Figure 218) to the outer frame by securing the hinges H H. Fix a knob and fastener as shown in Figure 218, then fill in the nail-holes with putty, and rub up the outer frame,

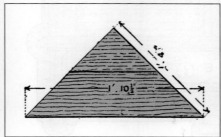

Figure 220—Top or Bottom of Corner Cupboard

Figure 221—Corner Cupboard Door

door frame, and panel with glasspaper; then the cupboard may be stained, sized, and varnished, or painted.

Household Tidy

A small cabinet should prove most useful. Figure 222 is part front elevation, part section. In Figure 223, which is a section on N N (Figure 222), A is one of a pair of vertical side pieces, into each of which a shelf B is housed. A top C is joined as shown in Figure 222. A back may be fitted in by grooving the inner edges of the sides, or by rebating and screwing on a strip as in Figure 224.

Two small doors hinged to the sides are framed up and paneled flush on the inside. Figure 225 shows a section of them at the meeting stiles. The piece D (Figure 223) should be screwed from behind. E (Figure 222) is a strip to carry hooks for keys, and F (Figure 223), to which this strip is fastened, is screwed to the shelf. A space G (Figure 222) is divided centrally, and is useful for holding small books, etc. The spaces H (Figure 222) may receive small drawers, which may be made of tobacco boxes with the lids removed, and with wooden fronts L (Figure 223) screwed through holes punched in the front of the boxes. Small compartments J (Figure 222) are closed by

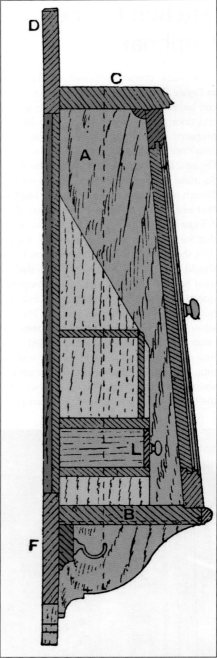

Figure 223—Cross Section of Tidy

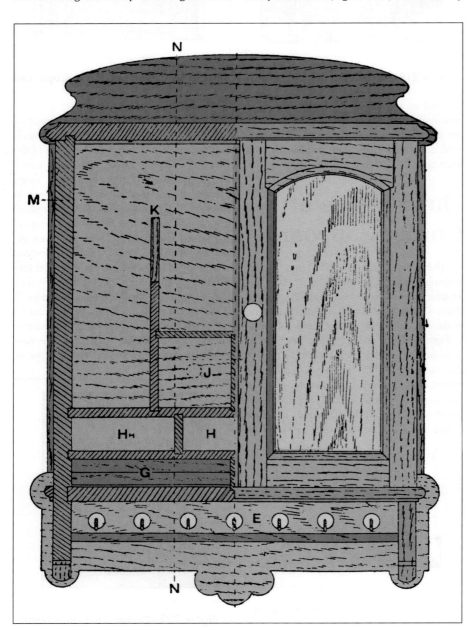

Figure 222—Part Elevation and Section of Tide

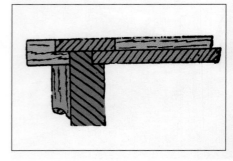

Figure 224—Back of Tidy Jointed to Side

Figure 225—Section through Door Meeting Stiles

Figure 226—Looped Leather Strip

sliding fronts pierced by center-bit holes, as indicated by the dotted circle. Each space may contain a ball of string, the hole being used help remove the front and admit one end of the string. The vertical portion K is carried well up so that a gum or paste-pot with a protruding brush may be protected if placed in the recess at the side. The inside of each door has a strip of leather near the top fastened transversely (Figure 226), and in the loops formed some such article as a hammer, screwdriver, or sardine-tin opener may be placed. You will find this very handy for many household tools. The cabinet may be hung to the wall by eye-plates, attached one on each side at about the level of M (Figure 222). The total width of the article is 17 inches, the height is 23 ¼ inches, and the depth is 6 inches.

Spice Box with Drawers

To make a spice box like the one in Figure 227, yellow pine or deal is the best material, but, in any case the wood must be thoroughly dry. The following pieces will be required:

1. Back of case, one piece 7¾ inches by 6½ inches by ¼ inch
2. Sides, two 7¾ inches by 3 inches by ¼ inch
3. Top and bottom, two 7¾ inches by 3¼ inches by ¼ inch
4. Shelves, four 6½ inches by 2⅝ inches by ³⁄₁₆ inch
5. Partition, one 6⅛ inches by 1 inch by ¼ inch
6. Ornament, one 7¼ inches by 1¼ inches by ¼ inch

Figure 227—Spice Box with Drawers

7. Sides of drawers, eighteen 2⅜ inches by 1⅛ inches by ³⁄₁₆ inch
8. Back of small drawers, eight 2⅝ inches by 1⅛ inches by ³⁄₁₆ inch
9. Front of small drawers, eight 3 inches by 1⅜ inches by ¼ inch
10. Bottom of small drawers, eight 3 inches by 2⅜ inches by ³⁄₁₆ inch
11. Back of bottom drawer, one 5⅞ inches by 1⅛ inches by ³⁄₁₆ inch
12. Front of bottom drawer, one 6¼ inches by 1⅜ inches by ¼ inch
13. Bottom of drawer, one 6¼ inches by 2⅜ inches by ³⁄₁₆ inch

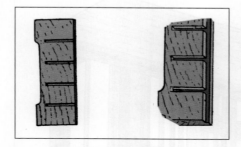

Figures 228 and 229—Spice Box Sides

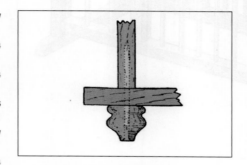

Figure 230—Spice Box Foot

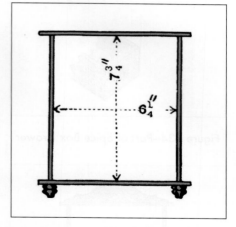

Figure 231—Vertical Outline of Spice Box

These are all finished sizes. The other materials required are nine small brass knobs, two bracket eyes, four wooden feet, and a handful of small nails. To make the case, cut the two side pieces to the shape shown by Figure 228. Cut four grooves (Figures 228 and 229) ⅛ inch deep and 2½ inches long, to fit in the shelves, and treat the back edge of these pieces in the same manner. The width of the grooves is just sufficient to allow the shelves to fit tightly into them. At equal distances from each other carefully mark out the positions for the grooves by dividing the side piece into five equal parts. The places for the grooves being determined, draw lines across representing the widths of the grooves; then cut these lines down to a uniform depth with a chisel, cutting downwards, or guided by a straightedge, draw it along to that it cuts into the wood. The bottoms of the grooves

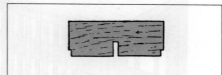

Figure 232—Spice Box Shelf

Figure 233—Spice Box Case

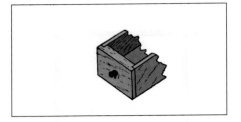

Figure 234—Part of Spice Box Drawer

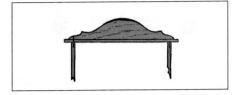

Figure 235—Top of Spice Box

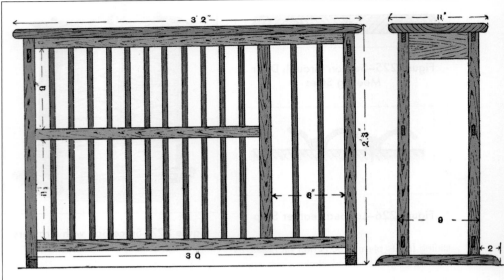

Figures 237 and 238—Front and End Elevations of Plate Rack

need not be absolutely smooth, as they are not seen when the parts are fitted together. Nail the top and bottom on to the sides. The feet can be put on at the same time by driving the nail first through the foot (see Figure 230). When you complete the case, it should measure inside 7¾ inches by 6 ¼ inches. Next take three of the four shelves and cut them as shown by Figure 232. The piece cut from the middle is 1 inch long and ¼ inch wide. The pieces cut from the

sides are ⅜ inch long, and ⅛ inch wide. The fourth and bottom shelf is cut similarly, only the middle piece is left in. Fit the shelves and partition into their respective places, the partition being nailed to the top of the case and to the bottom shelf. If the back is now fixed in its place, the case (Figure 233) may be considered

complete. All that's left to be done is to make the drawers. The front of the drawers should be cut so that the sides will fit into them as in Figure 234. After making the drawers, fix a knob on each to serve as a handle for pulling them out. If the remaining piece is cut to the shape shown in Figure 235, a passable ornament will be the result. It is fixed to the top by nails driven down into the two side pieces. By fixing two bracket eyes to the back, the box can either be hung against the wall or stood in any convenient place. To add a finish to the box it can either be stained or polished, or painted—according to your taste.

Pantry Safe or Cupboard Pantry

Figure 239 is a front elevation of a pantry safe which is rectangular in plan, and Figure 240 shows the end elevation. For the three pieces of framing, six stiles, 2 feet, 8 inches by 2 inches by 1 inch will be required. These pieces are mortised and tenoned together, the six top and bottom rails having haunched tenons, as shown in Figure 241. The other three rails have tenons and the pieces of framing are glued and wedged together. The two framings for the ends are rebated ½ inch each way on the back edges to receive the back, which is formed of 3-inch by ½-inch tongue and groove-jointed matchboarding. The front edge of the ends has a ¾-inch chamfer on the outside corners. The shelves and top should be got out of a wide board of

Figure 236—Plate Rack

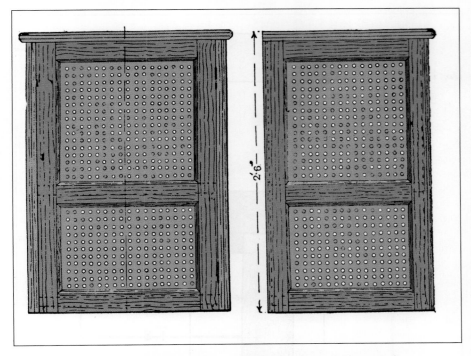

Figure 239 and 240—Front and End Elevations of Pantry Safe

pine or whitewood. The top, when finished, is 1 foot, 10½ inches long and 1 foot, 6 inches wide, which allows it to project ¾ inch over the front and ends. The projecting should either have a nosing worked on, as shown in Figures 239 and 240, or a chamfer. The shelves are 1 foot, 3¾ inches wide by 1 inch thick, supported by four fillets, 1 foot, 3¾ inches by ¾ inch by ½ inch, screwed to the sides of the safe as shown in Figures 243 and 344. In fixing together, the top may be secured to the ends with 1½-inch screws three through each

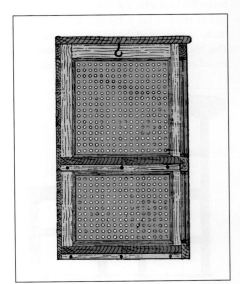

Figure 243—Cross Section of Pantry Safe

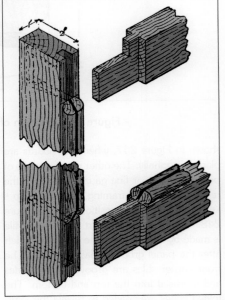

Figure 241—Top and Bottom Rail Joint
Figure 242—Middle Rail Joint

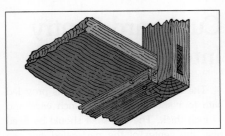

Figure 244—Bottom shelf of Pantry Safe

top rail. The bottom shelf is fixed with screws to the fillets, thus securing the lower ends of the sides as shown in Figure 244.

Fitting Together Pantry Safe

Before fixing the back, the framing should be squared, and a temporary lath fastened diagonally across the front to hold it square until the back is completed. Fit in the boards for the back and fasten with 1½-inch oval wire nails. Figure 245 shows the end rebated to receive the back. A 1½-inch screw through the top into the edge of the back will stiffen it considerably. Two or three hooks should be screwed into the top for meat to hang from (see Figure 243), and the middle shelf should be left loose, so that it can be removed when the hooks are used. The door may next be fitted in, and a ⅜-inch bead, worked down each side, will prevent the butt hinges looking

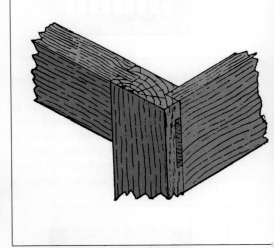

Figure 245—Back Corner of Pastry Safe

unsightly on the hanging side; 2-inch butt hinges are used, and are fixed 3 inch from each end, the whole of the hinge being let into the door stile. The spaces are covered with perforated zinc, which can be fastened on the inside of the safe with tacks, or may be secured with beads which is a neater and better method. About 60 feet of ½-inch beading will be required for the latter method, the beads being mitered at the corners, and fastened with 1-inch brads. The door may be fastened either with an ordinary lock, or with a special catch.

Cupboard Pantry Door

The door is 7/8 inch thick, with stiles and rails mortised, tenoned, and wedged together. You will need to mold and groove the edges of the framing to create a panel. You can substitute a square-edged framing and afterwards pin a molding to the opening, but this is not always the best option.

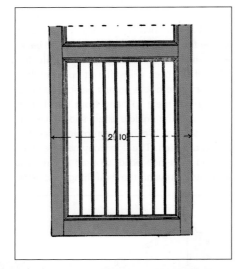

Figure 246—Door with Vertical Paneling

The panel should be made up of two or three widths of timber, butt-jointed and glued, or you can fill the framework as shown in Figure 246. Notice the grooved, tongued, and beaded boards 3 inches by 3/8 inch placed vertically in the frame. Another method is

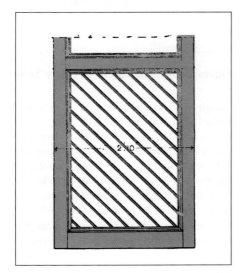

Figure 247—Door with Diagonal Paneling

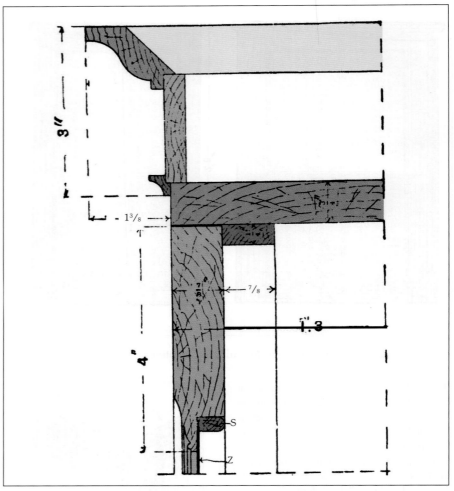

Figure 248—Section of Cornice and Door Rail

shown in Figure 247, where the boards are placed diagonally. The other panel consists of perforated zinc or fine gauze, and fits in the rebate of the door framing, a molded slip s keeping the zinc z in position, seen in Figure 248. The rebate in this part of the door stile is made by cutting away the inside piece left after the panel grooving has been run. The front corner stiles are grooved to the sides and mortised into the top and bottom. The folded edges are beaded to relieve the joint. Hang the door with strong brass butts, and furnish it with a knob, lock, and key.

Cupboard Pantry Interior

The interior of the cupboard is provided with four shelves, 1 foot, 1 inch wide and 1/2 inch thick. The top shelf should be fixed near the center of the perforated zinc, so that the two top spaces can be used for fresh meat. The lower shelves can be fitted as needed. The cornice has a small molding to cover the joint with the top (see Figure 248), and is rebated for a frieze panel 3/8 inches thick, surmounted by a cornice molding. The moldings and panel are mitered and keyed at the corners, and strengthened with

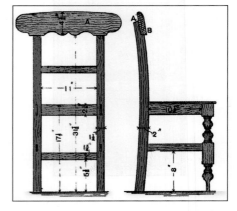

Figures 249 and 250—Back and Side Elevations of a Basic Kitchen Chair

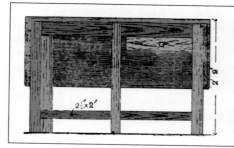
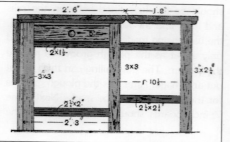

Figures 251 and 252—Side and Front Elevations of Gate-legged Table

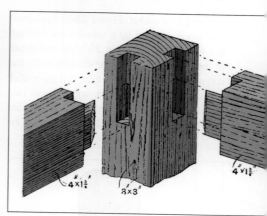

Figure 255—Joint of Top Rails and Leg

blocks glued at the angles. When finished, the cornice forms a separate piece from the carcass, so that it can be removed when not in use. The four feet are turned in the lathe, and have dowel ends fitting into the carcass bottom and fixed with wedges. A plinth finish can be adapted if you prefer. This cupboard pantry can be easily converted to a wardrobe if desired by removing the zinc panel and inserting a bevel panel corresponding with the lower one, so that it is desirable to finish the interior neatly.

Gate-Legged Kitchen Table

Figures 251 to 254 illustrate a gate-legged kitchen table which, when closed, is 2 feet, 6 inches wide by 4 feet, 4 inches long. The leading points in the making are as follow: The legs and rails should be planed up square to sizes, and the legs set out for mortising. The mortises for the lower rails are of a simple character, the tenons being stubbed inches. For the long top rails the mortises and tenons at one end should be as shown at A (Figure 255) and at the other end

as shown at B (Figure 256). Each end of the top rail C (Figure 254) is of the form shown at C (Figure 255). The two rails F (Figure 254) for the drawer have the joints illustrated at D and E (Figure 256), the upper rail being dovetailed in the top of the legs. The two gate legs and their rails are stub-mortised and tenoned together. Figures 256 and 258 show how the gate-leg stile is jointed to rails. After the joints are made the whole of the framing should be carefully fitted, and the joints numbered; then the framing should be separated, and the internal parts smoothed off. After you glue and fit the joints, cramp them in position until the glue is dry; odd strips of wood, with a block nailed on each end and a wedge inserted, may be used for this purpose. The top, including the flaps, should be formed of 1½-inch boards, ploughed, tongued, grooved, and glued together. The top and flaps should be planed off true, and the top secured to the rails of the framing by 2½-inch screws driven obliquely from the inside of the rails. The flaps may be attached to the top by 2½-inch wrought-iron back-flap hinges as shown in Figure 253. The top and flaps should be strengthened by 2-inch by 1-inch thick fillets, which are screwed on as indicated at Figures 252 and 253. The stiles of the gate legs should be fixed at the bottom end by a pin

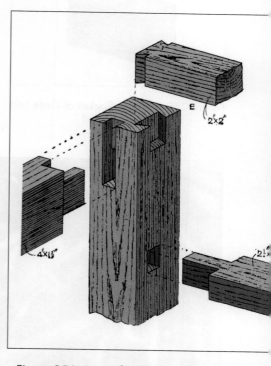

Figure 256—Joint of Rails and Leg at Drawer End

Figure 253—Underneath Plan of Gate-legged Table

Figure 254—Gate-legged Table Framing

Figure 257—Jointing Gate Leg Stile and Rail

working in a socket (Figure 258), the upper end being secured by a screw sunk through the top rail as in Figure 257. As the depth of the side rails will not be enough to fix the runners of the drawer, pieces G (Figure 251 and 254) should be added, and to these two runners can be fixed, and also a cross rail; see H and K (Figure 254). The drawer front should be carefully fitted between the rails and legs, and the sides and back prepared, the back being made wide enough to extend only as far as the plough groove to receive the bottom. The dovetailing can then be set out and made. After this the plough grooves for the bottom should be made. Next some ½-inch boards should be glued up for the bottom, the edges being chamfered to fit into the plough grooves. To secure the bottom it should be nailed into the lower edge of the back, and have strips underneath fixed to the bottom and the sides, these being secured with glue and planed off flush. A knob or handle should be provided and fixed to the front of the drawer.

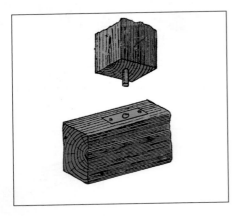

Figure 258—Pin and Socket of Gate Leg Stile and Rail

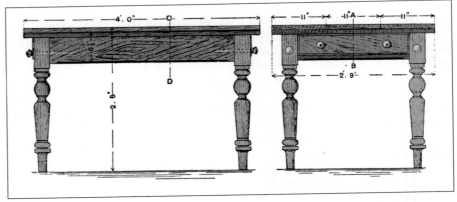

Figures 259 and 260—Side and End Elevations of Table with Turned Legs

Bedroom Furniture

Shaving Cabinet

The shaving cabinet illustrated in Figure 261 is arranged so that when you open the doors the light is reflected on each side of the face. You should begin by making the back of the cabinet. It is of ½-inch material, 2 feet, 7 inches by 2 feet, 5 inches, and is either jointed up, as shown in section by Figure 262, or framed, a ¼-inch board forming its center. Cut it out with a bow-saw; the ornamental panels at the top have the outlines cut in about ⅛ inch, the centers being beveled down and punched with the point of a wire nail. The top and bottom boards at the back are ½ inch thick and 2 feet, 6½ inches by 7½ inches The ends form an angle of 45° with the back. The front edge has a length of 1 foot, 3½ inches These boards are 1 foot apart, and are fixed with screws through the back as at a (Figure 263). Frame up the front 1 foot, 3 inches by 1 foot, the rails and stiles being 1¼ inches by ¾ inch, rebated for the glass, and the back edge of the stiles being beveled off at an angle of 67.5°. Cut and fit the brackets as shown in Figure 262. They are glued and screwed through the bottom board and back. Get out the ornamental rails at the top and bottom, miter them, and glue and brad them in place, ½ inch within each edge, and they will then cover up the screws used to secure the front, as at c (Figure 263). Fasten a slip cut to the section of B in each angle at the back; with these slips the doors will come flush. Frame the doors 1 foot by 9⅛ inches of similar section to the front, with the hinge stile beveled to 67.5°, and hand the door in place. This can be fitted either with locks or with hanging handles and catches. Stain and polish, or otherwise finish to your taste, and, finally, insert the mirrors. These should be securely wedged like at D (Figure 263), and covered at the back with thin wood to protect them from scratches.

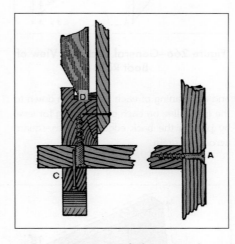

Figure 263—Joints of Shaving Cabinet

Boot and Shoe Rack

Figures 264 and 265 show a rack for boots, shoes, and slippers, which you will find useful in almost any part of your house. The rack can be made of ¾-inch yellow pine or sound red deal, and requires two ends, 2 feet, 9 inches by 11 inches by ¾ inch; one top, 4 feet by 11¾ inches by ¾ inch two braces, 3 feet, 6 inches by 3 inches by ½ inch; six rails for shelves, 3 feet, 7 inches by 2 inches by 1 inch; and one molding, 6 feet by 1½ inches by ⅞ inch. First prepare the rails, then plane them square, and take off the sharp corner to form a slight chamfer. Shoulder and tenon each end ready for the ends, which can then be prepared with the mortises. Space them out as shown in Figure 266, the top two rails for children's boots and slippers. After you cut the mortises, gently drive on the ends and wedge the rail tenons. Across the back, screw two braces B (Figure 266), and then screw on the top from the under-side. Make sure to gouge the ends and braces for the screws. At the front, immediately below the top, place a 5/16-inch diameter iron rod I R (Figure 266) suspended by brass hooks, on which to fix a curtain c (Figures 264 and 266) to cover the boots. The finishing molding along the top can be fixed next, and the rack is then often ready for painting, or staining and varnishing. Rails are preferable to solid shelves, as they admit of a free current of air to dry the soles of the boots.

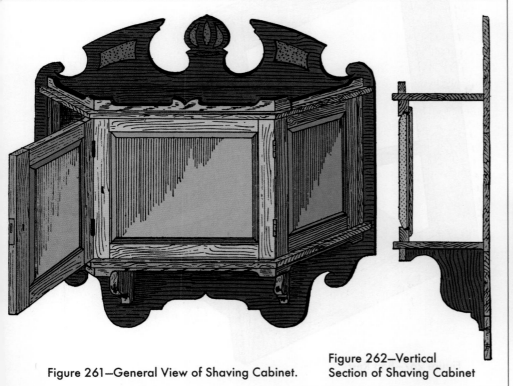

Figure 261—General View of Shaving Cabinet.

Figure 262—Vertical Section of Shaving Cabinet

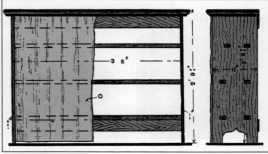

Figures 264 and 265—Elevations of Boot and Shoe Rack

Bed Rest

For the bed rest illustrated by Figure 267 the following materials will be required: For the frame A (Figure 267) two pieces 2 feet, 1 inch by 3 inches by 1 inch, planed to 2¾ inches by ⅞ inch. For the frame B, two pieces 1 foot, 4½ inches, planed to 1¾ inches by 1⅛ inches. For the frame c, two pieces 10½ inches by 1¾ inches by 1¼ inches, planed to 1½ inches by 1 inch; and one piece 11½ inches by 1¾ inches by 1¼ inches, planed to 1½ inches by 1 inch. There will also be required between 5 yd. and 6 yd. of webbing, one pair of 1¼-inches back flaps, some ¾-inch screws, and two coach-screws, 3 inches by ¼ inch. The wood must first be planed to true. Begin by planing true the face side and face edge of each piece of wood, afterwards gauging to the several widths and thickness, taking care that in each measurement the gauge is set quite 1/32 inch full to allow for smoothing off when glued up. When this is true, lay it on the bench, face up, and as the other pieces are planed lay them on the prepared piece, and if each piece does not lie perfectly flat, plane off the part where it touches, that being the highest part. The face edge of each piece can be done in a similar manner. To test with the eye alone, hold up a piece of wood to your line of sight, then slowly turn up the further edge into your line of sight. Notice which corner comes into view first—that will be the highest part. Plane this side down, and the opposite corner of the diagonal. This is because if one corner curls up ⅛ inch, taking 1/16 inch off the opposite corner will make it equally true and the material will hold thicker that if ⅛ inch was planed off one corner. Having planed all the material on the face sides and edges, set the gauge for the width. As each piece is gauged, plane down to the line, taking care that the edge is exactly square before the line is reached; if two gauges are to hand, then the other gauge can be set to the thickness,

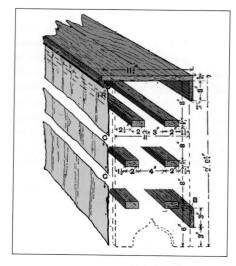

Figure 266—General Sectional View of Boot Rack

and the planing of each piece; plan down to the gauge line on each edge, testing for level by placing the back edge of the try-square across the grain.

Setting out Bed Rest

Now proceed to set out, remembering that in setting out work the face sides and face edges must always pair. For example, take the frame marked A (Figure 267). Take one of the sides, 2 feet, 1 inch long, 2¾ inches by ⅞ inch, and, keeping the face side to the worker and the face edge uppermost, find the center of length; mark off 1 foot (half the height) on each side of this, and square lines across the edge. Then from these lines mark 2¾ inches inwards, being the width of rail at top and bottom of frame, and set out 1⅝ inches width of tenon. Note that the tenon does not go through, only entering 1⅝ inches, which is quite sufficient, the joint being fastened by fox-wedging, which is here shown by Figure 268, but which has already been described. Thus, setting-out lines will only be needed on the face edge. Having set out one side of frame a, take the other side and place it beside the one set out, so that

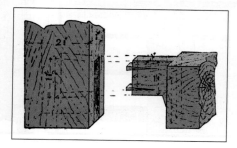

Figure 268—Joint at Corner of Bed Rest Frame

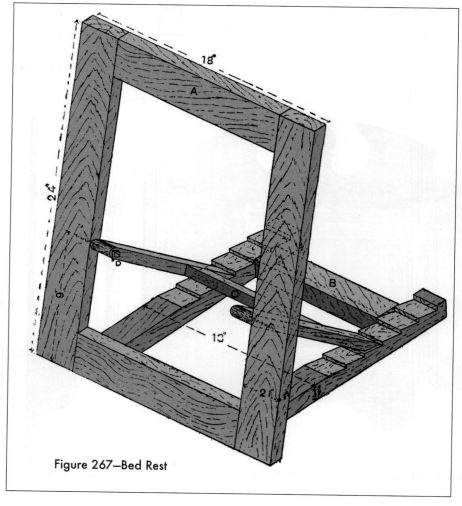

Figure 267—Bed Rest

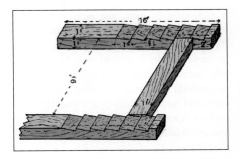

269—Horizontal Frame of Bed Rest

Figures 272 and 273—Side and Back Views of Corner of Bed Rest Frame

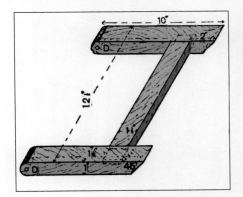

Figure 270—Bed Rest Strut

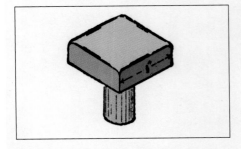

Figure 271—Head of Coach-screw

the two sides pair—that is, both face sides are outside and both face edges uppermost. Having squared lines across, set the mortise gauge to the chisel that is nearest one-third the thickness of material, adjusting the one teeth of the mortise gauge so that the chisel just rests comfortably between the points of the teeth, then set the gauge head so that the teeth are exactly in the center, trying first one side and then the other by pricking in the points of the teeth until the points coincide; then tighten up, and holding the stock of the gauge close to the face side, mark the mortises and continue gauge lines for the haunching, running the lines on to the end. Next proceed to set out top and bottom rail of frame A (Figure 267). Take one rail, find the center of its length, and set off half-width of frame less 2¾ inches (width of stile) on each side of center. Square all these lines,

taking great care, for it is very important that these, being shoulder lines, should be marked true. Note that the try-square should always be applied to either a face side or face edge. The lines must next be squared to the other rail, it having been first noted in placing the rails together that they pair. then with the mortise gauge mark the tenons all round the end from shoulder to shoulder, Now set out frame b (Figure 267). The material should be ½ inch longer than the finished lengths. Set out one side, starting ¼ inch from one end, and marking off 1 foot, 4 inches, thus leaving a little waste to be sawn off, and enabling a square end to be cut, which can also be shot just before the gluing together. From the first line set off 2 inches (Figure 269). On the face side of one of the 1-foot, 6-inches pieces, commencing ¾ inch from the rail end, set out successively six spaces of 1½ inches each; place the other side so that they pair. Square lines across the two pieces be means of the try-square; you should also square down each edge a little way, and with a marking gauge set to ¼ inch just make a mark on the line; mark the slope with a striking knife, using the edge of the try-square as a straight-edge; then set out the rails 9 ½ inches between shoulders, the tenons 1 inch longer, prepared for fox-wedging, as in frame A. Frame c (Figure 267) comes next, taking one side and proceeding to set it out as in the case of frame B. For dimensions of c, see Figure 270. Note that one end of frame c is semicircular, and that at the center of this portion a ⁵⁄₁₆ inch hole must be bored through each piece (see D D, Figure 270), so as to enable the coach-screw (the head of which is shown in Figure 271) to pass through for the purpose of hanging frame c to frame A, and allowing it to hinge easily. The other ends of the frame c must be cut at an angle of 45°, to obtain which it is not necessary to have a bevel, for if the thickness of the stuff is set off from the end and a line joined diagonally from this point to the top edge, the angle will be 45°. The cross-rail

is set out similarly to frame B; it is 9 ½ inches between the shoulders, and the tenons are 1 inch longer. The setting-out is not complete.

Making a Bed Rest

Begin the actual work by mortising the sides for frame a (Figure 267). The mortises are 1 ⁵⁄₈ inches, and are widest at bottom for fox-wedging (see Figure 268); note also that the haunching (see Figure 268) tapers from nothing at the end to ³⁄₈ inch at the tenon. Great care must be used in mortising, because if the hole is not perpendicular it will cause the frames to twist, as the tenons will follow the holes, and so tend to take the stuff with it. After mortising, cut the tenons, ripping down first, then sawing the shoulders, afterwards marking width of tenons 1⁵⁄₈ inches from face edges, and the haunched part of tenon ³⁄₈ inch to nothing (see Figure 268). Next make two cuts in each tenon about ¼ inch from the side and 1 inch long, and prepare and insert in each cut a small wedge of the same thickness as the tenon, ¾ inch long, and tapered ³⁄₁₆ inch to nothing, leaving the wedge projecting at the tenon end. On driving or cramping the frame together the wedges are driven in, and the end of each tenon is widened so that, with the glue, it is impossible to pull

Figure 274—Webbing on Bed Rest Frame

it out. A hole should now be bored in the center of each face edge of stiles 9 inches from the bottom, the hole to be a little less than ¼-inch diameter for the coach-screw to fasten in tight when hanging. It is best done before gluing up, for when together it would be found awkward to bore except with a rather large gimlet. Smooth the edges of the frame so as to leave it clean, and the frame is ready for gluing. The frames B and C can be served in a similar manner. In gluing up the frames, be sure the glue is hot. Get the cramp or cramps together before you begin, and make sure the fox-wedges are right. When the frames are glued up, test them with a large square while applying

the cramp; a slight movement of the cramp will correct any error. In frame A, if the diagonals are the same length it is square. If an iron cramp is not to hand, a wooden one can be constructed by screwing cleats of wood on a piece of board, 2 inches or 3 inches wide, and allowing for a pair of wood between the frame and the hammer to avoid dents. When the glue is set, the frames can be smoothed over and the fixing together can begin. A pair of 1¼-inch back flaps and ¾-inch screws will be required to hinge frame B to A (see Figures 272 and 273); about half the knuckle of the hinge projects. The drawings will fully explain the position of the hinges. Two coach-screws,

3 inches by ¼ inch, are needed to hang frame C in position in frame A (see Figure 267); these can be turned in with a pair of pincers. The front should now be covered with webbing. The lengths are cut to the dimensions of frame A, about ¾ inch at each end being turned under and fastened by two ¾-inch clout nails at each end. The spaces between webs should be no more than the width of the webbing, as shown in Figure 274, where it will be noticed that the webbing is interlaced; 5 yd. will suffice, though a little more would certainly make the whole a little stiffer. Finish with a light stain and a coat of varnish.

Figures 275 and 276—Elevations of Linen Chest

Linen Chest

Deal is the best material probably with which to make the linen chest illustrated by Figure 275. The elevations and inside view of the chest, which is 2 feet, 3 inches high, 4 feet long, and 1 foot 10 inches wide, are represented by Figures 275 to 277. The two uprights at the front corners are strips of 1½-inch material, 2 feet, 3 inches long by 4 inches wide. Figure 278 shows a front and Figure 279 a side view of that to the right hand. In both figures, a denotes where the upper 4 inches is cut away from the front to a depth of ½ inch to receive the end of the upper front rail; while b is a similar cut to take the end of the lower front rail. At c and D are other cuts for the end rails, all being ½ inch deep. A side view of the right-hand back upright is shown by Figure 284. Its length and breadth are those of its fellow at the front, but it is of ¾-inch board only; 1½-inch wood may be used, to make the job stronger, but one half will have to be sawn away as far as e to receive the ends of the back-boards. Below E, where the double thickness is needed to form the leg, a second piece of ¾-inches material should be screwed on. At F and G this upright is cut away ½ inch to receive the ends of the rails. The rails are all of ¾-inches board and 4 inches wide. The front ones are 4 feet long. The ends of these are cut away behind for a distance of 4 inches and to the depth of ¼ inch to fit the

corresponding cuts in the uprights. To these they are fastened with round-headed screws, as shown. The stiles are also of ¾-inches material, 1 foot, 4 inches long and 3 inches wide, with the exception of the middle one of the front, which is 4 inches. Their tops and bottoms are halved in front for 1½ inches (See H H, Figure 281), corresponding cuts being made in the rails above and below to receive them. Their front edges, as well as those of the rails against the panels, are chamfered off. The paneling of the front is done with ½-inches board, cut to 1-foot, 7-inch lengths. These reach from the bottom of the lower rail to the middle of the upper one, above which is a longitudinal strip of the same board. This may be seen in Figure 277, which shows the inner side of the chest front. This arrangement will make a neater finish than could be gained by carrying the upright boards to the top. In Figure 277 the position of the stiles is shown by dotted lines, and the joints of the paneling should be set up so that it is covered by them.

The panels which have been thus made are 1 foot, 1 inch by 7½ inches at sight; and, before leaving them, it will be well to finish them by adding the small ornamental spandrils, one of which is illustrated by Figure 282. These are of ¼-inch board, and the curved edge of each is worked to a hollow molding with the gorge. The length

of the end rails is 1 foot, 9½ inches The backs of their ends are cut away ¼ inch deep to fit the openings in the uprights—the front end for 1 inch and the back end for 1½ inches The end of the chest (Figure 276) has a 3-inches stile, and is paneled like the front. The back is of ¾-inches boarding, the pieces being 4 feet long and (together) 1 foot, 9 inches wide. At the corners, openings will have to be cut, 4 inches by ½ inch, to admit the ends of the rails. The back-boards are screwed to the uprights, and should also be doweled

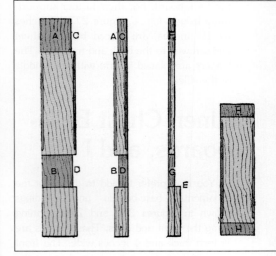

Figures 278 and 279—Front Corner Upright of Linen Chest
Figure 280—Back Corner Upright
Figure 281—Front Middle Stile

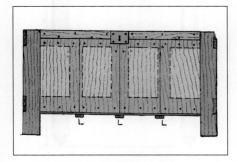

Figure 277—Inside View of Chest Front

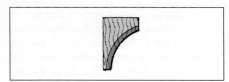

Figure 282—Spandril for Linen Chest Panels

Figure 283—Oak Bedstead

together. For the bottom ½-inches match-boarding will be best, the pieces being 4 feet long and (together) 1 foot, 10 inches wide. Openings are cut at the corners, 4 inches by 1½ inches, to admit the uprights. These boards are screwed to the lower end rails, also to the lower front rail and to the back-board. For their further support, three ledgers (L L, L, Figure 277), 2 inches by 1½ inches, are placed beneath them and screwed to the back and front rail. The ledgers are placed in line with the middle of the stiles.

Linen Chest Base-boards, and Lids

You may prefer to add to the chest the ornamental base-boards and moldings shown in Figures 275 and 276, running along the front and ends. These boards are ¾ inch thick and 4 inches wide. The front one is 4 feet, 1½ inches long; those at the ends are 1 foot, 10¾ inches long. They are mitered where they meet at the corners, and their lower edges are so shaped as to hide the ends of the ledgers Their tops overlap the lower rails 1 inch, and they are screwed to the uprights. Place a strip of molding on these, ¾ inches square, with its front edges simply rounded off. These strips, like the boards below, are mitered at the corners; the front strip is 4 feet, 3 inches long, the end strips 1 foot, 11½ inches. The lid is made of ¾-inch boards, 4 feet, 6 inches long and (together) 2 feet wide. These boards are

screwed down to four ledgers, two of which appear in Figures 275 and 276; the others are within the chest, and are placed at equal distances from the outer ones and from each other. As a prop for the lid, have a lath about 1 foot, 4 inches long, working on a screw driven into the inner side of one of the end rails near the front. The chest may be stained to a dark oak color.

Oak Bedstead

Figure 283 shows a beautiful oak bedstead. Leading dimensions are given on Figures 284 and 285, but these dimensions can be varied to your own requirements. Alternative methods of connecting the sides to the foot and head as shown, Figures 286 and 287 illustrating the old

bed-screw method of connecting by means of nuts and bolts. In this case a hole A (Figure 286) is bored from the end of the tenon longitudinally in the rail; a mortise is made from the inside of the rail, and a nut B (Figure 287) is inserted, the bolt-head being hidden by the turned wooden button C. A method that is not being generally used is illustrated in Figure 289, the principle being that used on iron bedsteads. A dovetail piece with flanges is screwed to the post, the flanges being let in, and a corresponding dovetail socket piece is fitted and screwed to each end of the rails. The illustration shows the construction fully. The pieces of timber having been cut to their several sizes, each should be planed to finished dimensions. Then the posts and rails should be set out and the mortising and tenoning done. The forms of joints for connecting the stiles and rails of the head and foot are clearly shown at D and E (Figure 288), and the top rails of

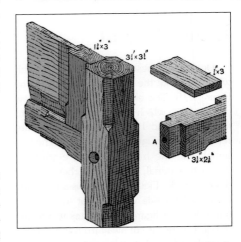

Figure 286—Part of Bedstead Post and Panel

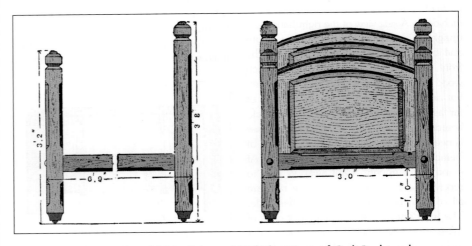

Figures 284 and 285—Side and End Elevations of Oak Bedstead

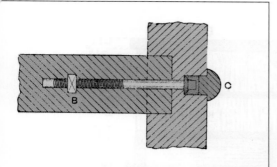

Figure 287—Bed-screw Joint

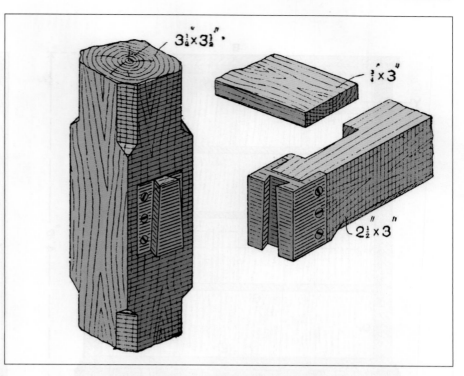

Figures 289—Metal Dovetail for Bed Rails

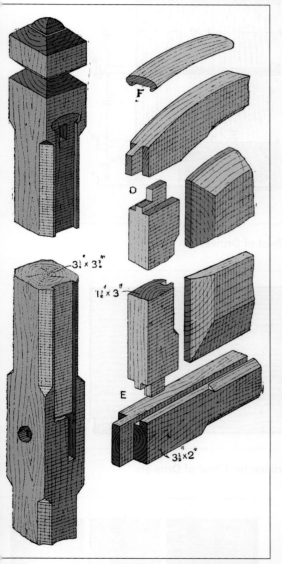

288—Jointing Bedstead Framework

splayed margins. Cut an oval-sectioned piece F (Figure 288) to proper sweep and sink out the under-side to fit on to the curved rail, the posts being recessed so as to receive the ends of the capping. After you fit the joints properly, the posts, rails, and stiles need to be stop-chamfered and the tops and bottoms of the posts worked to shape. The joints of the head and foot framings can then be glued together, and when the glue is dry the stiles and rails can be smoothed off. Then the top and bottom rails and stiles are ready to be fit into the posts, which can also be glued together. Four castors should be fixed on the bottom of each post to complete the bedstead.

Framework of Chest of Drawers

The chest of drawers illustrated by Figure 290 is 4 feet wide, 4 feet, 6 inches high, and 2 feet deep outside. The two sides A, 1¼ inches thick, 1 foot, 10¾ inches wide, and 3 feet, 11⅛ inches high, support the top B, which is 1½ inches thick, 2 feet wide, and 4 feet long, the grooves being ½ inch deep. Two inner pieces D, 1 foot, 10 ¼ inches wide and 3 feet, 5 ¼ inches long, rest on piece C (see Figures 291 and 292). The top piece F and the two

lower partitions G are 5 ½ inches wide and 3 feet, 5 ¼ inches long, the shorter partitions H being 5 ½ inches wide and 10½ inches long. The partition J is 1 foot, 4 ¾ inches wide and 3 feet, 5 ¼ inches long, and the vertical partitions K are 1 foot, ¾ inch deep by 1 foot, 4 ¾ inches long, cut to the shape shown by Figure 293, to rest in ⅜-inch grooves in the pieces B, F, and J. Cut all the partitions, including F and bottom E, as shown in Figure 294, and rest, as shown in Figure 291, in ⅜-inch grooves

Figure 290—Chest of Drawers

the head and foot are curved for a better appearance. The curved top rail has a stub-tenon at each end to enter corresponding mortises in each post. The stiles and rails are ploughed to receive a panel, which is shown in Figures 286 and 288, with

Figures 291 and 292—Longitudinal and Cross Sections of Chest of Drawers

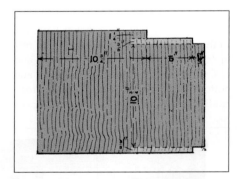

Figures 293—Vertical Partition for Chest of Drawers

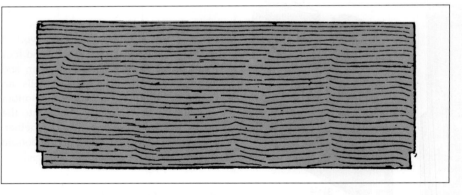

Figures 294—Horizontal Partition for Chest of Drawers

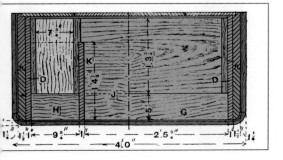

Figure 295—Horizontal Section of Chest of Drawers

in the partitions D and K, are put in at each side of the drawer spaces for the drawers to run on (see Figures 292 and 295). A ¼-inch boarding separates the drawers, as seen by the black lines in Figure 292. The plinth is 2½ inches thick and 5 inches deep, and should be cut out to receive the bottom and sides, as shown. The whole is raised slightly from the ground by means of buttons, 1 inch thick, screwed to the plinth. The framework should be strengthened where necessary by triangular blocks.

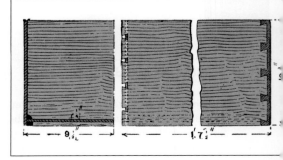

Figures 296 and 297—Drawer Details

The Drawers

Fit the drawers together as shown in Figures 296 and 297, but the bottoms of the long drawers are strengthened with two battens. Make the drawer fronts 1 inch thick, the sides ³⁄₈ inch thick, and the bottom and back ¼ inch thick; the smaller drawers can be made of thinner material.

The clear spaces for the drawers are as follows:

1. Bottom drawer, 11 inches deep, 3 feet, 4 ½ inches wide, by 1 foot, 9¾ inches

2. Bonnet drawer, 1 foot, 10¾ inches deep, 1 foot, 7 inches wide, by 1 foot, 7½ inches;

3. Four short drawers, 5 inches deep, 9¾ inches wide by 1 foot, 7½ inches

Drawer stops should be fixed on the runners. You can make the handles of either wood or brass. The back should be ½ inch thick, and is strengthened vertically by two battens, and rests in a groove in the top B (see Figure 292). You can make the bottom long drawer deeper than the rest, and, instead of the four small drawers and center bonnet drawer, two short drawers can be put in if you desire. Figure 295 is a section on Y Y (Figure 291).

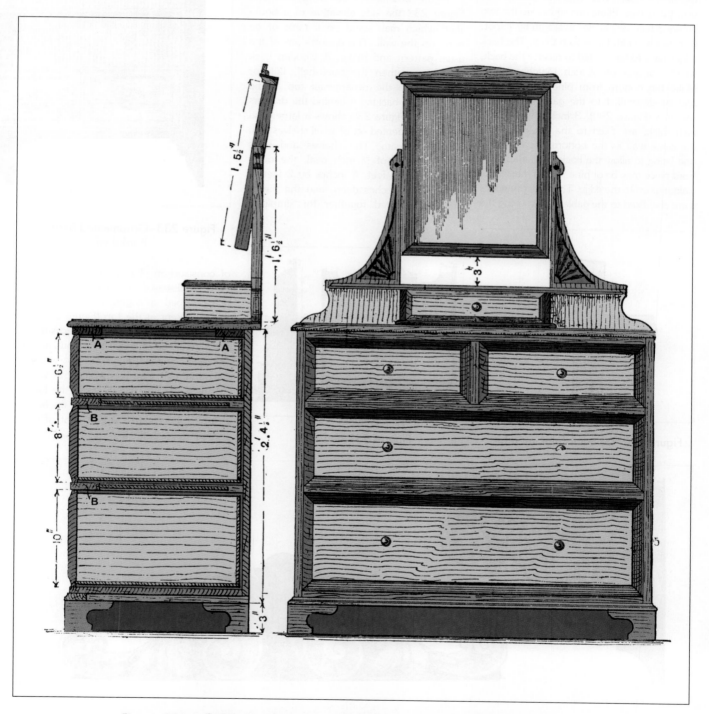

Figures 298 and 299—Section and Elevation of Chest of Drawers and Dressing Glass

Chest of Drawers and Dressing Glass

Figure 298 shows the front elevation of a chest of drawers, with a dressing glass and jewel drawer at the top. This chest contains four drawers, one 10 inches deep, one 8 inches deep, and two short drawers 6½ inches deep. The chest measures 3 feet over the gables. Plane and square up the two gables 2 feet, 3½ inches in length by 1 foot, 6 inches in width by ⅞ inch thick. The back edges have to be rebated to receive the back, which consists of ⅝-inch matchboarding. Make the bottom from pine ¾ inch thick and lap-dovetail it to the gable ends. Two pieces A (Figure 299), 3 inches wide by ¾ inch thick, are fixed to the gable ends in the same way as the bottom, the object of one being to allow the back to overlap. The front piece may be of pine, with a thin slip of walnut glued to the edge. The two fore-edges B are also fixed to the gables by grooving the latter to form a dovetail (Figure 230). The fore-edges are slid in from the back. Stop the grooves at the front, and rebate the fore-edges to bring them flush there.

Bedroom Bookshelves

Figure 231 shows the front view and Figure 232 the side elevation of a bookshelf which may stand on a table or be hung on the wall. The drawers are of the usual pattern and make. A wooden back would strengthen the bookshelf, but for ordinary use the ornamental top A and the piece of backing B behind the drawers is enough. Figure 233 shows a larger and highly ornamented set of small shelves with lockers beneath. The shelves and sides can be made of ⅝-inch deal, the outer frame being 3 feet, 4 inches by 2 feet, 4 inches by 6 inches deep, and the pieces may be screwed together for simplicity

Figure 233—Ornamented Bedroom Bookshelf

of construction. The back and top should be of well-seasoned ½-inch boards, glued and clamped up together. The outlines of the ornament at the top (Figure 234) and at the lower ends of the side pieces (Figure 235) should be roughly shaped out, and the ornament applied in gesso. The whole of the ornament may be applied this way or it can be carved. The doors of the lockers should be paneled and hinged at the bottom to open downwards. Add a coat of white enamel to finish the work.

Figure 230—Dovetail Joint in Gable

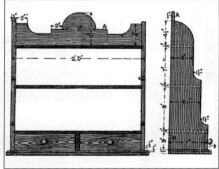

Figures 231 and 232

Figures 234 and 235—Gesso Ornaments for Bookshelf

Appendix 1
Alternative Energy

Solar Thermal Energy   412

Installing a Passive Solar Space
Heater   413

Make Your Own Solar
Cooking Oven   413

Windmills   414

Selecting and Installing a
Geothermal Heat Pump
System in Your Home   417

Appendix 1

Solar Thermal Energy

Solar thermal (heat) energy is used most often for heating swimming pools, heating water to be used in homes, and heating specific spaces in buildings. Solar space heating systems are either passive or active.

Passive Solar Space Heating

Passive space heating is what happens in a car on a sunny summer day—the car gets hot inside. In buildings, air is circulated past a solar heat surface and through the building by convection—less dense, warm air tends to rise while the denser, cooler air moves downward. No mechanical equipment is needed for passive solar heating.

Passive solar space heating takes advantage of the warmth from the sun through design features, such as large, south-facing windows and materials in the floors and/or walls that absorb warmth during the day and release it at night when the heat is needed most. Sunspaces and greenhouses are good examples of passive systems for solar space heating.

Passive solar systems usually have one of these designs:

1. Direct gain—This is the simplest system. It stores and slowly releases heat energy collected from the sun shining directly into the building and warming up the materials (tile or concrete). It is important to make sure the space does not become overheated.
2. Indirect gain—This is similar to direct gain in that it uses materials to hold, store, and release heat. This material is generally located between the sun and the living space, usually in the wall.
3. Isolated gain—This collects solar energy separately from the primary living area (a sunroom attached to a house can collect warmer air that flows through the rest of the house).

Active Solar Space Heating

Active heating systems require a collector to absorb the solar radiation. Fans or pumps are used to circulate the heated air or the heat-absorbing fluid. These systems often include some type of energy storage system.

There are two basic types of active solar heating systems. These are categorized based on the type of fluid (liquid or air) that is heated in the energy collectors. The collector is the device in which the fluid is heated by the sun. Liquid-based systems heat water or an antifreeze solution in a hydronic collector. Air-based systems heat air in an air collector. Both of these systems collect and absorb solar radiation, transferring solar heat to the interior space or to a storage system, where the heat is then distributed. If the system cannot provide adequate heating, an auxiliary or backup system provides additional heat.

Liquid systems are used more often when storage is included and are well suited for radiant heating systems, boilers with hot water radiators, and absorption heat pumps and coolers. Both liquid and air systems can adequately supplement forced air systems.

Active solar space heating systems are comprised of collectors that absorb solar radiation combined with electric fans or pumps to distribute the solar heat. These systems also have an energy-storage system that provides heat when the sun is not shining.

Another type of active solar space heating system, the medium temperature solar collector, is generally used for solar space heating. These systems operate in much the same way as indirect solar water heating systems but have a larger collector area, larger storage units, and much more complex control systems. They are usually configured to provide solar water heating and can provide between 30 and 70 percent of residential heating requirements. All active solar space heating systems require more sophisticated design, installation, and maintenance techniques than passive systems.

Passive Solar Water Heaters

Passive solar water heaters rely on gravity and on water's natural tendency to circulate as it is heated. Since these heaters contain no electrical components, passive systems are more reliable, easier to maintain, and work longer than active systems. Two popular types of passive systems are:

1. Integral-collector storage systems— These consist of one or more storage tanks that are placed in an insulated box with a glazed side facing the sun. The solar collectors are best suited for areas where temperatures do not often fall below freezing. They work well in households with significant daytime and evening hot-water needs but they do not work as efficiently in households with only morning hot-

water draws as they lose most of the collected energy overnight.

2. Thermospyhon systems—These are an economical and reliable choice particularly in newer homes. These systems rely on natural convection of warm water rising to circulate the water through the collectors and into the tank. As water in the collector heats, it becomes lighter and rises to the tank above it and the cooler water flows down the pipes to the bottom of the collector. In freeze-prone climates, indirect thermosyphons (using glycol fluid in the collector loop) can be installed only if the piping is protected.

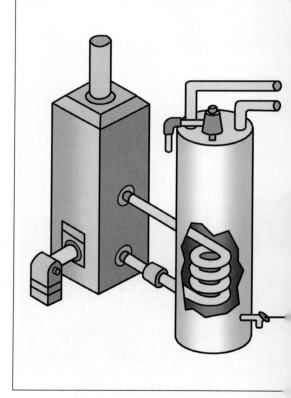

⌃ A combination of an indirect water heater and a highly efficient boiler can provide a very inexpensive method of water heating.

Active Solar Water Heaters

Active solar water heaters rely on electric pumps and controllers to circulate the water (or other heat-transfer fluids). Two types of active solar water heating systems are:

1. Direct circulation systems—These use pumps to circulate pressurized potable water directly through the collectors.

These systems are most appropriate for areas that do not have long freezes or hard/acidic water.

2. Indirect circulation systems— These pumps heat transfer fluids through the collectors. These heat exchangers then transfer the heat from the fluid to potable water. Some of these indirect circulation systems have overheat protectors so the collector and glycol fluid do not become superheated. Common indirect systems include antifreeze, in which the heat transfer fluid is usually a glycol-water mixture, and drainback, in which pumps circulate the water through the collectors and then the water in the collector loop drains back into a reservoir tank when the pump stops.

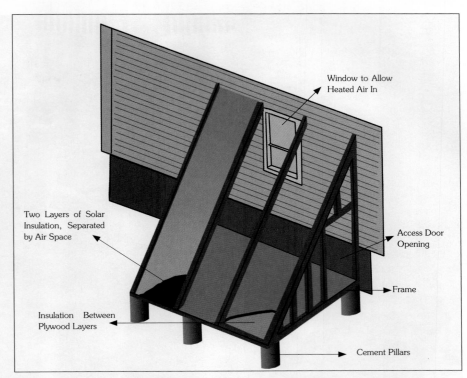

≫ A passive solar space heater.

Installing a Passive Solar Space Heater

A passive solar space heater works when the sun shines through the solar panels to heat the air inside a box. As the air heats up in the box, it rises and moves into the house. Cool air moves into the box and out of the house—in this way, the house is heated without the use of a mechanized heating system. Using a passive solar heater works best if you have a house that faces south and has both basement and first floor windows on that side of the house. If your house meets these requirements (and there aren't too many obstructions that would impede the sun from shining on the heater), then you can begin construction.

The passive solar space heater is made up of a floor and two triangular end walls, all of which can be made simply out of plywood. In between the open space, insulation can be placed. A lid can also be added to cover the heater in the summer.

To build such a solar space heater, first decide where on the southern wall your collector will be located. If you can place the heater in between windows, that is the best option. You may need to cut through the wall near a window to allow for the proper ventilation but if you don't want to do this, you can also purchase a detachable plywood "chimney" to move the heated air into the house. Next, find the studs that will support the fiberglass panel and find a panel that will be of the appropriate size.

Next, make the base for your solar heating system. The base can be made of ⅜-inch plywood board. Nail the board to a 2 x 4 and level it. Next, add insulation (the kind found on rolls is best), nailing it to the plywood. Then, nail the whole board to the side of the house. Make sloping supports out of 2 x 4s. Make sure the end wall studding is nailed in, and then attach the outside panel to it.

Under the shingles, install flashing or something else that will keep water out of the top of the solar heater. Then, install the fiberglass panels, making sure the edges are caulked so no water can come in. Enclose the edges of the fiberglass with small strips of plywood. Then, install the outer fiberglass panel so that it is flush with the top surface and caulk it. To finish up, paint the inside of the plywood surfaces black to absorb the heat. The inside of the cover panel should be painted white to reflect the light.

Make Your Own Solar Cooking Oven

This type of simple, portable solar oven is perfect for camping trips or if you want to do an outdoor barbeque with additional cooked foods in the summer. This homemade solar oven can reach around 350°F when placed in direct sunlight.

Supplies

- A reflective car sunshade or any sturdy but flexible material (such as cardboard) covered with tin foil and cut to the notched shape of a car sunshade
- Velcro
- A bucket
- A cooking pot
- A wire grill
- A baking bag

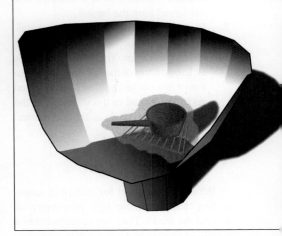

≫ A simple solar oven.

⌃ Solar ovens can be fashioned in a variety of ways. The goal is to have as much surface area as possible reflecting the sun toward your food.

Directions

1. Place the car sunshade on the ground. Cut the Velcro into three separate pieces and stick on half of each piece onto the edge near the notch. Then, test the shade to see if the Velcro pieces, when brought together, form a funnel. Place the funnel atop the bucket.

2. Place the cooking pot on the wire grill. Put this all in the baking bag and put it inside the funnel. The rack should now be lying on top of the bucket. Now place the whole cooker in direct sunlight and angle the funnel in the direction of the sun. Adjust the angle as the sun moves.

Windmills

Windmills are used for pumping water, milling, and operating light machinery all around the world. They are constructed in a variety of shapes and some are quite picturesque. When set up properly, windmills cost nothing to operate and if the wheel is made well, it will last for many years without need for major repairs. To make a windmill requires a good understanding of carpentry and workmanship but it is not incredibly difficult or expensive to do.

Constructing a Windmill

Windmills can be of all sizes, though the larger the windmill, the more power it can generate. This windmill and tower can be easily constructed out of wood, an old wheel, and a few iron fittings you may be able to find at a hardware store or home center. Constructing the windmill in sections is the easiest way to create this structure. Simply follow these directions to make your own energy-producing windmill:

The Tower (Fig. 2)

1. The tower is the first part to be built and should be constructed out of four spruce sticks that are 16 feet long and 4 inches square, in a configuration that

measures 30 inches square at the top and 72 inches square at the base.

2. The deck should be 36 inches square and should project 2 inches over the top rails.

3. The rails and cross braces can be spruce or pine strips and should measure 4 inches wide and ⅞ inch thick. Attach these to the corner posts with steel-wire nails.

4. Embed the corner posts 2 feet into the ground, leaving 14 feet above the surface.

The rail at the bottom, which is attached to the four posts, should measure 3 feet above the ground. Midway between this and the top rail of the deck, run a middle rail around the post. Make sure that where your wheel will be attached, this point rises at least 2 feet above any obstructions (buildings, trees, etc.) so it can have access to the blowing wind.

5. The cross braces should be beveled at the ends so they fit snugly against the corner.

6. The posts, rails, and braces should be planed so they present a nice appearance at the end of the building. A ladder can also be constructed at one side of the tower to allow easy access to the mill.

7. Nail a board across two of the rails halfway up the tower. Secure the lower end of a trunk tightly here if you are constructing a pumping mill. However, if a wooden mill is what you are after, you can use an old wheel from a wagon and six blades of wood.

The Turntable

1. The turntable holds the wheel and tail. It should be built of 2½ x 2-inch timber and 2-inch galvanized wrought iron "water" tube and flanges.

2. The upper flange supports the timber framing. It should be countersunk, using a half-round file, and screwed tightly onto the tube as far as possible. The end of the tube should project just slightly beyond the face of the flange so that it can be riveted over to fill the countersink.

3. Bolt the two loose flanges to the framework of the tower. Use them with 2-inch pipe with the thread filed away so they may slide freely onto the tube. The upper loose flange should form a footstep bearing and the lower flange a guide for the turntable.

4. Now mount the turntable on the ball bearing to make sure the mill head can turn freely. Screw on two back nuts to guard against any possibility of the turntable being lifted out of place by a strong wind.

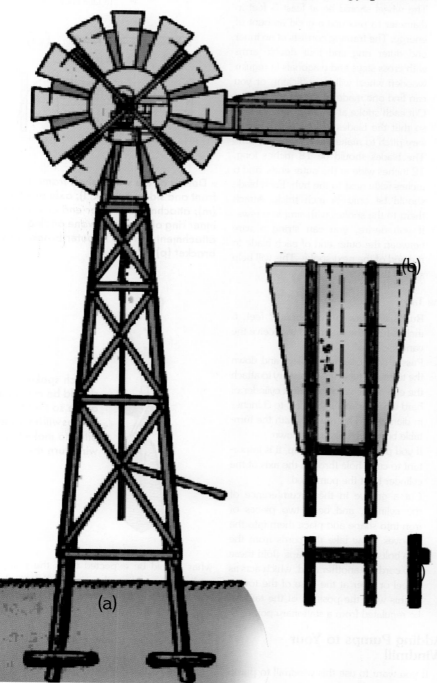

(b)

(a)

⩔ Details of the windmill. Figure (a) shows a general view with the tail turned to "off" position. Figure (b) shows details of the tail, and (c) shows a cross-piece of the tail.

⩔ Beveled cross braces fit snugly against the corners.

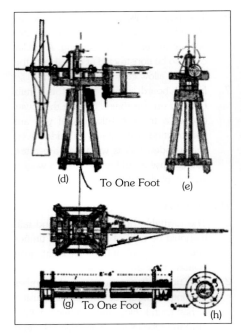

(d) To One Foot (e)

(g) To One Foot (h)

≈ The windmill turntable holds the wheel and tail. The head carries the wheel spindle.

The Head

1. This is the part that will carry the wheel spindle.
2. Notch the joints and secure them with 2-inch bolts.
3. The upright, which carries a bolt or pin for the spur-wheel to revolve upon, is kept in place in the front and at the sides by a piece of hoop iron.
4. The tail vane swivel is a piece of 5-inch bore tube with back nuts and washers. Pass an iron bolt or other piece of iron through this, screw it to each end, and fit it with four nuts and washers.

The Wheel Shaft

1. Use wrought-iron tubing and flanges to create the wheel shaft. The bore of the tube is at least 5 inches, and the outside diameter should be roughly 1½ inches. Both the tube and the fittings should be of good quality and a thick gauge (steam quality is preferred).
2. If lathe is available, lightly skim it over the tubing. However, if it's not, a careful filing will do just as well to smooth down the edges.
3. Screw the tube higher up on one end to receive the flanges forming the hub. Screw these on and secure them on one side with back nuts and on the other with a distance piece made out of a 1½-inch bore tube. Fit a cap to close the open front end of the tube.

4. Grease two plummer blocks with some form of lubrication. These will be the bearings for the shaft.
5. A pinion is needed of at least 2½ inches in diameter at the pitch circle. Bore it to fit the wheel shaft. A spur wheel of 7 inches in diameter should follow that (gear wheels from a lawn mower can be used if available).

The Wheel

1. The wheel should be at least 5 feet in diameter to produce a good amount of energy. The framing consists of an inner and outer ring and four double arms with cross stays and diagonals (a regular wooden wheel will be sufficient, or you can find one made of galvanized steel).
2. Cut each spoke at an angle on one side so that the blades will have the necessary pitch to make the wind turn them.
3. The blades should be 18 inches long, 12 inches wide at the outer ends, and 6 inches wide next to the hub. Each blade should be only ¾ inch thick. Attach them to the spokes with simple screws.
4. If you desire, you can string a wire between the outer end of each blade to the end of the next spoke. This will help steady the blades.

The Tail

1. Run a fine saw cut up about 2 feet, 6 inches from the outer end to receive the vane (optional).
2. Pass a cord over two pulleys and down the turntable tube. It is necessary to attach the end of the cord to a short cylinder of hard wood or metal (about 2 to 3 inches in diameter). This revolves with the turntable but can be slid up or down.
3. If you plan on using a pump, it is important to cut a hole through the axis of the cylinder to fit the pump rod.
4. Cut a groove in the circumference of the cylinder, and bend two pieces of iron into shape and place them into the grooves. Now take the cords from the two bolts, untying the straps. Join these two cords to another cord, which acts as a reel or lever at the base of the tower. In this way, the position of the tail can be regulated from a stationary point.

Adding Pumps to Your Windmill

If you want to use this windmill to pump water, then you may need to do some experimenting with different lengths of pump stroke. Below is a table indicating

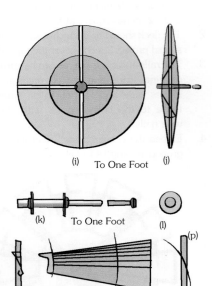

(i) To One Foot (j)

(k) To One Foot (l)

(m) (n) (o) (p)

≈ Details of the wheel shaft frame (i, j); front and side views, (k, l); axle of wheel (m); attachment of inner end of vane to inner ring of frame (n); vane on rings (o); attachment of vane to outer brackets by bracket (p).

« Each spoke should be cut at an angle so that the blades will have the pitch to make the wind turn them.

what should be expected from the pump, and also providing the size of the single-action pump suitable for a given lift (using a ratio of 1 to 3).

Make sure that your pump is not too large; otherwise, it may not start in a light wind or breeze.

The pump is driven by a pin screwed into the side of the spur wheel and is secured with a lock nut. Drill and tap three or four holes at different distances from the center

Total Lift	Gallons per Hour	Bore of Pump	Approximate Stroke
26 ft	100	2 in.	3½ in.
60 ft	50	2 in.	1½ in.
100 ft	25	1½ in.	1½ in.

of the wheel so the length of the stroke can be adjusted. If the spokes on the wheel are too thin for drilling, you can use a clamp with a projecting pin instead.

A pump rod—a continuous wooden rod about 1 inch square and thicker at the top end—can be used in connecting the bottom end (by bolting) to the "bow" supplied with the pump. Intermediate joints, if needed, can be fashioned with 1 x ½-inch fish plates roughly 6 inches long. If the pump is no more than 12 feet below the crank pin, one guide will be adequate. The pump rod must be able to revolve with the head and will be need to be thickened up in a circular section where it passes through the guide. Make the guide in two halves and screw or bolt it to a bar running across the tower.

Final Touches

When construction is finished, paint all of the woodwork any color that complements your yard or property and, if desired, lacquer it to protect the wood from rain and snow. A windmill of this size will create at least a one quarter horsepower in a 15 mph wind.

Selecting and Installing a Geothermal Heat Pump System in Your Home

The heating efficiency of commercial ground-source and water-source heat pumps is indicated by their coefficient of performance (COP)—the ratio of heat provided in Btu per Btu of energy input. The cooling efficiency is measured by the energy efficiency ratio (EER)—the ratio of heat removed to the electricity required (in watts) to run the unit. Many geothermal heat pump systems are approved by the U.S. Department of Energy as being energy efficient products and so, if you are thinking of purchasing and installing this type of system, you may want to check to

see if there is any special financing or incentives for purchasing energy efficient systems.

Evaluating Your Site

Before installing a geothermal heat pump, consider the site that will house the system. The presence of hot geothermal fluid containing low mineral and gas content, shallow aquifers for producing the fluid, space availability on your property, proximity to existing transmission lines, and availability of make-up water for evaporative cooling are all factors that will determine if your site is good for geothermal electric development. As a rule of thumb, geothermal fluid temperature should be no less than 300°F.

In the western United States, Alaska, and Hawaii, hydrothermal resources (reservoirs of steam or hot water) are more readily available than the rest of the country. However, this does not mean that geothermal heat cannot be used throughout the country. Shallow ground temperatures are relatively constant throughout the United States and this means that energy can be tapped almost anywhere in the country by using geothermal heat pumps and direct-use systems.

To determine the best type of ground loop systems for your site, you must assess the geological, hydrological, and spatial characteristics of your land in order to choose the best, most effective heat pump system to heat and cool your home:

1. Geology—This includes the soil and rock composition and properties on your site. These can affect the transfer rates of heat in your particular system. If you have soil with good heat transfer properties, your system will require less piping to obtain a good amount of heat from the soil. Furthermore, the amount of soil that is available also contributes to which system you will choose. For example, areas that have hard rock or shallow soil will most likely benefit from a vertical heat pump system instead of a system requiring large and deep trenches, such as the horizontal heat pump system.

2. Hydrology—This refers to the availability of ground or surface water, which will affect the type of system to be installed. Factors such as depth, volume, and water quality will help determine if surface water bodies can be used as a source of water for an open-loop heat pump system or if

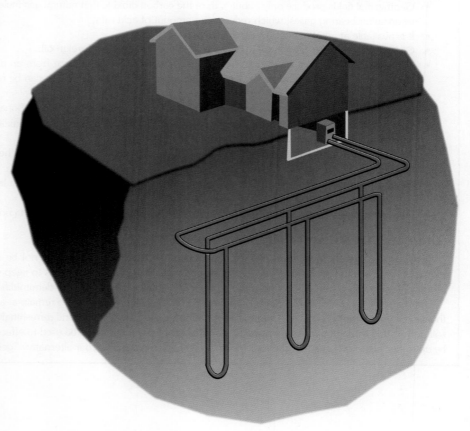

≫A vertical closed-loop geothermal system

they would work best with a pond/lake system. Before installing an open-loop system, however, it is best to determine your site's hydrology so potential problems (such as aquifer depletion or groundwater contamination) can be avoided.

3. Available land—The acreage and layout of your land, as well as your landscaping and the location of underground utilities, also play an important part in the type of heat pump system you choose. If you are building a new home, horizontal ground loops are an economical system to install. If you have an existing home and want to convert your heat and cooling to geothermal energy, vertical heat pump systems are best to minimize the disturbance to your existing landscaping and yard.

Installing the Heat Pumps

Geothermal heat pump systems are somewhat difficult to install on your own—though it can certainly be done. Make sure, before you begin any digging, to contact your local utility company to make sure you will not be digging into gas pipes or electrical wires.

The ground heat exchanger in a geothermal heat pump system is made up of closed- or open-loop pipe—depending on which type of system you've determined is best suited for your site. Since most systems employed are closed-loop systems, high density polyethylene pipe is used and buried horizontally at 4 to 6 feet deep or vertically at 100 to 400 feet deep. These pipes are filled with an environmentally friendly antifreeze/water solution that acts as a heat exchanger. You can find this at your local home store or contact a contractor to see where it is distributed. This solution works in the winter by extracting heat from the earth and carrying it into the building. In the summertime, the system reverses, taking heat from the building and depositing it into the ground.

Air delivery ductwork will distribute the hot or cold air throughout the house's ductwork like traditional, conventional systems. An air handler—a box that contains the indoor coil and fan—should be installed to move the house air through the heat pump system. The air handler contains a large blower and a filter, just like standard air conditioning units.

Cost-Efficiency of Geothermal Heat Pump Systems

By installing and using a geothermal heat pump system, you will save on the costs of operating and maintaining your heating and cooling system. While these systems are generally a bit pricier to install, they prove to be more efficient and thus save you money on a monthly and yearly basis. Especially in the colder winter months, geothermal heat pump systems can reduce your heating costs by about half. Annual energy savings by using a geothermal heat pump system range from 30 to 60 percent.

Benefits of Using Geothermal Energy

- It is clean energy. Geothermal energy does not require the burning of fossil fuels (coal, gas, or oil) in order to produce energy.
- Geothermal fields produce only about ⅙th of the carbon dioxide that natural gas-fueled power plants do. They also produce little to no sulfur-bearing gases, which reduces the amount of acid rain.
- It is available at any time of day, all year round.
- Geothermal power is homegrown, which reduces dependence on foreign oil.
- It is a renewable source of energy. Geothermal energy derives its source from an almost unlimited amount of heat generated by the earth. And even if energy is limited in an area, the volume taken out can be reinjected, making it a sustainable source of energy.
- Geothermal heat pump systems use 25 to 50 percent less electricity than conventional heating and cooling systems. They reduce energy consumption and emissions between 44 and 72 percent and improve humidity control by maintaining about 50 percent relative humidity indoors (GHPs are very effective for humid parts of the country).
- Heat pump systems can be "zoned" to allow different parts of your home to be heated and cooled to different temperatures without much added cost or extra space required.
- Geothermal heat pump systems are durable and reliable. Underground piping can last for 25 to 50 years and the heat pumps tend to last at least 20 years.
- Heat pump systems reduce noise pollution since they have no outside condensing unit (like air conditioners).

Alternate "Geothermal" Cooling System

True geothermal energy systems can be very expensive to install and you may not be able to use one in your home at this time. However, here is a fun alternative way to use the concepts of geothermal systems to keep your house cooler in the summer and your air conditioning bills lower. All you need are a basement, small window fan, and dehumidifier.

Your basement is a wonderful example of how the top layers of earth tend to remain at a stable temperature throughout the year. In the winter, your basement may feel somewhat warm; in the summer, it's nice and refreshingly cool. This is due to the temperature of the soil permeating through the basement walls. And this cool basement air can be used to effectively reduce the temperature in your home by up to five degrees during the summer months. Here are the steps to your alternative "geothermal" cooling system:

1. Run the dehumidifier in your basement during the night, bringing the humidity down to about 60 percent.

2. Keep your blinds and curtains closed in the sunniest rooms in your home.

3. In the morning, when the temperature inside the house reaches about 77°F, open a small window in your basement, just a crack, and open one of the upstairs windows, placing a small fan in it and directing the room air out of the window.

4. With all other windows and outside doors closed, the fan will suck the cool basement air through your home and out the open window. Doing this for about an hour will bring down the temperature inside your home, buying you a couple of hours of reprieve before switching on the AC.

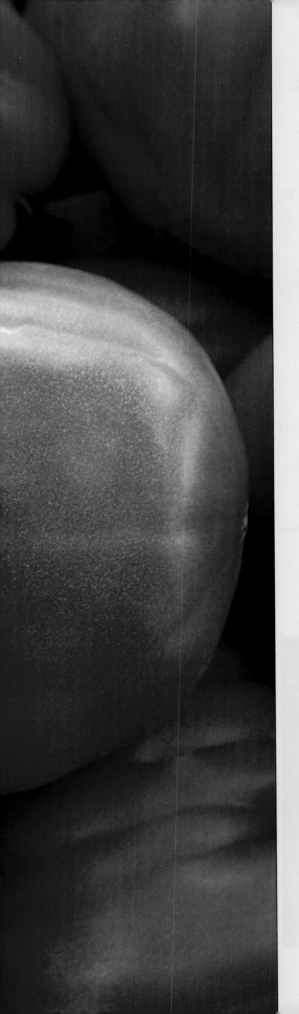

Appendix 2

Food Co-op Directory

Appendix 2

Food Co-ops and Distributors by State

Alabama

Grow Alabama
2301 Finley Boulevard
Birmingham, Alabama 35202
(205) 991-0042
info@growalabama.com
www.growalabama.com

Alaska

Organic Alaska
3404 Willow Street
Anchorage, Alaska 99517
(907) 306-3931
organic@alaska.com
www.organicalaska.com

Arizona

Food Conspiracy Co-op
412 North 4th Avenue
Tucson, Arizona 85705
(520) 624-4821
www.foodconspiracy.org

Arkansas

Ozark Natural Foods Co-op
1554 North College Avenue
Fayetteville, Arkansas 72703
(479) 521-7558
www.ozarknaturalfoods.com

California

Briar Patch Community Market
290 Sierra College Dr, Ste A
Grass Valley, California 95945
(530) 272-5333
info@briarpatch.coop
www.briarpatch.coop

Co-opportunity Natural Foods
1525 Broadway
Santa Monica, California 90404
(310) 451-8902
service@coopportunity.com
www.coopportunity.com

Davis Food Co-op
620 G Street
Davis, California 95616
(530) 758-2667
www.davisfood.coop

Isla Vista Food Co-op
6575 Seville Road
Isla Vista, California 93117
(805) 968-1401
gm@islavistafood.coop
www.islavistafood.coop

Kresge Food Co-op
600 Kresge Court
Kresge College UCSC
Santa Cruz, California 95064
(831) 426-1506
www.kresge.ucsc.edu/activities/coops/food-coop.html

North Coast Co-op, Arcata
811 I Street
Arcata, California 95521
(707) 822-5947
co-oparc@northcoastco-op.com
www.northcoastco-op.com

North Coast Co-op, Eureka
25 4th Street
Eureka, California 95501
(707) 443-6027
www.northcoastco-op.com

Ocean Beach People's Organic Food Market
4765 Voltaire Street
San Diego, California 92107
(619) 224-1387
www.obpeoplesfood.coop

Other Avenues Community Food Store
3930 Judah Street
San Francisco, California 94122
(415) 661-7475
info@otheravenues.org
www.otheravenues.org

Quincy Natural Foods Co-op
269 Main Street
Quincy, California 95971
(530) 283-3528
www.qnf.coop

Rainbow Grocery Co-op
1745 Folsom Street
San Francisco, California 94103
(415) 863-0620
general@rainbow.coop
www.rainbow.coop

Sacramento Natural Foods Co-op
1900 Alhambra Boulevard
Sacramento, California 95816
(916) 455-2667
www.sacfoodcoop.com

Santa Rosa Community Market & Café
1899 Mendocino Avenue
Santa Rosa, California 95401
(707) 546-1806
www.srcommunitymarket.com

Ukiah Natural Foods
721 South State Street
Ukiah, California 95482
(707) 462-4778
www.ukiahcoop.com

Colorado

Fort Collins Food Co-op
250 East Mountain Avenue
Fort Collins, Colorado 80524
(970) 484-7448
info@ftcfood.coop
www.ftcfoodcoop.com

High Plains Food Co-op
5655 South Yosemite Street,
Suite 400
Greenwoods Village,
Colorado 80111
(785) 626-3640
info@highplainsfood.org
http://highplainsfood.org

Connecticut

Willimantic Food Co-op
91 Valley Street
Willimantic, Connecticut 06226
(860) 456-3611
www.willimanticfood.coop

Delaware

Newark Natural Foods Cooperative
280 East Main Street, Market East Plaza
Newark, Delaware 19711

(302) 368-5894
www.newarknaturalfoods.com

Florida

Ever'man Cooperative
315 West Garden Street
Pensacola, Florida 32502
(850) 438-0402
info@everman.org
www.everman.org

Homegrown Organic Local Food Co-op
2310 North Orange Avenue
Orlando, Florida 32804
(407) 895-5559
info@homegrowncoop.org
www.homegrowncoop.org

Sunseed Food Co-op, Inc.
6615 North Atlantic Avenue
Cape Canaveral, Florida 32920
(321) 784-0930
www.sunseedfoodcoop.com

Georgia

Daily Groceries Food Co-op
523 Prince Avenue
Athens, Georgia 30601
(706) 548-1732
info@dailygroceries.org
www.dailygroceries.org

Life Grocery Natural Food Co-op & Café
1453 Roswell Road
Marietta, Georgia 30062
(770) 977-9583
www.lifegrocery.com

Sevananda Food Co-op
467 Moreland Avenue NE
Atlanta, Georgia 30307
(404) 681-2831
info@sevananda.com
www.sevananda.coop

Hawaii

Kokua Market Natural Foods Co-op
2643 South King Street
Honolulu, Hawaii 96826
(808) 941-1922
info@kokua.coop
www.kokua.coop

Idaho

Boise Consumer Co-op
888 West Fort Street
Boise, Idaho 83702
(208) 472-4500
www.boisecoop.com

Moscow Food Co-op
121 East 5th Street
Moscow, Idaho 83843
(208) 882-8537
www.moscowfood.coop

Illinois

Common Ground Food Co-op
300 South Broadway Suite 166
Urbana, Illinois 61801
(217) 352-3347
info@commonground.coop
www.commonground.coop

Duck Soup Co-op
129 East Hillcrest Drive
DeKalb, Illinois 60115
(815) 756-7044
ducksoupcoopgm@gmail.com
www.ducksoupcoop.com

Neighborhood Co-op Grocery
1815 West Main Street
Carbondale, Illinois 62901
(618) 529-3533
info@neighborhood.coop
www.neighborhood.coop

South Suburban Food Co-op
208 Forest Boulevard
Park Forest, Illinois 60466
(708) 747-2256
info@southsuburbanfoodcoop.com
www.southsuburbanfoodcoop.com

Stone Soup Ashland
4637 North Ashland Avenue
Chicago, Illinois 60640
(773) 669-7687
www.stonesoupcoop.org

West Central Illinois Food Cooperative
176 North Farnham Street
Galesburg, Illinois 61401

Indiana

Bloomingfoods Kirkwood
419 East Kirkwood Avenue
Bloomington, Indiana 47408
(812) 336-5300
www.bloomingfoods.coop

Bloomingfoods East
3220 East Third Street
Bloomington, Indiana 47401
(812) 336-5400
www.bloomingfoods.coop

Bloomingfoods Near West Side
316 West Sixth Street

Bloomington, Indiana 47404
(812) 333-7312
www.bloomingfoods.coop

Clear Creek Food Co-op
710 East Main Street
Richmond, Indiana 47374
(765) 939-4390

Lost River Market & Deli
26 Library Street
Paoli, Indiana 47454
(812) 723-3735
www.lostrivercoop.com

Maple City Market
314 South Main Street
Goshen, Indiana 46526
(574) 534-2355
info@maplecitymarket.com
www.maplecitymarket.com

River City Food Co-op
116 Washington Avenue
Evansville, Indiana 47713
(812) 401-7301
www.rivercityfoodcoop.org

Three Rivers Food Co-op's Natural Grocery
1612 Sherman Boulevard
Fort Wayne, Indiana 46808
(260) 424-8812
gm@3riversfood.coop
www.3riversfood.coop

Iowa

New Pioneer Co-Op Coralville
1101 2nd Street
Coralville, Iowa 52241
(319) 358-5513
www.newpi.com

New Pioneer Co-op Iowa City
22 South Van Buren Street
Iowa City, Iowa 52240
(319) 338-9441
www.newpi.com

Oneota Community Food Co-op
312 West Water Street
Decorah, Iowa 52101
(563) 382-4666
customerservice@oneotacoop.com
http://oneotatestsite.com

Wheatsfield Cooperative Grocery
413 Northwestern Avenue
Ames, Iowa 50010
(515) 232-4094

shop@wheatsfield.coop
www.wheatsfield.coop

Kansas

Community Market & Deli
901 Iowa Street
Lawrence, Kansas 66044
(785) 843-8544
themerc@themerc.coop
www.themerc.coop

Prairieland Market
138 South 4th
Salinas, Kansas 67401
(785) 827-5877

Southeast Kansas Buying Club
11th & Walnut
Independence, Kansas 67301
(620) 205-7095
http://seksbuyingclub.wordpress.com

Topeka Natural Food Coop
503 Southwest Washburn Avenue
Topeka, Kansas 66606
(785) 235-2309
http://topekafoodcoop.wordpress.com

Kentucky

Good Foods Market & Café
455 Southland Drive
Lexington, Kentucky 40503
(859) 278-1813
goodfoods@goodfoods.coop
www.goodfoods.coop

Otherworld Food Co-op, Inc.
1865 Celina Road
Burkesville, Kentucky 42717
(270) 433-7400
otherworld@duo-county.com
www.duo-county.com/~otherworld

Louisiana

New Orleans Food Co-op
2372 St. Claude Avenue, Suite 110
New Orleans, Louisiana 70117
(504) 264-5579
www.nolafood.coop

Maine

Belfast Co-op
123 High Street
Belfast, Maine 04915
(207) 338-2532
www.belfast.coop

Blue Hill Co-op Community Market & Cafe
4 Ellsworth Road, P.O. Box 1133
Blue Hill, Maine 04614-1133

(207) 374-2165
info@bluehill.coop
http://bluehill.coop

Fare Share Market
443 Main Street
Norway, Maine 04268
(207) 743-9044
www.faresharecoop.org

Rising Tide Co-op
323 Main Street
Damariscotta, Maine 04543
(207) 563-5556
customercare@risingtide.coop
www.risingtide.coop

Maryland

Common Market Co-op
5728 Buckeystown Pike, Unit B1
Frederick, Maryland 21704
(301) 663-3416
www.commonmarket.coop

Glut Food Co-op
4005 34th Street
Mt. Rainier, Maryland 20712
(301) 779-1978
www.glut.org

Maryland Food Co-op
B-0203 Stamp Student Union
University of Maryland, College Park, Maryland 20742
(301) 314-8089
http://thestamp.umd.edu/food/md_food_co-op

Takoma Park Silver Spring Co-op
201 Ethan Allen Avenue
Takoma Park, Maryland 20912
(301) 891-2667
www.tpss.coop

Massachusetts

Berkshire Co-op Market
42 Bridge Street
Great Barrington, Massachusetts 01230
(413) 528-9697
community@berkshire.coop
www.berkshirecoop.org

Green Fields Market
144 Main Street
Greenfield, Massachusetts 01301
(413) 773-9567
www.greenfieldsmarket.coop

Harvest Co-op Markets
580 Massachusetts Avenue
Cambridge, Massachusetts 02139

(617) 661-1580
www.harvest.coop

Harvest Co-op Markets
57 South Street
Jamaica Plain, Massachusetts 02130
(617) 524-1667
www.harvest.coop

River Valley Market
330 North King Street
Northampton, Massachusetts 01061
(413) 584-2665
info@rivervalleymarket.coop
www.rivervalleymarket.coop

Michigan

Brighton Food Cooperative
2715 West Coon Lake Road
Howell, Michigan 48843
(517) 546-4190
bfc@brightonfoodcoop.com
http://www.brightonfoodcoop.com

Dibbleville Food Cooperative
106 East Elizabeth Street
Fenton, Michigan 48430
(810) 629-1175
contact@dibbleville.com
www.dibbleville.com

East Lansing Food Co-op
4960 Northwind
East Lansing, Michigan 48823
(517) 337-1266
info@elfco.coop
www.elfco.coop

Grain Train Natural Food Market
220 East Mitchell Street
Petoskey, Michigan 49770
(231) 347-2381
www.graintrain.coop

Ionia Natural Food Co-op
6070 David Highway
Saranac, Michigan 48881
infc_mc@hotmail.com
https://www.facebook.com/IoniaNatural FoodsCooperative/info

Keweenaw Food Co-op
1035 Ethel Avenue
Hancock, Michigan 49930
(906) 482-2030
info@keweenaw.coop
www.keweenaw.coop

Marquette Food Co-op
502 West Washington Street
Marquette, Michigan 49855

(906) 225-0671
info@marquettefood.coop
www.marquettefood.coop

Oryana Food Cooperative
260 East 10th Street
Traverse City, Michigan 49684
(231) 947-0191
www.oryana.coop

People's Food Co-op & Café Verde
216 North 4th Avenue
Ann Arbor, Michigan 48104
(734) 994-9174
info@peoplesfood.coop
www.peoplesfood.coop

People's Food Co-op of Kalamazoo
507 Harrison Street
Kalamazoo, Michigan 49007
(269) 342-5686
outreach@peoplesfoodco-op.org
www.peoplesfoodco-op.org

Simple Times Farm Market & Buying Club
9044 Gale Road
Goodrich, Michigan 48438
(810) 280-2143
www.simpletimesfarm.com

West Michigan Co-op
1475 Northeast Michigan Street
Grand Rapids, Michigan 49503
(616) 951-3287
help@wmcoop.com
www.westmichigancoop.com

Ypsilanti Food Co-op and River Street Bakery
312 North River Street
Ypsilanti, Michigan 48198
(734) 483-1520
info@ypsifoodcoop.org
www.ypsifoodcoop.org

Minnesota

Bluff Country Co-op
121 West 2nd Street
Winona, Minnesota 55987
(507) 452-1815
bccoop@bluff.coop
www.bluff.coop

Cook County Whole Foods Co-op
20 East 1st Street
Grand Marais, Minnesota 55604
(218) 387-2503
info@cookcounty.coop
www.cookcounty.coop

Crow Wing Food Co-op
720 Washington Street
Brainerd, Minnesota 56401
(218) 828-4600
cwfoodco-op@brainerd.net
www.crowwingcoop.com

Eastside Food Co-op
2551 Central Avenue
Minneapolis, Minnesota 55418
(612) 788-0950
luna@eastsidefood.coop
www.eastsidefood.coop

Hampden Park Food Co-op
928 Raymond Avenue
St. Paul, Minnesota 55114
(651) 646-6686
www.hampdenparkcoop.com

Harmony Food Co-op
302 Irvine Avenue
Bemidji, Minnesota 56601
(218) 751-2009
www.harmonycoop.com

Just Food Co-op
516 South Water Street
Northfield, Minnesota 55057
(507) 650-0106
www.justfood.coop

Lakewinds Natural Foods
435 Pond Promenade
Chanhassen, Minnesota 55317
(952) 697-3366
www.lakewinds.coop

Lakewinds Natural Foods
17501 Minnetonka Boulevard
Minnetonka, Minnesota 55345
(952) 473-0292
www.lakewinds.coop

Linden Hills Food Co-op
3815 Sunnyside Avenue
Minneapolis, Minnesota 55410
(612) 922-1159
info@lindenhills.coop
www.lindenhills.coop

Mississippi Market Food Co-op
622 Selby Avenue
St. Paul, Minnesota 55104
(651) 310-9499
info@msmarket.coop
www.msmarket.coop

Mississippi Market Natural Foods Co-op
1500 West 7th Street

St. Paul, Minnesota 55105
(651) 690-0507
info@msmarket.coop
www.msmarket.coop

Natural Harvest Whole Food Co-op
505 North 3rd Street
Virginia, Minnesota 55792
(218) 741-4663
www.naturalharvestfoodcoop.com

People's Food Co-op
519 1st Avenue Southwest
Rochester, Minnesota 55901
(507) 289-9061
www.pfc.coop

Pomme De Terre Food Co-op
613 Atlantic Avenue
Morris, Minnesota 56267
(320) 589-4332
www.pdtfoods.org

River Market Community Co-op
221 North Main Street
Stillwater, Minnesota 55082
(651) 439-0366
info@rivermarket.coop
www.rivermarket.coop

Seward Community Co-op
2823 East Franklin Avenue
Minneapolis, Minnesota 55404
(612) 338-2465
www.seward.coop

St. Peter Food Co-op & Deli
228 Mulberry Street
St. Peter, Minnesota 56082
(507) 934-4880
www.stpeterfood.coop

Valley Natural Foods Co-op
13750 County Road 11
Burnsville, Minnesota 55337
(952) 891-1212
info@valleynaturalfoods.com
www.valleynaturalfoods.com

Wedge Community Co-op
2105 Lyndale Avenue South
Minneapolis, Minnesota 55405
(612) 871-3993
www.wedge.coop

Whole Foods Co-op
610 East 4th Street
Duluth, Minnesota 55805
(218) 728-0884
info@wholefoods.coop
http://wholefoods.coop

Mississippi

Rainbow Co-op
2807 Old Canton Road
Jackson, Mississippi 39216
(601) 366-1602
www.rainbowcoop.org

Missouri

City Food Co-op
2639 Cherokee Street
St. Louis, Missouri 63118
(314) 771-7213
info@cityfoodcoopstl.com
www.cityfoodcoopstl.com

Montana

Community Food Co-op
908 West Main Street
Bozeman, Montana 59715
(406) 587-4039
info@bozo.coop
www.bozo.coop

Nebraska

Open Harvest Natural Foods Co-op
1618 South Street
Lincoln, Nebraska 68502
(402) 475-9069
harvest@openharvest.com
www.openharvest.com

Nevada

Great Basin Community Food Coop
240 Court Street
Reno, Nevada 89501
(775) 324-6133
info@greatbasinfood.coop
http://greatbasinfood.coop

New Hampshire

Concord Cooperative Market
24 South Main Street
Concord, New Hampshire 03301
(603) 225-6840
info@concordfoodcoop.coop
www.concordfoodcoop.coop

Co-op Community Food Market
43 Lyme Road
Hanover, New Hampshire 03755
(603) 643-2725
comment@coopfoodstore.com
www.coopfoodstore.com

Co-op Food Stores: Hanover
45 South Park Street
Hanover, New Hampshire 03755
(603) 643-2667
comment@coopfoodstore.com
www.coopfoodstore.com

Co-op Food Stores: Lebanon
12 Canterra Parkway
Lebanon, New Hampshire 03766
(603) 643-4889
comment@coopfoodstore.com
www.coopfoodstore.com

New Jersey

George Street Co-op
89 Morris Street
New Brunswick, New Jersey 08901
(732) 247-8280
www.georgestreetcoop.com

Purple Dragon Co-op
289 Washington Street
Glen Ridge, New Jersey 07028
(973) 429-0391
info@purpledragon.com
www.purpledragon.com

Sussex County Food Co-op
30 Moran Street
Newton, New Jersey 07860
(973) 579-1882
info@sussexcountyfoods.org
www.sussexcountyfoods.org

New Mexico

La Montañita Co-op and Food Market
226 West Coal Avenue
Gallup, New Mexico 87301
(505) 863-5383
http://lamontanita.coop

Silver City Food Co-op
520 North Bullard Street
Silver City, New Mexico 88061
(575) 388-2343
www.silvercityfoodcoop.com

New York

Abundance Cooperative Market
62 Marshall Street
Rochester, New York 14607
(585) 454-2667
www.abundance.coop

Flatbush Food Cooperative
1415 Cortelyou Road
Brooklyn, New York 11226
(718) 284-9717
info@flatbushfoodcoop.com
www.flatbushfoodcoop.com

4th Street Food Co-op
58 East 4th Street
New York, New York 10003-8914
(212) 674-3623
www.4thstreetfoodcoop.org

GreenStar Cooperative Market
701 West Buffalo Street
Ithaca, New York 14850
(607) 273-9392
www.greenstar.coop

High Falls Food Coop
1398 State Road 213
High Falls, New York 12440
(845) 687-7262
www.highfallsfoodcoop.com

Honest Weight Food Coop
100 Watervliet Ave
Albany, New York 12206
(518) 482-2667
www.honestweight.coop

Lexington Co-operative Market
807 Elmwood Avenue
Buffalo, New York 14222
(716) 866-2667
board@lexington.coop
www.lexington.coop

Park Slope Food Co-op
782 Union Street
Brooklyn, New York 11215
(718) 622-0560
www.foodcoop.com

Potsdam Consumer Co-op
24 Elm Street
Potsdam, New York 13676
(315) 265-4630
mail@potsdamcoop.com
www.potsdamcoop.com

Syracuse Real Food Co-op
618 Kensington Road
Syracuse, New York 13210
(315) 472-1385
www.syracuserealfood.coop

The Cambridge Food Co-op
1 West Main Street
Cambridge, New York 12816
(518) 677-5731
http://cambridgefoodcoop.com

North Carolina

Chatham Marketplace
480 Hillsboro Street, Suite 320
Pittsboro, North Carolina 27312
(919) 542-2643
mary@chathammarketplace.coop
www.chathammarketplace.coop

Company Shops Market
268 East Front Street
Burlington, North Carolina 27215

(336) 223-0390
info@companyshopsmarket.coop
www.companyshopsmarket.coop

Deep Roots Market
3728 Spring Garden Street
Greensboro, North Carolina 27407
(336) 292-9216
www.deeprootsmarket.com

French Broad Food Co-op
90 Biltmore Avenue
Asheville, North Carolina 28801
(828) 255-7650
info@frenchbroadfood.coop
www.frenchbroadfood.coop

Hendersonville Community Co-op
715 South Grove Street
Hendersonville, North Carolina 28792
(828) 693-0505
www.hendersonville.coop

Tidal Creek Cooperative Food Market
5329 Oleander Drive, Suite 100
Wilmington, North Carolina 28403
(910) 799-2667
mail@tidalcreek.coop
www.tidalcreek.coop

West Village Market & Deli
771 Haywood Road
Asheville, North Carolina 28806
(828) 225-4949
www.westvillagemarket.com

North Dakota

Amazing Grains Natural Food Market
214 De Mers Avenue
Grand Forks, North Dakota 58201
(701) 775-4542
www.amazinggrains.org

Ohio

Clintonville Community Market
200 Crestview Road
Columbus, Ohio 43202
(614) 261-3663
info@communitymarket.org
www.communitymarket.org

Kent Natural Foods Co-op
151 East Main Street
Kent, Ohio 44240
(330) 673-2878
http://kentnaturalfoods.org

Oregon

Ashland Food Co-op
237 North 1st Street
Ashland, Oregon 97520
(541) 482-2237
www.ashlandfood.coop

Astoria Cooperative
1355 Exchange Street
Astoria, Oregon 97103
(503) 325-0027
store@astoria.coop
www.astoria.coop/wp

Brookings Natural Food Co-op
630 Fleet Street
P.O. Box 8051
Brookings, Oregon 97415
(541) 469-9551

First Alternative Natural Foods Co-op
2855 Northwest Grant Avenue
Corvallis, Oregon 97330
(541) 452-3115
cs_north@firstalt.coop
www.firstalt.coop

First Alternative Natural Food Co-op
1007 Southeast 3rd Street
Corvallis, Oregon 97333
(541) 753-3115
cs_south@firstalt.coop
www.firstalt.coop

Food Front Cooperative Grocery
2375 Northwest Thurman Street
Portland, Oregon 97210
(503) 222-5658
info@foodfront.coop
http://foodfront.coop

Oceana Natural Foods Co-op
159 Southeast 2nd Street
Newport, Oregon 97365
(541) 265-8285
www.oceanafoods.org

People's Food Co-op
3029 Southeast 21st Avenue
Portland, Oregon 97202
(503) 232-9051
info@peoples.coop
www.peoples.coop

Pennsylvania

East End Food Co-op
7516 Meade Street
Pittsburgh, Pennsylvania 15208
(412) 242-3598
www.eastendfood.coop

Swarthmore Co-op
341 Dartmouth Avenue
Swarthmore, Pennsylvania 19081
(610) 543-9805
generalmanager@swarthmore.coop
www.swarthmore.coop

Weavers Way Cooperative Association
559 Carpenter Lane
Philadelphia, Pennsylvania 19119
(215) 843-2350
contact@weaversway.coop
www.weaversway.coop

Whole Foods Co-op
1341 West 26th & Brown Avenue
Erie, Pennsylvania 16508
(814) 456-0282
www.wholefoodscoop.org

Rhode Island

Alternative Food Co-op
357 Main Street
Wakefield, Rhode Island 02879
(401) 789-2240
www.alternativefoodcoop.com

South Carolina

Upstate Food Co-op
404 John Holiday Road
Six Mile, South Carolina 29682
(864) 868-3105
info@upstatefoodcoop.com
www.upstatefoodcoop.com

South Dakota

The Co-op Natural Foods
2504 South Duluth Avenue
Sioux Falls, South Dakota 57104
(605) 339-9506
www.coopnaturalfoods.com

Tennessee

Marketplace Buying Club Co-op
3511 Belmont Boulevard
Nashville, Tennessee 37214
www.marketplaceco-op.org

Morningside Buying Club
215 Morningside Lane
Liberty, Tennessee 37095
(615) 563-2353
www.morningsidefarm.com

Three Rivers Market
1100 North Central Street
Knoxville, Tennessee 37917
(865) 525-2069
info@threeriversmarket.coop
www.threeriversmarket.coop

Appendix 2

Texas

Central City Co-op
2515 Waugh Drive
Houston, Texas 77006
info@centralcityco-op.com
www.centralcityco-op.com

Wheatsville Food Co-op
3101 Guadalupe Street
Austin, Texas 78704
(512) 478-2667
gm@wheatsville.com
www.wheatsville.com

Utah

The Community Food Co-op of Utah
1726 South 700 West
Salt Lake City, Utah 84104
(801) 746-7878
general@thecommunitycoop.org
https://thecommunitycoop.com

Vermont

Brattleboro Food Co-op
2 Main Street
Brattleboro, Vermont 05301
(802) 257-0236
adminbfc@sover.net
www.brattleborofoodcoop.com

City Market–Onion River Co-op
82 South Winooski Avenue
Burlington, Vermont 05401
(802) 861-9700
info@citymarket.coop
www.citymarket.coop

Hunger Mountain Co-op
623 Stone Cutters Way
Montpelier, Vermont 05602
(802) 223-8000
www.hungermountain.com

Putney Food Co-op
8 Carol Brown Way
Putney, Vermont 05346
(802) 387-5866
ptnycoop@sover.net
www.putneycoop.com

St. J. Food Co-op
490 Portland Street
St. Johnsbury, Vermont 05819
(802) 748-9498
info@stjfoodcoop.com
www.stjfoodcoop.com

Virginia

Eats Natural Foods Co-op
708A North Main Street
Blacksburg, Virginia 24060
(540) 552-2279
eatsnatural@gmail.com
www.eatsnaturalfoods.com

Healthy Foods Co-op
110 West Washington Street
Lexington, Virginia 24450
(540) 463-6954
healthyfoods@embarqmail.com
http://healthyfoodscoop.org

Roanoke Natural Foods Co-op
1319 Grandin Road Southwest
Roanoke, Virginia 24015
(540) 343-5652
info@roanokenaturalfoods.coop
www.roanokenaturalfoods.coop

Washington

Community Food Co-op
1220 North Forest Street
Bellingham, Washington 98225
(360) 734-8158
info@communityfood.coop
www.communityfood.coop

The Food Co-op
414 Kearney Street
Port Townsend, Washington 98368
(360) 385-2883
info@foodcoop.coop
www.foodcoop.coop

Madison Market/Central Co-op
1600 East Madison Street
Seattle, Washington 98122
(206) 329-1545
www.centralcoop.coop

Puget Consumers' Co-op–Fremont
600 North 34th Street
Seattle, Washington 98103
(206) 632-6811
www.pccnaturalmarkets.com

Puget Consumers' Co-op–Greenlake
7504 Aurora Avenue North
Seattle, Washington 98103
(206) 525-3586
www.pccnaturalmarkets.com

Puget Consumers' Co-op–Issaquah
1810 12th Avenue Northwest
Issaquah, Washington 98027
(425) 369-1222
issaquah.storems@pccsea.com
www.pccnaturalmarkets.com

Puget Consumers' Co-op–Kirkland
10718 Northeast 68th Street
Kirkland, Washington 98033
(425) 828-4622
www.pccnaturalmarkets.com

Puget Consumers' Co-op–Seward Park
5041 Wilson Avenue South
Seattle, Washington 98118
(206) 723-2720
www.pccnaturalmarkets.com

Puget Consumers' Co-op–View Ridge
6514 40th Street Northeast
Seattle, Washington 98115
(206) 526-7661
www.pccnaturalmarkets.com

Puget Consumers' Co-op–West Seattle
2749 California Avenue Southwest
Seattle, Washington 98116
(206) 937-8481
www.pccnaturalmarkets.com

Sno-Isle Natural Foods Co-op
2804 Grand Avenue
Everett, Washington 98201
(425) 259-3798
info@snoislefoods.coop
www.snoislefoods.coop
Yelm Food Co-op
17835 State Route 507
Yelm, Washington 98597
(360) 400-2210
yelmfoodcoop@gmail.com
http://yelmfarmersmarket.com

West Virginia

Mountain People's Market
1400 University Avenue
Morgantown, West Virginia 26505
(304) 291-6131
http://mountainpeoplescoop.com

Wisconsin

Basic Cooperative
1711 Lodge Drive
Janesville, Wisconsin 53545
(608) 754-3925
www.basicshealth.com

Kickapoo Exchange Food Co-op
209 Main Street
Gays Mills, Wisconsin 54631
(608) 735-4544
kickapooexchange@yahoo.com

Mega Pik N Save
1201 South Hastings Way
Eau Claire, Wisconsin 54701
(715) 839-5200
www.megafoods.com

Menomonie Market Food Co-op
521 2nd Street East
Menomonie, Wisconsin 54751
(715) 235-6533
info@mmfc.coop
www.mmfc.coop

Nature's Bakery Co-op
1019 Williams Street
Madison, Wisconsin 53703
(608) 257-3649
mail@naturesbakery.coop
www.naturesbakery.coop

Outpost Natural Foods Co-op
100 East Capital Drive
Milwaukee, Wisconsin 53212
(414) 961-2597
www.outpostnaturalfoods.coop

Outpost Natural Foods Co-op: Wauwatosa
7000 West State Street
Wauwatosa, Wisconsin 53213
(414) 778-2012
www.outpostnaturalfoods.coop
People's Food Co-op
315 South 5th Avenue
La Crosse, Wisconsin 54601
(608) 784-5798
www.pfc.coop

Riverwest Co-op Grocery & Café
733 North East Clarke Street
Milwaukee, Wisconsin 53212
(414) 264-7933
www.riverwestcoop.org

Viroqua Food Cooperative
609 North Main Street
Viroqua, Wisconsin 54665
(608) 637-7511
info@viroquafood.coop

Willy Street Co-op
1221 Williamson Street
Madison, Wisconsin 53703
(608) 251-6776
www.willystreet.coop

Yahara River Grocery Cooperative
229 East Main Street
Stoughton, Wisconsin 53589
(608) 877-0947
info@yaharagrocery.coop
www.yaharagrocery.coop

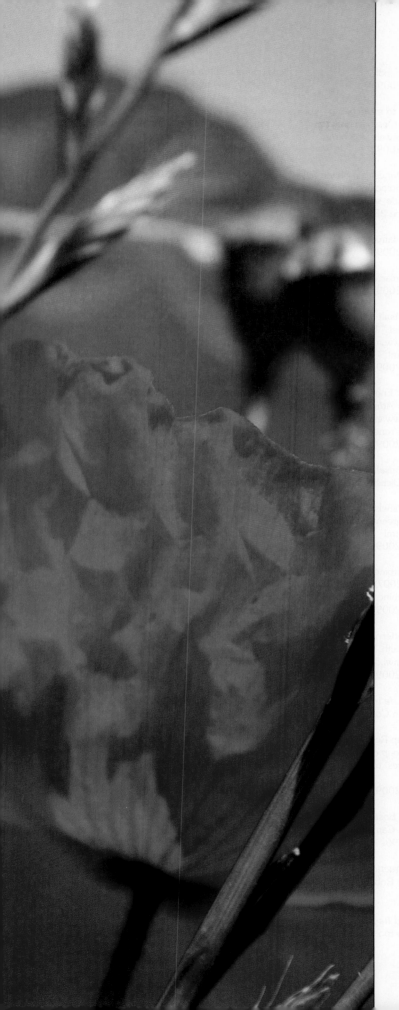

Sources

Adams, Joseph H. *Harper's Outdoor Book for Boys*. New York: Harper & Brothers, 1907.

American Heart Association. *How Can I Manage Stress?*. american-heart.org/downloadable/heart/1196286112399ManageStress.pdf (accessed June 24, 2009).

American Wind Energy Association. *Wind Energy Fact Sheet*. www.awea.org/pubs/factsheets/HowWindWorks2003.pdf (accessed June 22, 2009).

Andersen, Bruce and Malcolm Wells. *Passive Solar Energy Book*. Build It Solar (2005). www.builditsolar.com/Projects/Solar-Homes/PasSolEnergyBk/PSEbook.htm (accessed June 23, 2009).

Anderson, Ruben. "Easy homemade soap." *Treehugger: A Discovery Company*. treehugger.com/files/2005/12/easy_homemade_s.php (accessed June 24, 2009).

Andress, Elizabeth L. and Judy A. Harrison, ed. *So Easy to Preserve,* 5th ed. Athens, GA: The University of Georgia Cooperative Extension, 2006.

Anthony, G.A., F.G. Ashbrook, and Frants P. Lund. U.S. Department of Agriculture. Farmers' Bulletin: "Pork on the Farm: Killing, Curing, and Canning," Issues 1176-1200. Washington: Government Printing Office, 1922.

Autumn Hill Llamas & Fiber. "Llama Fiber Article." *Autumn Hill Llamas & Fiber*. autumnhillllamas.com/llama_fiber_article.htm (accessed June 24, 2009).

Bailey, Henry Turner, ed. *School Arts Book*, vol. 5. Worcester, MA: The Davis Press, 1906.

Beard, D.C. *The American Boy's Handy Book*. NH: David R. Godine, 1983.

Beard, Linda and Adelia Belle Beard. *The Original Girl's Handy Book*. New York: Black Dog & Leventhal Publishers Inc., 2007.

Bell, Mary T. *Food Drying with an Attitude*. New York: Skyhorse Publishing, Inc., 2008.

Bellows, Barbara. "Solar Greenhouse Resources." *ATTRA: National Sustainable Agriculture Information Service* (2009). attra.ncat.org/attra-pub/solar-gh.html (accessed June 24, 2009).

Ben. "My Inexpensive 'Do It Yourself' Geothermal Cooling System." *Trees Full of Money*. www.treesfullofmoney.com/?p=131 (accessed June 29, 2009).

Sources

Benton, Frank. U.S. Department of Agriculture. *The Honey Bee: A Manual of Instruction in Apiculture*. Washington: Government Printing Office, 1899.

Brooks, William P. *Agriculture vol. III: Animal Husbandry, including The Breeds of Live Stock, The General Principles of Breeding, Feeding Animals; including Discussion of Ensilage, Dairy Management on the Farm, and Poultry Farming*. Springfield, MA: The Home Correspondence School, 1901.

Bower, Mark. "Building an inexpensive solar heating panel." *Mobile Home Repair* (Aberdeen Home Repair, 2007). www.mobilehomerepair.com/article17c.htm (accessed June 22, 2009).

Boy Scouts of America. *Handbook for Boys*. New York: The Boy Scouts of America, 1916.

"Build a Solar Cooker." *The Solar Cooking Archive*. www.solarcooking.org/plans/default.htm (accessed June 22, 2009).

Burgdorf, David, Thomas Cogger, Glenn Lamberg, Rick Knysz, and Paul Johnson. "Community Garden Guide Season Extension: Hoop Houses." U.S. Department of Agriculture, Natural Resources Conservation Service. http://plant-materials.nrcs.usda.gov/pubs/mipmcarcgghoophouse.pdf (accessed July 1, 2010).

California Integrated Waste Management Board. "Compost—What Is It?" ciwmb.ca.gov/organics/CompostMulch/CompostIs.htm (accessed June 24, 2009).

California Integrated Waste Management Board. "Home Composting." ciwmb.ca.gov/Organics/HomeCompost (accessed June 24, 2009).

Call Ducks: Call Duck Association UK. callducks.net (accessed June 24, 2009).

"Candle making." *Lizzie Candles Soap*. lizziecandle.com/index.cfm/fa/home.page/pageid/12.htm (accessed June 24, 2009).

Comstock, Anna Botsford. *How to Keep Bees; A Handbook for the Use of Beginners*. New York: Doubleday, 1905.

Cook, E.T., ed. *Garden: An Illustrated Weekly Journal of Horticulture in all its Branches*, vol. 64. London: Hudson & Kearns, 1903.

Corie, Laren. "Building a Very Simple Solar Water Heater." *Energy Self Sufficiency Newsletter* (Rebel Wolf Energy Systems, September 2005). www.rebelwolf.com/essn/ESSN-Sep2005.pdf (accessed June 22, 2009).

"Craft instructions: how to make hemp jewelry." *Essortment*. essortment.com/hobbies/makehempjewelr_sjbg.htm (accessed June 24, 2009).

Dahl-Bredine, Kathy. "Windshield Shade Solar Cooker." *Wikia*. solarcooking.wikia.com/wiki/Windshield_shade_solar_funnel_cooker (accessed June 22, 2009).

Dairy Connection Inc. dairyconnection.com (accessed June 24, 2009).

Danlac Canada Inc. danlac.com (accessed June 24, 2009).

Davis, Michael. "How I built an electricity producing Solar Panel." *Welcome to Mike's World*. www.mdpub.com/SolarPanel/index.html (accessed June 22, 2009).

Department of Energy. "Energy Kid's Page." *Energy Information Administration*, November 2007. www.eia.doe.gov/kids/energyfacts/sources/renewable/solar.html (accessed June 26, 2009).

Dickens, Charles, ed. *Household Worlds*, vol. 1. London: Charles Dickens & Evans, 1881.

"DIY Home Solar PV Panels." *GreenTerraFirma*. greenterrafirma.com/home-solar-panels.html (accessed June 23, 2009).

"Do-It-Yourself Wind Turbine Project." *GreenTerraFirma* (2007). greenterrafirma.com/DIY_Wind_Turbine.html (accessed June 23, 2009).

Druchunas, Donna. "Pattern: Fingerless Gloves for Hand Health." *Subversive Knitting*. sheeptoshawl.com (accessed June 24, 2009).

Earle, Alice M. *Home Life in Colonial Days*. New York: Macmillan Company, 1899.

"Easy Cold Process Soap Recipes for Beginners." *TeachSoap.com: Cold Process Soap Recipes*. teachsoap.com/easycpsoap.html (accessed June 24, 2009).

Farmer, Fannie M. *The Boston Cooking-School Cook Book*. Cambridge: University Press, 1896.

Farrington, Edward I. *Practical Rabbit Keeping*. New York: Robert M. McBride and Co.,1919.

Flach, F., ed. *Stress and Its Management*. New York: W.W. Norton & Co., 1989.

"Fun-Panel." *Wikia*. solarcooking.wikia.com/wiki/Fun-Panel (accessed June 22, 2009).

Gegner, Lance. "Llama and Alpaca Farming." *Appropriate Technology Transfer for Rural Areas (ATTRA)*. attra.ncat.org/attra-pub/llamaalpaca.html (accessed June 24, 2009).

Glengarry Cheesemaking and Dairy Supply Ltd. glengarrycheesemaking.on.ca (accessed June 24, 2009).

"Guide to Herbal Remedies." *Natural Health and Longevity Resource Center*. all-natural.com/herbguid.html (accessed June 24, 2009).

Hall, A. Neely and Dorothy Perkins. *Handicraft for Handy Girls: Practical Plans for Work and Play*. Boston: Lothrop, Lee & Shepard Company, 1916.

Hasluck, Paul N. *The Handyman's Book*. London: Cassell and Co., 1903.

Hill, Thomas E. *The Open Door to Independence: Making Money from the Soil*. Chicago: Hill Standard Book Company, 1915.

"Homemade Solar Panel." pyronet.50megs.com/RePower/Homemade%20Solar%20Panels.htm (accessed June 24, 2009).

"Homemade Teat Dip & Udder Wash Recipe." *Fias Co Farm*. fiascofarm.com/goats/teatdip-udderwash.html (accessed June 24, 2009).

"How to Build a Composting Toilet." *eHow, Inc.* www.ehow.com/how_2085439_build-composting-toilet.html (accessed June 28, 2009).

"How to Knit a Hat." *Knitting for Charity: Easy, Fun and Gratifying*. knittingfor-charity.org/how_to_knit_a_hat.html (accessed June 24, 2009).

"How to Knit a Scarf for Beginners." *AOK Coral Craft and Gift Bazaar*. aokcorral.com/how2oct2003.htm (accessed June 24, 2009).

"How to Make Hemp Jewelry." *Beadage: All About Beading!* beadage.net/hemp/index.shtml (accessed June 24, 2009).

"How to Make Taper Candles" *How To Make Candles.info*.howtomakecandles.info/cm_article.asp?ID=CANDL0603 (accessed June 24, 2009).

"How to Milk a Goat." *Fias Co Farm*. fiascofarm.com/goats/how_to_milk_a_goat.htm (accessed June 24, 2009).

"How to Sell Your Crafts on eBay." *Craft Marketer: DIY Home Business Ideas.* craftmarketer.com/sell-your-crafts-on-ebay-article.htm (accessed June 24, 2009).

J.G. "The Fragrance of Potpourri." *Good Housekeeping*, January 1917. New York: Hearst Corp., 1916.

Junket: Making Fine Desserts Since 1874. junketdesserts.com (accessed June 24, 2009).

Kellogg, Scott and Stacy Pettigrew. *Toolbox for Sustainable City Living: A Do-It-Ourselves Guide.* Cambridge, MA: South End Press, 2008.

Kendall, P. and J. Sofos. "Drying Fruits." *Nutrition, Health & Food Safety.* Colorado State University Cooperative Extension: No. 9.309). uga.edu/ nchfp/how/dry/csu_dry_fruits.pdf (accessed June 24, 2009).

Kleen, Emil, and Edward Mussey Hartwell. *Handbook of Massage.* Philadelphia: P. Blakiston Son & Co.,1892.

Kleinheinz, Frank. *Sheep Management: A Handbook for the Shepherd and Student,* 2nd ed. Madison, WI: Cantwell Printing Company, 1912.

Ladies' Work-Table Book, The: Containing Clear and Practical Instructions in Plain and Fancy Needlework, Embroidery, Knitting, Netting and Crochet. Philadelphia: G.B. Zeiber & Co., 1845.

Lambert, A. *My Knitting Book.* London: John Murray, 1843.

Lamon, Harry M. and Rob R. Slocum. *Turkey Raising.* New York: Orange Judd Publishing Company, 1922.

"Learn to Make Beeswax Candles." *MyCraftBook.* mycraftbook.com/ Make_Beeswax_Candles.asp (accessed June 24, 2009).

Lindstrom, Carl. *Greywater.* www.greywater.com (accessed June 25, 2009).

Llucky Chucky Llamas. llamafarm.com/ welcome.html (accessed June 24, 2009).

Lynch, Charles. *American Red Cross Abridged Text-book on First Aid: General Edition, A Manual of Instruction.* Philadelphia: P. Blakiston's Son & Co., 1910.

"Make Your Own Paper." *Environmental Education for Kids!* dnr.wi.gov/org/caer/ce/eek/cool/paper.htm (accessed June 24, 2009).

"Marketing your homemade crafts." *Essortment.* essortment.com/all/crafts-marketing_mfm.htm (accessed June 24, 2009).

McGee-Cooper, Ann. *You Don't Have to Go Home From Work Exhausted!: The energy engineering approach.* Dallas, Texas: Bowen & Rogers, 1990.

Moore, Donna. "Shear Beauty." *International Lama Registry.* lamaregistry.com/ ilreport/2005May/shear_beauty_may.html (accessed June 24, 2009).

Moorlands Cheesemakers: Suppliers of Farm and Household Dairy Equipment. cheesemaking.co.uk (accessed June 24, 2009).

Mountain, Johnny. "Raising Rabbits." http://www.thefarm.org/charities/i4at/lib2/rabbits.htm (accessed July 1, 2010).

Morais, Joan. "Beeswax Candles." *Natural Skin and Body Care Products.* naturalskinandbodycare.com/2008/12/beeswax-candles.html

Murphy, Karen. "How to make beeswax candles." *SuperEco.* supereco.com/how-to/how-to-make-beeswax-candles (accessed June 24, 2009).

Natural Skin and Body Care Products. naturalskinandbodycare.com (accessed June 24, 2009).

N., Beth. "How to Make Taper Candles." *Associated Content.* associatedcontent.com/article/360786/how_to_make_taper_candles.html?cat=24 (accessed June 24, 2009).

National Ag Safety Database. "Basic First Aid: Script." *Agsafe.* nasdonline.org/docs/d000101-d000200/d000105/d000105.html (accessed June 24, 2009).

National Center for Complementary and Alternative Medicine. "Herbal Medicine." *MedlinePlus: Trusted Health Information for You.* nlm.nih.gov/ medlineplus/herbalmedicine.html (accessed June 24, 2009).

National Center for Complementary and Alternative Medicine. *Herbs at a Glance.* nccam.nih.gov/health/herbsataglance.htm (accessed June 24, 2009).

National Center for Complementary and Alternative Medicine. *Massage Therapy: An Introduction.* nccam.nih.gov/health/massage/#1 (accessed June 24, 2009).

National Center for Home Food Preservation. "Drying: Herbs." uga.edu/ nchfp/how/dry/herbs.html (accessed June 24, 2009).

National Center for Home Food Preservation. "General Freezing Information." uga.edu/nchfp/how/freeze/dont_freeze_foods.html (accessed June 24, 2009).

National Center for Home Food Preservation. "USDA Publications: USDA Complete Guide to Home Canning, 2006." uga.edu/nchfp/publications/publications_usda.html (accessed June 24, 2009).

National Institutes of Health: Office of Dietary Supplements. "Botanical Dietary Supplements: Background Information." *Office of Dietary Supplements.* ods.od.nih.gov/factsheets/BotanicalBackground.asp (accessed June 24, 2009).

National Renewable Energy Laboratory. "Wind Energy Basics." Learning About Renewable Energy. www.nrel.gov/learning/re_wind.html (accessed June 24, 2009).

New England Cheesemaking Supply Company. cheesemaking.com (accessed June 24, 2009).

Nissen, Hartvig. *Practical Massage in Twenty Lessons.* Philadelphia: F.A. Davis Company, 1905.

Nucho, A. O. *Stress Management: The Quest for Zest.* Illinois: Charles C. Thomas, 1988.

Nummer, Brian A. "Fermenting Yogurt at Home." National Center for Home Food Preservation: uga.edu/nchfp/publications/nchfp/factsheets/ yogurt.html (accessed June 24,2009).

Ostrom, Kurre Wilhelm. *Massage and the Original Swedish Movements: Their Application to Various Diseases of the Body.* 6th ed. Philadelphia: P. Blakiston's Son & Co., 1905.

Ponder, T. *How to Avoid Burnout.* Mountainview, CA: Pacific Press Publishing Association, 1983.

Reyhle, Nicole. "Selling Your Homemade Goods." *Retail Minded.* retailminded.com/blog/2009/01/selling-your-homemade-goods (accessed June 24, 2009).

Retail Minded. retailminded.com/blog (accessed June 24, 2009).

Sanford, Frank G. *The Art Crafts for Beginners.* New York: The Century Co., 1906.

Sell, Randy. "Llama" *Alternative Agriculture Series*, no. 12. ag.ndsu.edu/pubs/alt-ag/llama.htm (accessed June 24, 2009).

Sheep to Shawl. sheeptoshawl.com (accessed June 24, 2009).

Sources

Sherlock, Chelsa C. Care and Management of Rabbits. Philadelphia: David McKay Co.,1920.

Singleton, Esther. *The Shakespeare Garden.* New York: The Century Co., 1922.

Smith, Kimberly. "Where to Sell Your Homemade Crafts Offline." *Associated Content.* associatedcontent.com/ article/1678550/ where_to_sell_your_ homemade_crafts.html (accessed June 24, 2009).

"Soap making–General Instructions." *Walton Feed, Inc.* waltonfeed.com/old/old/ soap/ soap.html (accessed June 24, 2009).

"Soy candle making." *Soya–Information about Soy and Soya Products.* soya.be/ soy-candle-making.php (accessed June 24, 2009).

Swenson, Allan A. *Foods Jesus Ate and How to Grow Them.* New York: Skyhorse Publishing, Inc., 2008.

Szykitka, Walter. *The Big Book of Self-Reliant Living: Advice and Information on Just About Everything You Need to Know to Live on Planet Earth,* 2nd ed. Guilford, CT: The Lyons Press, 2004.

Taylor, George Herbert. *Massage: Principles and Practice of Remedial Treatment by Imparted Motion.* New York: John B. Alden, 1887.

Thompson, Nita Norphlet and Sue McKinney-Cull. "Soothing Those Jangled Nerves: Stress Management." *ARCH Factsheet,* no. 41. archrespite.org/archfs41.htm (accessed June 24, 2009).

U.S. Department of Agriculture: Food Safety and Inspection Service. *Fact Sheets: Egg Products Preparation.* fsis.usda.gov/ Factsheets/ Focus_On_Shell_Eggs/index. asp (accessed June 24, 2009).

U.S. Department of Agriculture: Food Safety and Inspection Service. *Fact Sheets: Poultry Preparation.* fsis.usda.gov/Fact_ Sheets/ Chicken_Food_Safety_Focus/ index.asp (accessed June 24, 2009).

U.S. Department of Agriculture: Natural Resources Conservation Service. "Backyard Conservation: Composting." nrcs. usda.gov/feature/ backyard/compost.html (accessed June 24, 2009).

U.S. Department of Agriculture: Natural Resources Conservation Service. "Backyard Conservation: Nutrient Management." nrcs.usda.gov/ feature/backyard/nutmgt. html (accessed June 24, 2009).

U.S. Department of Agriculture: Natural Resources Conservation Service. "Composting in the Yard." nrcs.usda.gov/ feature/backyard/ compyrd.html (accessed June 24, 2009).

U.S. Department of Agriculture: Natural Resources Conservation Service. "Home and Garden Tips: Composting." nrcs.usda. gov/feature/ highlights/homegarden/compost. html (accessed June 24, 2009).

U.S. Department of Agriculture: Natural Resources Conservation Service. "Home and Garden Tips: Lawn and Garden Care." nrcs.usda.gov/ feature/highlights/ homegarden/lawn.html (accessed June 24, 2009).

U.S. Department of Agriculture: National Agricultural Library. "Organic Production." afsic.nal.usda.gov/nal_display/index. php?info_center= 2&tax_level=1&tax_ subject=296 (accessed June 24, 2009).

U.S. Department of Energy. "Active Solar Heating." *Energy Efficiency and Renewable Energy: Energy Savers.* www.energysavers.gov/your_home/space_heating_ cooling/index.cfm/mytopic=12490 (accessed June 26, 2009).

U.S. Department of Energy. "Benefits of Geothermal Heat Pump Systems." *Energy Efficiency and Renewable Energy: Energy Savers.* www.energysavers.gov/your_ home/space_heating_cooling/index.cfm/ mytopic=12660 (accessed June 25, 2009).

U.S. Department of Energy. "Energy Efficiency and Renewable Energy." *Wind and Hydropower Technologies Program.* www1.eere.energy.gov/windandhydro/ (accessed June 24, 2009).

U.S. Department of Energy. "Energy Technologies." *Efficiency and Renewable Energy: Solar Energy Technologies Program.* www1.eere.energy.gov/solar/ want_pv.html (accessed June 26, 2009).

U.S. Department of Energy. "Geothermal Heat Pumps." *Energy Efficiency and Renewable Energy: Energy Savers.* www.energysavers.gov/your_home/ space_heating_cooling/index.cfm/ mytopic=12650 (accessed June 26, 2009).

U.S. Department of Energy. "Heat Pump Water Heaters." *Energy Efficiency and Renewable Energy: Energy Savers.* www.energysavers.gov/your_home/ water_heating/index.cfm/mytopic=12840 (accessed June 26, 2009).

U.S. Department of Energy. "Hydropower Basics." *Energy Efficiency and Renewable Energy: Wind and Hydropower Technologies Program.* www1.eere.energy.gov/ windandhydro/hydro_basics.html (accessed June 26, 2009).

U.S. Department of Energy: National Renewable Energy Laboratory. "Direct Use of Geothermal Energy." *Office of Geothermal Technologies.* www1.eere. energy.gov/geothermal/pdfs/directuse.pdf (accessed June 26, 2009).

U.S. Department of Energy: National Renewable Energy Laboratory. "Wind Energy Myths." *Wind Powering American Fact Sheet Series.* www.nrel.gov/docs/fy05osti/37657. pdf (accessed June 26, 2009).

U.S. Department of Energy. "Renewable Energy." *Energy Efficiency and Renewable Energy: Energy Savers.* www.energysavers.gov/renewable_energy/ solar/index.cfm/mytopic=50011 (accessed June 26, 2009).

U.S. Department of Energy. "Selecting and Installing a Geothermal Heat Pump System." *Energy Efficiency and Renewable Energy: Energy Savers.* www.energysavers.gov/your_home/ space_heating_cooling/index.cfm/ mytopic=12670 (accessed June 25, 2009).

U.S. Department of Energy. "Technologies." *Energy Efficiency and Renewable Energy: Geothermal Technologies Program.* www1.eere.energy.gov/geothermal/faqs. html (accessed June 25, 2009).

U.S. Department of Energy. "Solar." *Energy Sources.* www.energy.gov/energysources/ solar.htm (accessed June 26, 2009).

U.S. Department of Energy. "Toilets and Urinals." *Greening Federal Facilities,* second edition. www.eere.energy.gov/ femp/pdfs/29267-6.2.pdf (accessed June 29, 2009).

U.S. Department of Energy. "Your Home." *Energy Efficiency and Renewable Energy: Energy Savers.* www.energysavers.gov/ your_home/space_heating_cooling/index. cfm/mytopic=12300 (accessed June 26, 2009).

U.S. Environmental Protection Agency. "Composting Toilets." *Water Efficiency Technology Fact Sheet.* www.epa.gov/ owm/mtb/comp.pdf (accessed June 29, 2009).

U.S. House of Representatives. United States Department of Agriculture. *Report of the Commissioner of Patents for the Year 1831: Agriculture.* 37th congress, 2nd sess., 1861.

University of Maryland. *National Goat Handbook*. uwex.edu/ces/ cty/richland/ag/ documents/national_goat_handbook.pdf (accessed June 24, 2009).

"Where to sell crafts? Consider these often overlooked alternative markets…" *Craft Marketer: DIY Home Business Ideas*. craftmarketer.com/ where_to_sell_crafts. htm (accessed June 24, 2009).

Whipple, J. R. "Solar Heater." *J. R. Whipple & Associates*. www.jrwhipple.com/sr/ solheater.html (accessed June 23, 2009).

Wickell, Janet. *Quilting*. Teach Yourself Books. Chicago: NTC Publishing Group, 2000.

Williams, Archibald. *Things Worth Making*. New York: Thomas Nelson and Sons, Ltd., 1920.

"Wind Energy Basics." *Wind Energy Development Programmatic EIS*, windeis.anl.gov/ guide/basics/index.cfm (accessed June 25, 2009).

Wolok, Rina. "How to Build a Composting Toilet." *Greeniacs*, June 15, 2009. greeniacs.com/GreeniacsGuides/How-to-Build-a-Composting-Toilet.html (accessed June 29, 2009).

Woods, Tom. "Homemade Solar Panels." *Forcefield* (2003). www.fieldlines.com/ story/2005/1/5/51211/79555 (accessed June 24, 2009).

Woolman, Mary S. and Ellen B. McGowan. *Textiles: A Handbook for the Student and the Consumer*. New York: The Macmillan Company, 1921.

Worcester Polytechnic Institute. "A Passive Solar Space Heater for Home Use." *Solar Components Corporation*. www.solar-components.com/SOLARKAL. HTM#doityourself (accessed June 22, 2009).

Young Ladies' Journal, The: Complete Guide to the Work-Table. London: E. Harrison, 1885.

Acknowledgments

"Not what we say about our blessings, but how we use them, is the true measure of our thanksgiving."

—W.T. Purkiser

"As we express our gratitude, we must never forget that the highest appreciation is not to utter words, but to live by them."

—John Fitzgerald Kennedy

Sincere thanks to my Skyhorse team; you are the ones who make this work so fulfilling. In particular, thanks to those of you who worked to imagine, edit, and design these pages and to get them into the hands of readers:

Tony Lyons
Bill Wolfsthal
Ann Treistman
Kathryn Mennone
Adam Bozarth
Julie Matysik
Kaylan Connally
Anna Gorovoy

This book would never have been completed without the help of several researchers and writers:

Michael Coulter
Sara Kitchen
Laura Ofstad
Kirsten Rischert-Garcia
Melanie Trice

Timothy Lawrence, thank you for everything you are and do.

Index

A

agave, 132
Aphids, 78
apiculture. *See* beekeeping
apple cider doughnuts, 122
asparagus, 132

B

baking tips, 118
basketweaving
    basket with triple twist, 273
    birch bark basket, 274
    coiled basket, 273–274
    small reed basket, 272
bath salts and scrubs, 268
beans, 6
bedroom furniture
    bed rest, 400
    bedroom bookshelves, 408
    boot and shoe rack, 399
    chest of drawers, 405–406
    linen chest, 403–404
    making a bed rest, 401–402
    oak bedstead, 404–405
    setting out bed rest, 400–401
    shaving cabinet, 399
beech, 132
beekeeping
    bee pastures, 313
    beeswax, 314
    hives
        building, 312–313
        making, 315
        situating, 313
    honey extraction, 314
    reasons for, 312
    stings, avoiding, 312
    swarming, 313–314
    tips, 315
    winter, preparing bees for, 313–314
berry ink, 249
birds, butterflies, and bees
    plant species for, 93
blackberries, 133
bleeding heart plants, 12
bookbinding, 285–286
boomerangs, 259
burdock, 133
butter, 145

C

cakes
    applesauce cake, 111
    banana layer cake, 111
    boiled butterscotch frosting, 114
    cherry-nut cake, 111
    chocolate cake, 110–111
    Christmas cupcakes, 112
    fruit cake, 111
    gingerbread, 112–113
    seafoam icing, 114–115
    spice cake, 113–114
    troubles with, 113
    yellow layer cake, 110
candles, 261
    essential oils for mosquito repellants, 265
    floating, 264
    gourd votives, 264
    jarred soy candles, 264–265
    natural dyes for, 270
    rolled beeswax candle, 262
    taper candles, 262, 264
canning
    altitude, 162–163
    apple butter, 177
    apple juice, 176
    applesauce, 177–178
    apricots, 179
    beans
        baked, 200–201
        green, 201
        shelled or dried, 199–200
    beets, 202
    benefits of, 156–157
    berries, 179
    berry syrup, 181
    bread-and-butter pickles, 213–214
    canners, 168
        boiling-water canners, 169
        pressure canners, 170–171
    carrots, 202–203
    controlling headspace, 166
    corn
        cream style, 203–204
        whole kernel, 204
    dill pickles, 210–211, 214
    ensuring high-quality, 163
    food acidity and processing methods, 162
    fruit purées, 181–182
    glossary, 158–160
    grape juice, 182
    hot packing, advantages, 164, 166
    how canning preserves food, 157–158
    identifying and handling spoiled canned food, 173
    jams, jellies, and other fruit spreads, 189
        blueberry-spice jam, 196
        grape-plum jelly, 196–197
        jam without added pectin, 194
        jellies without added pectin, 189, 191–192
        lemon curd, 193–194
        peach-pineapple spread, 197
        pear-apple jam, 195
        refrigerated apple spread, 198
        refrigerated grape spread, 198
        remaking soft jellies, 198
        strawberry-rhubarb jelly, 196
    jars and lids
        cleaning, 166
        cooling, 171
        lid selection, preparation, and use, 167–168
        seals, 172
        sterilization of empty jars, 166–167
    maintaining color and flavor, 163–164
    marinated peppers, 212–213
    meat stock (broth), 208–210
    mixed vegetables, 204–205
    peaches, 182–184
    pears, 184
    peas, 205
    piccalilli, 213
    pickle relish, 214
    pickled horseradish sauce, 212
    pickled three-bean salad, 212
    pie fillings
        apple pie, 186–187
        blueberry pie, 187–188
        cherry pie, 188
        mincemeat pie, 188–189
    potatoes, sweet, 206–207
    processing time, 168, 174–185
    pumpkin, 207
    rhubarb, 185
    sauerkraut, 211
    soups, 208
    special diets, 173
    spiced apple rings, 178
    storing canned food, 172
    succotash, 207–208
    sweet pickles, quick, 215
    sweet pickles, reduced-sodium sliced, 215, 217
    syrups, 174, 175
    tomatoes
        chile salsa (hot tomato-pepper sauce), 221
        crushed, 218
        juice, 217–218
        ketchup, 221
        sauce, 218–219
        spaghetti sauce without meat, 220–221
        whole or halved, packed in water, 219–220
    what not to do, 163
cattails, 134
cheese, 147
    cheddar, 149–150
    cheese press, 149
    mozzarella, 149
    preparation, 147
    queso blanco, 148
    ricotta cheese, 148
    yogurt cheese, 147
chickens
    breed selection, 292–293
    chicks
        hatching, 294, 296
    eggs from, 296
    feeding, 293–294
    free-range, 293
    housing, 292
        coops, 292
    meat from, 296
    slaughtering, 303
chicory, 134
co-ops. *See* food co-ops
community gardens
    beginning your own, 84, 86
        school gardens, 84–86, 88–89
    benefits of, 88
    things to consider, 87–88
    tools needed, 86–87
companion planting, 10
composting
    barrels used for, 19
    Compost Lasagna, 18
    making your own, 18
    materials to use, 19
    preparation of, 17–18
    problems, 20
    types of
        cold or slow, 19
        vermicomposting, 19–20
    uses for, 20
containers
    flowers and, 64
    herbs and, 63
    preserving plants in, 64–66
    things to consider, 63

Index

vegetables and, 62–63
cookies
  brazil nut shortbreads, 122
  chocolate brownies, 119
  chocolate pinwheel cookies, 120
  date bars, 119–120
  German Christmas cookies, 121
  ginger cookies, 119
  raisin penny-pinchers, 120–121
  Swedish heirloom cookies, 122
corn, heirloom, 36
cornhusk dolls, 270–271
cows
  breeding, 328
  breeds, 327
  butchering, 335–339
  calf rearing, 328, 330
  diseases, 330
  feeding and watering, 328
  grooming, 327–328
  housing, 327
crackers, 149
cranberry, 134
Cutworms, 78

D
dandelion, 134
diagrams, 8
dried plants, 239
drying foods
food dehydrators, 222
  fruit leathers, 225–227
  herbs, 227, 229–230
  jerky, 230–231
  procedures, 223, 225
  pumpkin leather, 227
  tomato leather, 227
  vegetable leathers, 227
  woodstove dehydrator, 227
ducks
  breeds
    Aylesbury ducks, 297
    Black Cayuga ducks, 297
    Saxony ducks, 228
  diseases, 300
  ducklings
    caring for, 299–300
    hatching, 299
  feeding and watering, 299
  slaughtering, 303

E
eggs
  blown, 238–239
  leaf- or flower-stenciled, 240
  natural dyes for Easter, 240–241
elderberry, 134

F
farmers' markets
  benefits of, 90
  getting involved in, 90–91
feather pens, 249
flour, 98
flowerpots, mosaic, 242
flowers
  cutting, 46
  digging, 45–46
  planning and, 44
  purchasing seeds or plants, 46
  small gardens, 44
  things to consider, 46
  watering, 46
flowers, preserving
  natural wax flowers, 254
  pressed flowers and leaves, 253–254
food co-ops
  becoming a member, 128–129
  buying club, 127
  by state, 422–429
  co-op grocery store, 127
  definition of, 127
  starting your own, 128–129
freezing foods, 232–235
  foods that do not freeze well, 234
  fruits, 234–235
  meat, 235
  vegetables, 234
fruit bushes and trees
  blueberries, 52
  brambles and bush fruits, 50, 52
  currants and gooseberries, 52–53
  fruit growing chart, 50
  fruit trees, 53
  grapes, 54
  strawberries, 50

G
gardening tools, 10
gardening, choosing a site for
  elevation, 10
  irrigation, 9
  proximity, 9
  rain, 9
  soil quality, 9
  sunlight and, 9
  water availability, 9
gates and fences
  barways, 384
  field gate, 381–382
  hanging field gates, 382
  pole fencing, 384
  repairing field gates, 382–383
  wicket gates, 383
  wood for fencing, 383

geothermal heat pump, 417–419
germination temperatures, 23
goats
  breeds, 316
    angora, 319
  cheese, 318–319
  diseases, 316
  feeding, 316
  milk, 316, 318
grains
  growing and threshing, 55–57
  grains, milling, 104
  granola, 123

H
hammocks, 247–249
hams
  country-style ham, 141
  regular-cut ham, 141
  virginia ham
    aging, 142
    carving, 144
    cooking, 143
    curing, 142
    deboning, 144
    pests, 142–143
    preparing, 143
harvesting, 81–83
hay barns, 385–386
hazelnut, 134
Heather, 40
heirloom corn, 36
heirloom tomatoes, 36
hemp jewelry, 283
home gardening, 12
Homeschool Hints
  microorganisms, 32
  Root Vegetable "Magic," 42
  sprouting seeds, 23
horses
  breeding, 310
  breeds, 307–308
  diseases, 310
  feeding and watering, 310
  grooming, 308, 310
  housing, 308
houses, runs, and coops for poultry
  fowl-house with semicircular roof, 379–380
  hen coop with chicken run, 371
  interior arrangements of fowl-house, 372
  lean-to poultry house and run, 373
  outdoor chicken rearer, 378
  portable fowl-house and run, 371–371
  roosting shed of lean-to house, 372–375
  runs for chicken rearer, 379

sitting box for broody hens, 373
hydroponics
  aggregate culture, 73–74
  benefits and drawbacks of, 71
  nutrient solutions, 74
  plant nutrient deficiencies, 75
  pre-mixed chemicals, 75
  water culture, 72–73

I
ice cream, 150
insects. See pests

J
Japanese Beetles, 76
Junior Homesteader
  bread in a bag, 109
  compost lasagna, 18
  composting, 20
  family taste celebration, 108
  foods we eat, 6
  garden parties, 46
  gardens, 12
  grain game, 100–102
  how soap works, 268
  plant playhouses, 7
  starting a mini garden, 42
  treehouses, 376—377
juniper, 134–135

K
kitchen furniture
  cupboard pantry door, 396
  cupboard pantry interior, 396–397
  fitting together pantry safe, 395
  gate-legged kitchen table, 397–398
  household tidy, 392–393
  kitchen corner cupboard, 391
  outer frame of corner cupboard, 391–392
  pantry safe or cupboard pantry, 394–395
  spice box with drawers, 393–394
kites
  butterfly kite, 258
  fish kite, 258
  frog kite, 257
  Japanese square kite, 259
  shield kite, 258
knots
  bowline, 255
  clove hitch, 255
  halter, 255
  sheepshank knot, 255
  slip knot, 255
  square/reef knot, 256

timber hitch, 256
two half hitches, 256

L
labeling rows, 41
lamp shades, 274
lima beans, 42
llamas
cria, 325
diseases, 325
feeding, 324
fibers from, 325–326
housing, 324
shearing, 324–325
toenail trimming, 324
locally grown food, 125
log huts or cabins, 389–390
lotus, 135

M
maple syrup, 130
marsh marigold, 135
mint, 250
mulberry, 135
mulch
applying, 31
benefits of using, 29
where to find, 30–31
wood chips, 29
mushrooms (wild)
chanterelles, 138
coral fungi, 138
coral mushrooms, 139
morels, 138
pufballs, 138
shaggy mane mushrooms, 139

N
nettle, 135

O
oak, 135
open sheds, 385
organic gardens, 124
how to start your own, 32–34
saving seeds, 35–36
things to consider, 34
what it entails, 32

P
palmetto palm, 135
paper, handcrafted, 284
Periwinkle, 39
persimmon, 135–136
pests and disease
management, 10
disease identification, 78
powdery mildew leaf
disease, 80
identifying the problem, 76
insects and mites, 76
Aphids, 78
beneficial insects, 77

Cutworms, 78
Japanese Beetles, 76
Integrated Pest
Management (IPM), 76–78
natural repellants, 78
practices, 80
pies
apple crumb pie, 117
blueberry pie (no bake), 115–116
chocolate chiffon pie, 116
coconut bavarian pie, 116
lemon meringue pie, 115
peach chiffon pie, 116
pie crust, easy, 115
pie crusts, prebaking, 115
pumpkin chiffon honey pie, 116–117
pigs
breeds, 332–333
butchering, 335–339
diseases, 334
feeding, 333–334
housing, 333
terminology, 332
pine, 136
pine cone bird feeder, 255
plant a row for the hungry
(PAR) program, 151–152
plant nutrients
management of, 14
testing, 14
plantain, 136
planters
barrel plant holder, 47
rustic plant stand, 47
wooden window box, 47
plants
basic needs of
cold, 4
heat, 4–5
length of day, 4
light, 4
soil pH, 5
temperature, 4
water, 5
for decorative use, 239
glossary of, 5
shade-loving
bleeding heart plants, 12
flowering plants, 12
vegetable plants, 12
pokeweed, 136
portable workshop
with slant roof, 388
with span-roof, 387–388
potpourri, 250, 252
rose potpourri, 252
sachet potpourri, 253
"you choose" potpourri, 252

pottery
basic vase, 278
basics of, 276–277
bowls, 280
candlesticks, 280
decorating, 282
embellishments, 278–279
firing, 279
glazing, 282
jars, 280
vases, 282
wheel-working, 279
prickly pear cactus, 136
propogation rack, 22

Q
quick breads
basic quick bread recipe, 96
cinnamon bread, 96–97
cranberry coffee cake, 97
date muffins, 97
date-orange bread, 97
one-hour brown bread, 97
pineapple nut bread, 97
quilting, 275–276

R
rabbits
breeding, 316
breeds, 304
feeding, 316
health concerns, 316
housing, 314, 316
raised beds
how to make, 69–70
things to consider, 70
raspberries, 133
raw milk, 146
reindeer moss, 137
rooftop gardens
how to make, 67–68
suitable roofs for, 67
things to consider, 68
water leakage and, 67
root beer, 150
rugs, rag, 287

S
sachets, 253
sassafras, 137
sausage making
beef sausage, 140
bratwurst, cooked, 141
important considerations in, 140
polish sausage (kielbasa), 141
pork sausage, 140
saving seeds
hybrid vs. heirloom seeds, 36
scrapbooking, 286–287
seedlings, 23
sheep

breeds, 320
butchering, 335–339
diseases, 322
feeding, 320
housing, 320–321
milk, 322
shearing, 321
wool
carding, 322
spinning, 322
soaps
cold-pressed soap, 265–267
natural dyes for, 270
soap oils for, 267
soil
enriching
with compost, 16–20
with fertilizers, 15–16
nutrients
macronutrients, 14
management of, 14
micronutrients, 14
testing, 14, 16
pH, 5
quality of, 13–14
types of
clay, 24
loam, 24
sandy, 24
watering, 23–24
solar thermal energy, 412–413
spatterdock, 137
springtime wreath, 238
sprouting seeds, 22
strawberry, 137
sundials, 260

T
temperature zone map, 5
terracing, 38
building your own, 39–40
making use of slopes and, 40
materials needed for, 38
walls and, 38
terrariums, 241–242
thistle, 137
toboggans, 289
tomatoes, heirloom, 36
tools
classifying, 342
geometrical
bevels, 345
combination shooting
board, 349–350
compasses, dividers, and
calipers, 348
crenellated squares, 344–345
dividing a board with
rules, 344

marking and cutting gauges, 346–347

marking and scribing, 343

marking work for sawing, 345–346

miter blocks, 349

miter box, 349

miter templates, 350

rules, 343–344

spirit level, 351

squares and bevels, 344

testing surfaces with straight-edges, 343

try squares, 344

whitworth method of testing straight-edges, 343

holding tools

bench stops, 354

benches, 352–353

common bench screw vice, 354–355

cramping floor boards, 357

cramps, 356

pincers, 358

rope and block cramp, 356–357

sawing stools or trestles, 355–356

screw vice for kitchen table, 355

trees for shade or shelter

choosing, 58–59

maintenance of, 60–61

placement of, 59

planting, 59–60

pruning, 61

turkeys

breed selection, 301

chicks ("poults")

hatching, 302

raising, 302

diseases, 302

feeding, 301

housing, 301

slaughtering, 303

V

vegetables

hoophouses, 42–43

making your own, 41

things to consider, 41

venison or game sausage, 141

violets, 138

W

walnut, 137

water

conserving

choosing plants for low

water use, 25–26

rain barrels, 27–28

trickle irrigation systems, 26–27

water lily, 137

weddings

bouquets, 243–244

cake, 244–246

centerpieces, 245

edible flowers, 245

hanging flower pomander, 244

napkins, 246

programs for the ceremony, 246

sachets, 245

wild grapevine, 137–138

wild onion and garlic, 138

wild rose, 138

windmills, 414–417

wood sealers, 260

workshop furniture

bench with side and tail vices, 365–367

fixel sawing horse, 370

inside fittings of tool chest, 361

packing tool chest for transit, 362

portable folding bench, 367

portable sawing horses, 370

racks for saws and chisels, 360

sawing horse, 369–370

sawing stools and workshop trestles, 368–369

small tool chest with drawers, 362

strong stool or work bench, 365

tool cabinet, 363–364

tool chest partitions, 359–360

tool chest tills, 360–361

utilizing tool chest lids, 362–363

woodworker's tool chest, 359

worms, 19

Y

yeast bread

biscuits, 104

gluten-free bread, 105

making your own, 104

multigrain bread, 105

oatmeal bread, 105

raised buns (brioche), 105, 107

tips for making, 105

wheat and, 98, 101

yeast and, 97, 104

yogurt, 145–146

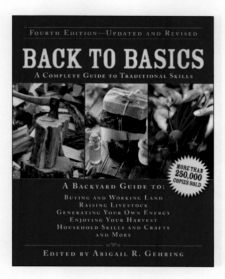

Back to Basics, Fourth Edition

A Complete Guide to Traditional Skills

Edited by Abigail R. Gehring

The classic guide to self-sufficiency, with more than 200,000 copies sold—now fully updated!

Anyone who wants to learn basic living skills—the kind employed by our forefathers—and adapt them for a better life in the twenty-first century need look no further than this eminently useful, full-color guide. Countless readers have turned to *Back to Basics* for inspiration and instruction, escaping to an era before power saws and fast-food restaurants and rediscovering the pleasures and challenges of a healthier, greener, and more self-sufficient lifestyle.

More than just practical advice, this is also a book for dreamers—even if you live in a city apartment, you will find your imagination sparked, and there's no reason why you can't, for example, make a loom and weave a rag rug. Complete with tips for old-fashioned fun (square dancing calls, homemade toys, and kayaking tips), this may be the most thorough book on voluntary simplicity available.

$27.95 Hardcover • ISBN 978-1-62914-369-9

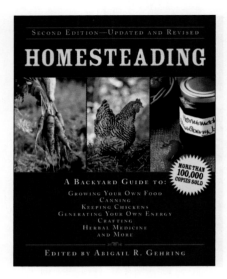

Homesteading, Second Edition

A Backyard Guide to Growing Your Own Food, Canning, Keeping Chickens, Generating Your Own Energy, Crafting, Herbal Medicine, and More

Edted by Abigail R. Gehring

The companion to the bestseller *Back to Basics* for country, urban, and suburban folks—now fully updated!

Who doesn't want to shrink their carbon footprint, save money, and eat homegrown food whenever possible? Even readers who are very much on the grid will embrace this large, fully illustrated guide on the basics of living the good, clean life. It's written with country lovers in mind—even those who currently live in the city.

Whether you live in the city, the suburbs, or even the wilderness, there is plenty you can do to improve your life from a green perspective. Got sunlight? Start container gardening. With a few plants, fresh tomato sauce is a real option with your own homegrown fresh tomatoes. Reduce electricity use by eating dinner by candlelight (using homemade candles, of course). Learn to use rainwater to augment water supplies. Make your own soap and hand lotion. Consider keeping chickens for the eggs. From what to eat to supporting sustainable restaurants to avoiding dry cleaning, this book offers information on anything a homesteader needs—and more.

$27.95 Hardcover • ISBN 978-1-62914-366-8

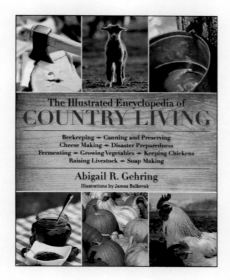

The Illustrated Encyclopedia of Country Living

Beekeeping, Canning and Preserving, Cheese Making, Disaster Preparedness, Fermenting, Growing Vegetables, Keeping Chickens, Raising Livestock, Soap Making

Abigail R. Gehring

Packed with step-by-step instructions, useful tips, time-honored wisdom, and both illustrations and photographs, this might just be the most comprehensive guide to back to basics living ever published. Fans of *Back to Basics*, *Homesteading*, and *Self-Sufficiency* have been asking for a one-stop resource for all the subjects covered in that successful series. In response, Gehring has compiled a massive, beautifully presented, single volume that covers canning and preserving, keeping chickens, fermenting, soap-making, how to generate your own energy, how to build a log cabin, natural medicine, cheese-making, maple sugaring, farm mechanics, and much, much more.

Whether you own one hundred acres or rent a studio apartment in the city, this book has plenty of ideas to inspire you. Learn how to build a log cabin or how to craft handmade paper; find out how to install a solar panel on your roof or brew your own tea from dried herbs; Cure a ham, bake a loaf of bread, or brew your own beer. This book has something for everyone.

$29.95 Hardcover • ISBN 978-1-61608-467-7

ALSO AVAILABLE

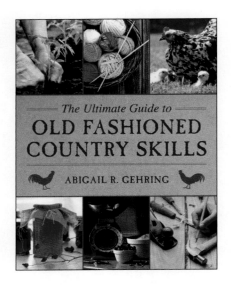

The Ultimate Guide to Old-Fashioned Country Skills

Abigail R. Gehring

Whether you're a suburbanite looking to live more simply or a die-hard homesteader interested in taking your garden to the next level, this guide is packed with step-by-step instructions, useful tips, vintage photographs and illustrations, and time-honored wisdom—creating one of the most comprehensive books on country skills available. This book is compiled of tested and practical experience passed down from generations of farmers and homesteaders.

Here readers can learn about:

- Creating a vegetable garden
- Canning and preserving
- Keeping poultry
- Natural medicine
- Bridge building
- Farm mechanics
- Crop rotation
- Cattle and dairying
- Foraging for wild food
- And much, much more!

Success comes to the person who works the most efficiently—not simply the person who works the hardest. Learn invaluable advice and tips for how to create a sustainable lifestyle and live off the land.

$24.95 Paperback • ISBN 978-1-62914-216-6

ALSO AVAILABLE

Mini Farming

Self-Sufficiency on ¼ Acre

Brett L. Markham

Start a mini farm on a quarter acre or less, provide 85 percent of the food for a family of four and earn an income.

Mini Farming describes a holistic approach to small-area farming that will show you how to produce 85 percent of an average family's food on just a quarter acre—and earn $10,000 in cash annually while spending less than half the time that an ordinary job would require. Even if you have never been a farmer or a gardener, this book covers everything you need to know to get started: buying and saving seeds, starting seedlings, establishing raised beds, soil fertility practices, composting, dealing with pest and disease problems, crop rotation, farm planning, and much more. Because self-sufficiency is the objective, subjects such as raising backyard chickens and home canning are also covered along with numerous methods for keeping costs down and production high. Materials, tools, and techniques are detailed with photographs, tables, diagrams, and illustrations.

$18.95 Paperback • ISBN 978-1-60239-984-6